Integrated River
Basin Development

Papers presented at the International conference on INTEGRATED RIVER BASIN DEVELOPMENT held at HR Wallingford, 13-16 September 1994. Organised and sponsored by HR Wallingford and the Institute of Hydrology, with co-sponsorship by the International Association for Hydraulic Research, the International Association for Hydrological Sciences, the British Hydrological Society, the UK Overseas Development Administration and the National Rivers Authority.

The valuable assistance of the UK Organising Committee, the International Organising Committee and Panel of Referees is gratefully acknowledged.

UK ORGANISING COMMITTEE

Dr W R White (chairman)	HR Wallingford
Dr J R Meigh	Institute of Hydrology
Dr J Gardiner	National Rivers Authority, Thames Region
Prof M J Hall	International Institute for Hydraulic and Environmental Engineeering, Delft
Dr R Herbert	British Geological Survey
Mr J Lazenby	Sir Alexander Gibb & Partners
Mr A Pepper	A T Pepper Engineering Consultancy Ltd
Dr R Bettess	HR Wallingford
Mr B Jackson	Overseas Development Administration

CONFERENCE ORGANISER

Jacqueline Watts	HR Wallingford

CONFERENCE EDITOR

Celia Kirby	Institute of Hydrology

INTERNATIONAL ORGANISING COMMITTEE

Dr J C Rodda	World Meteorological Organization, Switzerland
Prof Dr Ing W F Geiger	University of Essen, Germany
Prof M Nawalany	Warsaw University of Technology, Poland
Prof C S James	University of the Witwatersrand, South Africa

Integrated River Basin Development

Edited by

CELIA KIRBY
Instiutute of Hydrology, UK
W. R. WHITE
HR Wallingford, UK

JOHN WILEY & SONS
Chichester • New York • Brisbane • Toronto • Singapore

Published by John Wiley & Sons Ltd,
Baffins Lane, Chichester,
West Sussex PO19 1UD, England
Telephone   National   Chichester (0243) 779777
International   +44 243 779777

Other Wiley Editorial Offices

John Wiley & Sons, Inc., 605 Third Avenue,
New York, NY 10158-0012, USA

Jacaranda Wiley Ltd, 33 Park Road, Milton,
Queensland 4064, Australia

John Wiley & Sons (Canada) Ltd, 22 Worcester Road,
Rexdale, Ontario M9W 1L1, Canada

John Wiley & Sons (SEA) Pte Ltd, 37 Jalan Pemimpin #05-04,
Block B, Union Industrial Building, Singapore 2057

British Library Cataloguing in Publication Data

A catalogue record for this book is available from the British Library

ISBN 0-471-95361-X

Produced from camera-ready copy supplied by the Institute of Hydrology
Printed and bound in Great Britain by Bookcraft (Bath) Ltd

CONTENTS

PART II : INTEGRATED QUANTITY AND QUALITY OBJECTIVES

PART III : GROUNDWATER/SURFACE WATER INTERACTION

PART IV : MANAGEMENT OF LOW RIVER FLOWS

PART V : IMPACT OF RURAL LAND USE CHANGE

PART VI : EROSION AND SEDIMENTATION

List of Contributors

M.C. Acreman
IUCN, Rue Mauverney 28, CH-1196, Gland, Switzerland

N. Alpaslan
Dokuz Eylul University, Faculty of Engineering, Bornova 35100, Izmir,Turkey

S.R. Baconguis
Department of Environment and Natural Resources, Ecosystems Research and Development Bureau, College, Laguna 4031, The Philippines

T. Bagnati
CISE Tecnologie Innovative SpA, Via Reggio Emilia, 20090 Segrate, Milano, Italy

P.W. Balls
Marine Laboratory, Aberdeen AB9 2QJ, United Kingdom

J.R. Bevan
Department of Environment, University of Northumbria, Newcastle upon Tyne NE1 8ST, United Kingdom

S. Blazkova
T.G. Masaryk Water Research Institute, Podbabska 30, 160 62 Praha 6, Czech Republic

J.J. Bogardi
Wageningen Agricultural University, Nieuwe Kanaal 11 6709 PA, Wageningen, The Netherlands

P.A. Bradbury
HR Wallingford Ltd, Howbery Park, Wallingford OX10 8BA, United Kingdom

L.N. Braun
Libuse Bubenickova, Czech Hydrometeorological Institute, Prague, Czech Republic

A. Brookes
National Rivers Authority, Thames Region, Kings Meadow House, Kings Meadow Road, Reading RG1 8DQ, United Kingdom

R.L. Brown
*Binnie and Partners, Grosvenor House, 69 London Road, Redhill, Surrey RH1
1LQ, United Kingdom*

J.J. Burke
*Water Resources Branch, DESD/United Nations DC1-704, One United Nations
Plaza, New York NY 10017, USA*

G. Carnie
*Crouch Hogg Waterman, The Octagon, 25 Baird Street, Glasgow GE4 0EE, United
Kingdom*

F. Carvajal Monar
*HASKONING Royal Dutch Consulting Engineers & Architects, Barbarossastraat
35, P.O.Box 151, 6500 AD Nijmegen, The Netherlands*

M. Chalupka
*Research Centre of Agricultural and Forest Environment, Polish Academy of
Sciences, Bukowska 19, 60-809 Poznan, Poland*

M.J. Clark
*Geography Department and GeoData Institute, University of Southampton,
Southampton SO9 5NH, United Kingdom*

C.P. Crockett
*National Rivers Authority, Severn-Trent Region, Sapphire East, 550 Streetsbrook
Road, Solihull B91 1QT, United Kingdom*

B. de Bruine
*Ministry of Agriculture, Water and Rural Development, Private Bag 13193,
Windhoek, Namibia*

A. Dixit
Royal Nepal Academy of Science and Technology, Kathmandu, Nepal

P.W. Downs
*Department of Geography, University of Nottingham, University Park, Nottingham
NG7 2RD, United Kingdom*

J.A.L. Dunderdale
Silsoe College, Cranfield University, Silsoe, Bedford MK45 4DT, United Kingdom

A.C. Edwards
Macaulay Land Use Research Institute, Aberdeen AB9 2QJ, United Kingdom

J.B. Ellis
National Environment Research Council, Polaris House, North Star Avenue, Swindon SN2 1EU, United Kingdom

T.R. Franks
Development and Project Planning Centre, University of Bradford, West Yorkshire BD7 1DP, United Kingdom

J. Fürst
Universität für Bodenkultur, Institut für Wasserwirtschaft, Hydrologie und konstructiven Wasserbau, Nussdorfer Lande 11, A-1190 Wien, Austria

J. Gailhard
Electricité de France, Direction des Études et Recherches, Departement Environnement, 6 quai Watier, B.P. 49, 78401 Chatou, France

J. Gardiner
National Rivers Authority, Thames Region, Kings Meadow House, Reading RG1 8DQ, United Kingdom

C.N. Gibbins
Department of Environment, University of Northumbria, Newcastle upon Tyne NE1 8ST, United Kingdom

D.J.G. Gowing
Silsoe College, Cranfield University, Silsoe, Bedford MK45 4DT, United Kingdom

Ph. Gosse
Electricité de France, Direction des Études et Recherches, Departement Environnement, 6 quai Watier, B.P. 49, 78401 Chatou, France

R. Gray
W.S. Atkins Water, Woodcote Grove, Ashley Road, Epsom, Surrey KT18 5BW, United Kingdom

D. Gyawali
Royal Nepal Academy of Science and Technology, Kathmandu, Nepal

N.B. Harmancioglu
Dokuz Eylul University, Faculty of Engineering, Bornova 35100, Izmir, Turkey

G.L. Heritage
Centre for Water in the Environment, Department of Botany, University of the Witwatersrand, Private Bag 3, P.O. Wits 2050, South Africa

M.A. House
Urban Pollution Research Centre, Middlesex University, Bounds Green Road, London N11 2NQ, United Kingdom

G.E. Hollis
Department of Geography, University College, 26 Bedford Way, London WC1H 0AP, United Kingdom

C. Hottelot
Swiss Federal Institute of Technology ETH, Zurich, Switzerland

J.A. Hudson
Institute of Hydrology (Plynlimon), Staylittle, Llanbrynmair, Powys SY19 7DB United Kingdom

D.A. Hughes
Institute for Water Research, Rhodes University, P.O. Box 94, Grahamstown 6140, South Africa

A. Hututale
Ministry of Agriculture, Water and Rural Development, Private Bag 13193, Windhoek, Namibia

M.J. Jones
Coode Blizard Ltd, Consulting Engineers, Royal Oak House, Brighton Road, Purley, Surrey CR8 2BG, United Kingdom

P. Karki
BPC Hydroconsult, GPO Box 864, Kathmandu, Nepal

V. Kasimona
Department of Water Affairs, Lusaka, Zambia

J.M. Knowles
National Rivers Authority, North West Region, P.O. Box 12, Richard Fairclough House, Knutsford Road, Warrington WA4 1HG, United Kingdom

R. Kostal
Water Research Institute, Nabr. L. Svobodu 5, 812 49, Bratislava, Slovakia

S. Kumar
Centre for Water Resources Studies, (Patna University), Bihar College of Engineering, Patna 800 005, India

Zbigniew W Kundzewicz
Research Centre of Agricultural and Forest Environment, Polish Academy of Sciences, Bukowska 19, 60-809 Poznan, Poland

H. Laboyrie
HASKONING Royal Dutch Consulting Engineers & Architects, Barbarossastraat 35, P.O. Box 151, 6500 AD Nijmegen, The Netherlands

N.K. Lall
Royal Nepal Academy of Science and Technology, Kathmandu, Nepal

J.B.C. Lazenby
Sir Alexander Gibb & Partners, Earley House, London Road, Reading RG8 9EH, United Kingdom.

G.J.L. Leeks
Institute of Hydrology (Plynlimon), Staylittle, Llanbrynmair, Powys SY19 7DB, United Kingdom

Lin Xueyu
Institute of Applied Hydrogeology, Changchun University of Earth Sciences, 6 Xhimzhu Street, Changchun, Jilin 130026, China

J. Lindtner
Water Research Institute, Nabr. L. Svobodu 5, 812 49, Bratislava, Slovakia

A.M. Macdonald
Macaulay Land Use Research Institute, Craigiebuckler, Aberdeen AB9 2QJ, United Kingdom

K. Malatre
Electricité de France, Direction des Études et Recherches, Departement Environnement, 6 quai Watier, B.P. 49, 78401 Chatou, France

R.E. Manley
78 Huntingdon Road, Cambridge CB3 0HH, United Kingdom

D.W. Martin
National Rivers Authority, Severn-Trent Region, Sapphire East, 550 Streetsbrook Road, Solihull B91 1QT, United Kingdom

R.F. Merrix
National Rivers Authority, Northumberland and Yorkshire Region, Newcastle upon Tyne, NE3 3DU, United Kingdom

B.J. Middleton
Steffen, Robertson & Kirsten, P.O. Box 55291, Northlands 2116, South Africa

J. Morris
Silsoe College, Cranfield University, Silsoe, Bedford MK45 4DT, United Kingdom

K.A. Murdoch
Institute for Water Research, Rhodes University, P.O. Box 94, Grahamstown 6140, South Africa

K.D.W. Nanderlal
University of Peradeniya, Department of Civil Engineering, Perideniya, Sri Lanka

C. Neal
Institute of Hydrology, Crowmarsh Gifford, Wallingford OX10 8BB, United Kingdom

M.D. Newson
Centre for Land Use and Water Resources Research, University of Newcastle upon Tyne NE1 7RU, United Kingdom

L. Oyebande
Faculty of Environmental Sciences, University of Lagos, Box 160, Unilag Post Office, Akoka, Lagos 698, Nigeria

S. Parini
CISE Tecnologie Innovative SpA, Via Reggio Emilia, 200090 Segrate, Milano, Italy

M. Portapila
Department of Hydraulics and Sanitary Engineering, National University of Rosario, Argentina

N. Pouey
Departmento de Hidraulica e Ingenieria Sanitaria, Universidad Nacional de Rosario, Berutti 2353, 2000 Rosario, Sante Fe, Argentina

N. Prakash
Centre for Water Resources Studies, (Patna University), Bihar College of Engineering, Patna 800 005, India

T. Prasad
Centre for Water Resources Studies, (Patna University), Bihar College of Engineering, Patna 800 005, India

K.B. Pugh
North East River Purification Board, Aberdeen AB1 3RD, United Kingdom

E.P. Querner
The Winand Staring Centre for Integrated Land, Soil and Water Research (SC-DLO), Marijkeweg 11/22, P.O. Box 125, NL-6700 AC Wageningen, The Netherlands.

D. Ramsbottom
HR Wallingford, Howbery Park, Wallingford OX10 8BB, United Kingdom

B.R. Regmi
Royal Nepal Academy of Science and Technology, Kathmandu, Nepal

J. Rietjens
Electricité de France, Direction des Études et Recherches, Departement Environnement, 6 quai Watier, B.P. 49, 78401 Chatou, France

G. Roberts
Institute of Hydrology, Crowmarsh Gifford, Wallingford OX10 8BB, United Kingdom

C. Sabaton
Electricité de France, Direction des Études et Recherches, Departement Environnement, 6 quai Watier, B.P. 49, 78401 Chatou, France

I. Saccardo
ENEL/CRIS, C.so del Popolo 245 - 30172 Mestre, Venice, Italy

K. Sami
Institute for Water Research, Rhodes University, P.O. Box 94, Grahamstown 6140, South Africa

E. Saner
Dokuz Eylul University, Faculty of Engineering, Bornova 335100, Izmir, Turkey

K.D. Sharma
Central Arid Zone Research Institute, Pal Road, Jodhpur 342003, India

E. Schoeller
Department of Hydraulics and Sanitary Engineering, National University of Rosario, Argentina

K.J. Seymour
National Rivers Authority, North West Region, P.O. Box 12, Richard Fairclough House, Knutsford Road, Warrington WA4 1HG, United Kingdom

R. Slota
Water Research Institute, Nabr. L Svobodu 5, 812 49 Bratislava, Slovakia

B.J. Smith
Medway River Project, 3 Lock Cottages, Lock Lane, Sandling, Maidstone, Kent ME14 3AU, United Kingdom

H.A. Smithers
National Rivers Authority, North West Region, P.O. Box 12, Richard Fairclough House, Knutsford Road, Warrington WA4 1HG, United Kingdom

C. Soulsby
Department of Geography, University of Aberdeen AB9 2UF, United Kingdom

G. Spoor
Silsoe College, Cranfield University, Silsoe, Bedford MK45 4DT, United Kingdom

F.A. Stoffberg
Department of Water Affairs and Forestry, Private Bag X313, Pretoria 0001, South Africa

J.V. Sutcliffe
Sir Alexander Gibb & Partners, Earley House, London Road, Reading RG8 9EH, United Kingdom

B. Te Slaa
HASKONING Royal Dutch Consulting Engineers & Architects, Barbarossastraat 35, P.O. Box 151, 6500 AD Nijmegen, The Netherlands

F. Travade
Electricité de France, Direction des Études et Recherches, Departement Environnement, 6 quai Watier, B.P. 49, 78401 Chatou, France

D.A. Turnbull
Department of Environment, University of Northumbria, Newcastle upon Tyne NE1 8ST, United Kingdom

S. van Biljon
Department of Water Affairs and Forestry, Private Bag X313, Pretoria, South Africa

G. van Langenhove
Ministry of Agriculture, Water and Rural Development, Private Bag 13193,
Windhoek, Namibia

A.W. van Niekerk
Centre for Water in the Environment, Department of Civil Engineering,
University of the Witwatersrand, Private Bag 3, P.O. Wits 2050, South Africa

J.A. van Rooyen
Department of Water Affairs and Forestry, Private Bag X313, Pretoria 0001,
South Africa

F.C. van Zyl
Department of Water Affairs and Forestry, Private Bag X313, Pretoria 0001,
South Africa

A. Verdhen
Centre for Water Resources Studies, (Patna University), Bihar College of
Engineering, Patna 800 005, India

C. Weijun
Ministry of Water Resources, Beijing 100761, China

C. Woolhouse
National Rivers Authority, Thames Region, Aspen House, 97 Crossbrook Street,
Waltham Cross EN8 8HE, United Kingdom

E.P. Wright
The Baldons, Baker Street, Aston Tirrold, Blewbury OX11 9DD, United
Kingdom

Yang Yueso
Institute of Applied Hydrogeology, Changchun University of Earth Sciences, 6
Ximizhu Street, Changchun, Jilin 130026, China

E. Zimmerman
Department of Hydraulics and Sanitary Engineering, National University of
Rosario, Argentina

Preface

One in four people in the world have no access to safe water. The toll is high: illness from water-borne diseases still kills many people each year, especially young children. Water excess and scarcity, manifest as floods and droughts, also exact horrendous penalties in loss of life and human misery.

The importance of water as an essential element of life is no longer merely acknowledged as a fundamental truism, it is now an integral tenet of Agenda 21, the blueprint for sustainable management of the Earth, prepared at the UN Conference on Environment & Development at Rio de Janeiro in 1992. However, words alone will achieve nothing and action is required, not least discussion meetings of all kinds to explore what has been done and what can be done.

Our meeting in Wallingford in 1994, the papers from which are contained in this volume, was one such event. The response to our call for papers was considerable, and on the basis of the abstracts submitted, eight major themes emerged, on which were based the programme for the four-day conference and which also form the structure for this book:

- Sustainable resources
- Integrated quantity and quality objectives
- Groundwater/surface water interaction
- Management of low river flows
- Impact of rural land-use change
- Erosion and sedimentation
- Impact of urban and industrial development
- River basin management

By bringing together many different professional interests in water affairs from all round the world, we have attempted an objective look at the current state of progress in water resource management with sustainability as its main aim. We have tried to face up to the realities of the failures as well as acknowledging the successes so that sensible extrapolations can be made where relevent. We therefore hope that in some small way this volume will contribute to the tasks that lie ahead, to provide clean and wholesome water for all, to continue to meet the requirements for agricultural production, for energy supply and — whenever and wherever possible — mitigating the worst effects of climatic extremes.

Celia Kirby
W. R. White

Acknowledgements

We should like to acknowledge with much gratitude the hard work of John Griffin, Jane Gregory and Hilary Arnell in the preparation of the camera-ready copy and for their painstaking and meticulous proof-reading. A special vote of thanks is also due to Jo Manion for all her careful work on the illustrations to achieve clarity and consistency in style. Finally, we should like to thank Jacqueline Watts, the Conference Organiser, without whose enthusiasm and tireless efforts the event would not have taken place at all.

PART I

SUSTAINABLE RESOURCES

1 Sustainable integrated development and the basin sediment system: guidance from fluvial geomorphology

MALCOLM NEWSON

Centre for Land Use and Water Resources Research, University of Newcastle upon Tyne, UK

LAND AND WATER — THE BROAD CONTEXT

Downs, Gregory and Brookes (1991) have analysed the terms used to describe various levels of sector cohesion in river basin development. They reserve the word 'integrated' for an intermediate stage in which more than one sectoral interest is linked at both the operational and strategic levels. They reserve the word 'holistic' to describe schemes which approach the basin as an energy system or ecosystem. Mitchell (1990), however, sees holistic or comprehensive concepts as operating at the *strategic* level whilst integrated action is appropriate at the *operational* level.

It is impossible to concede that these couplings (holistic/integrated; strategic/operational) should not be part of all planned interventions in the river basin in the late Twentieth Century: the best way to take a system-wide, holistic approach to river basin development strategies is to strengthen a further coupling — that between land and water — in the minds of all politicians, planners, engineers and participants who live in the basin (Newson, 1992a). This paper sees the basin sediment system as part of the land-water coupling, worthy of consideration at the strategic stage but also of considerable relevance to operations (not only construction but maintenance — an essential aspect of sustainability). Because the basin's sediment system is its fundamental structural system (Figure 1), the foundation for development across its entire surface and the longer-term recipient of many of its impacts, its characteristic space and time signatures should be used to guide development itself.

STRATEGIC CONSIDERATIONS

Marchand & Toornstra (1986) present a simple ecosystem concept for the river basin; undisturbed, the basin shows spontaneous regulation functions (largely *storage* of water, sediment and nutrients) and extensive exploitation has little impact on these. Intensive exploitation, typical of rapid development, damages the spontaneous

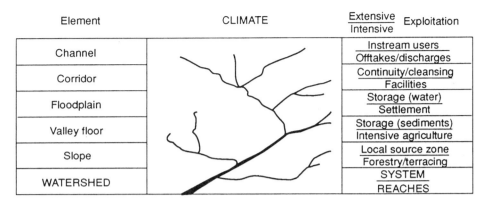

Element	CLIMATE	$\frac{\text{Extensive}}{\text{Intensive}}$ Exploitation
Channel		Instream users / Offtakes/discharges
Corridor		Continuity/cleansing / Facilities
Floodplain		Storage (water) / Settlement
Valley floor		Storage (sediments) / Intensive agriculture
Slope		Local source zone / Forestry/terracing
WATERSHED		SYSTEM / REACHES

Figure 1   Elements of the basin sediment system — extensive and intensive exploitation
(and "natural" controls)

regulation functions, which then have to be replaced by structural and legal/political analogues. Figure 1 develops these concepts.

Strategically, the basin sediment system from watershed to ocean should be seen as an important part of the spontaneous regulation function; the physical basis of the river basin is as a water and sediment transport system, connecting at the coast to the shelf transport system. Whilst recovery occurs from chemical pollution relatively quickly and campaigns of rehabilitation can achieve much in a short time, the span of timescales, from minutes to millennia, over which the sediment system operates its 'jerky conveyor belt' (Ferguson, 1981) means that physically destabilised basin systems can seldom be rectified within the lifetime of the project which damaged them. The sediment system creates, throughout its length, the morphology upon which the other functions of the system depend. It links closely to sectoral interests such as water storage, agricultural water use, flood defences and — via bonds between metals, nutrients and other pollutants — with sediment particles, it links to longer-term pollution control strategies.

'Control' of the sediment system can be illusory, even in such low-yield topo-climates as that of the UK; natural controls such as floodplains, wetlands and vegetation covers do a much better job for a longer period at cheaper cost than do revetments, groynes and traps. In general, our analogues of the spontaneous basin functions encourage *throughput*, not *storage*. Extensive exploitation systems work round the hazards of geomorphological change in an opportunistic, dynamic way. 'Surprise' is inevitable because the sediment system exercises intrinsic controls (Schumm, 1977) over its own dynamics through storage-dominated feedbacks; threshold phenomena are therefore common (Newson, 1992b). Predictions based solely on extrinsic variables like climate and flow are unlikely to be of use to a project designed around sustainability.

The lesson from these pointers is that the sediment system of each river basin deserves its own detailed environmental impact assessment before new development begins, using the key phases of EA: scoping of the range of impacts and their

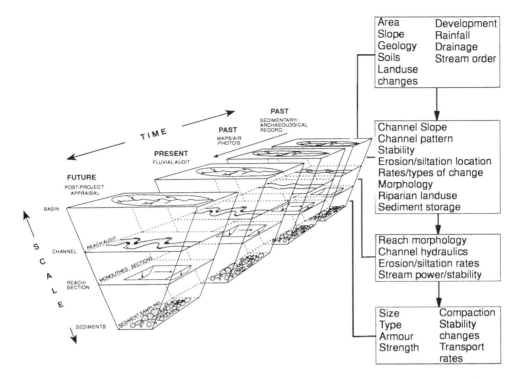

Figure 2 The geomorphological framework for rational river management

timescales, assessment of the magnitude and mitigation of impacts, and monitoring. Monitoring alone is insufficient and therefore wasteful, especially in view of the technical problems of measuring sediment fluxes (Bogen *et al.*, 1992), yet is often the agency panacea (e.g. World Bank/UNDP Hydrological Assessment, Sub-Saharan Africa) because it fits the conventional 'active science' component of project assessments to date. Fluvial geomorphology has other techniques to offer which tend to be more holistic, observational and historical (see Figure 2).

LOCATIONS IN THE RIVER SEDIMENT SYSTEM

At a very broad scale it is unfortunately the case that most of the less immediately hazardous river basins of the world have been developed, leaving those with substantial mountainous or drought-prone conditions to test our judgement. Basin development schemes now focus on them because of the potential contribution by dam construction to energy, flood control and food. Recent findings by Dedkov & Moszherin (1992) suggest that sediment yield in undeveloped mountainous basins is 27 times greater than in other regions but that this reduces to 3.2 times where development of the lowlands has occurred, doubling natural yields there. The other morphoclimatic region with high yields, especially after development, is the highly seasonal 'Mediterranean' zone; this finding contrasts with previous beliefs that semi-aridity generated the highest yields and illustrates continuing confusion about the precise role of vegetation in geomorphology. This is an active research area in the subject (Thornes, 1990) and must continue to be so if land/water couplings are to become a focus of sustainable development.

Turning to individual basins, despite geomorphology's insistence on continuity in space and time in the basin system, strategic ideas must become operational if we are not to be lost in transcendental research-level musings. We are at the point in world water development where the productivity of traditional extensive exploitation systems can no longer feed us (Falkenmark, 1986; Livovitch *et al.*, 1990). Thus intensive exploitation is inevitable and we must be specific in our location guidance to those planning dams, irrigation, flood protection and catchment management. Modern science has a fear of classification but operational classifications are becoming popular as a conceptual aid to auditable actions. In fluvial geomorphology the dynamics and stability of a site can usually be inferred from its location in the sediment system by field observation of signs and symptoms. Broadly, Schumm's classification of source, transfer and depositional zones educates us as to the important variables of sediment calibre and the intensity and duration of transport events. However, the sediment system has a long memory (the property of its huge storages). Slope deposition, floodplain deposition and channel storage as bars and bedforms account for 75% of the sediment output from headwaters (at the scale of most major intensive exploitation sites). It is therefore not sufficient to make locational or future planning decisions for development on the basis of simple patterns (Figure 3). Church and Slaymaker (1989) have shown how, in recently glaciated

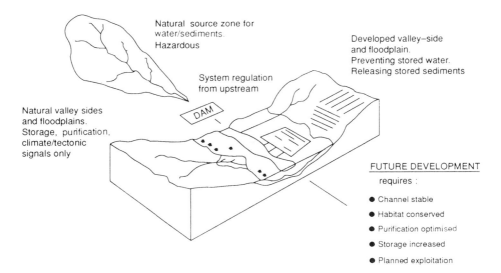

Figure 3   The disjointed river system under conventional development and the needs of sustainable development

landscapes, the source zone may now have moved down river, like an egg in a snake. Even in contemporarily active mountain regions the slope and channel systems exhibit transporting events of different magnitudes and frequencies, leading to storage delays of 20-25 years before a basin is 'clear' of debris flow sediments (Ergenzinger, 1992). The 'jerky conveyor belt' of coarse, channel-forming sediments leads to repetitions of the Schumm triplet and location of river structures and off-river developments must be guided by the current and future position of these features (Newson, 1992b). Only an experienced field geomorphologist is, however, likely to be able to decipher these patterns after a rapid survey.

At a reach scale the classification of channel type can be an important aid to strategic decisions; inherent in practical classifications are information on flow and sediment dynamics, their controls (which may also be controls on biological habitat) and their trends through time, i.e. their stability. Such classifications are becoming increasingly objective and sophisticated (e.g. Rosgen, *in press*), with the specific purpose of supporting and informing river and river corridor management decisions.

In order to advocate the use of geomorphological knowledge in the location strategy for river basin developments we must look at existing location decisions. The world-wide tradition in river engineering, dam construction and even ecology has been the reach length assessment, location-free in terms of systems. Location factors used by developers include geological (dam engineers), hydrological (evaporation *v.* precipitation) and the obvious socio-economic-political factors of 'need' or opportunity (including a domination of the interests of the urban elite). The addition of geomorphological assessment for strategies and projects will be a true test of the timescales over which impacts are being considered.

TIMING IN THE RIVER SEDIMENT SYSTEM

Beginning very broadly again, it is important to stress that 'bad times' for river basin development will inevitably be those of tectonic instability and climate shifts. We are therefore, globally, in a very difficult era. However, as many have pointed out, our 'incertitude' about future trends impacts more on how we make decisions (more attention to risk, subsidiarity) than on the evidence we use.

At the basin scale it is helpful, even if inaccurate, to structure a timescale model for the basin sediment system; this has been done theoretically by Statham (1977; Figure 4 here) and empirically from historical investigations by e.g. Trimble (1974). Such work points to the controlling role of storage in the basin, yet this phenomenon — the very substance of geomorphology — is seldom investigated (but see Swanson *et al.*, 1982). To the author's knowledge the only river basin to have a comprehensive research approach to its sediment storages is Redwood Creek in northern California (Madej, 1984); she divides, locates, surveys and estimates timings for the 720 km^2 basin (Figure 4). It is clear that, as well as being a 'jerky conveyor belt' in the linear sense, the river system is not unidirectional. Storage is both part of the feedback system leading to homeostasis *and* a potentially destabilising agency because of its

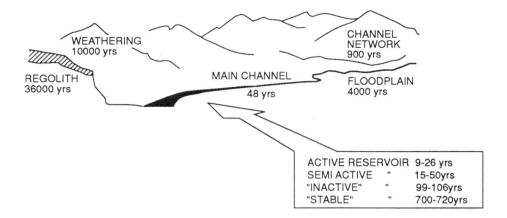

Figure 4   Statham's theoretical timings for the basin sediment system with (inset) the measurements of Madej (1984)

intrinsic geomorphological action as a landscape element — stored sediments have their own stability criteria and induce conditions for their own remobilisation. Whilst geomorphology and engineering science have been, to date, fascinated by lateral instability of rivers, the behaviour of storages in the sediment system normally produces vertical instability and much larger sediment outputs through feedback to slope conditions. The arroyo cycle and its impact on management of the Colorado is extensively documented by Graf (1985).

It is, however, inadequate to consider the temporal control over basin sediment yields as operating only at the scale of large storages; it is operational at scales from basin to particle and the interdependency of its components should be considered at both strategic and operational scales of development. As an example of particle-scale controls, Laronne & Reid (1993) have documented the relationship between flow duration, bed structuring and high sediment yields in the semi-arid environment; in the UK we are also becoming concerned about the habitat implications of flow management and engineering intervention on bed strength.

The time signatures of artifices in the channel system, such as dams, are also being calibrated by geomorphologists — Petts (1987) suggests a spectrum from five years (floodplain vegetation) to a century (channel form and substrate) for the impacts of river regulation.

Fortunately the methodologies of fluvial geomorphology are now extending to consider the multiplicity of controls on the time signature of the sediment system. At Newcastle, recent research on Holocene river development (Passmore *et al.*, 1993) indicates the breadth of evidence required by such ambitions: from geological exposures to tracing the individual particles of a cluster bedform and involving evidence from the channel, corridor, floodplain and valley floor. In a related operational study Sear & Newson (in press) have suggested how the scale of the evidence needed can be compressed if necessary by a scheme of 'fluvial auditing' which has an emphasis on river classification, field observation and archive searches.

The role of geomorphology in helping to set timescales of investment or sustainable development is about to be realised in the developed world but too often assessments of developing world projects rely heavily on conventional monitoring networks — the stock-in-trade of hydrology — leaving the sediment system to be calibrated from short records or flume-derived hydraulics. It is therefore reassuring that Thorne *et al.* (1993) show how, even at the vast scale of the Brahmaputra, complimentary use of field and archive information can be cost-effective in planning operations.

Three large difficulties remain to strategic and operational planners: the phenomenon of metastability (rivers will be difficult to manage during threshold changes of morphology and sediment load), the impact of major floods (which often initiate threshold behaviour) and the impact of major river structures (whose effects may mimic threshold behaviour).

LAND AND WATER AND LAND-CATCHMENT MANAGEMENT PLANNING

Environmental change will require new forms of risk-taking public decision-making; whilst existing conditions in many parts of the developing world militate against our concepts of planning by consultation it is important that plans are made and management follows them. In England and Wales the new system of catchment management planning has revealed a new willingness on the part of water professionals to examine the natural resources of a basin in their widest context as part of the consultation processes leading to catchment plans. At the much larger scale of problem basins in the developing world the same sorts of assessment would be required for proper implementation of e.g. the Helsinki Rules on resource allocation between nation states.

In terms of the geomorphologist's concept of the basin, it is clear that there is a land 'bid' to be made as part of the river management agenda; currently it may be phrased as affecting 'sensitive areas' or 'buffer strips' but there are good grounds for extending to floodplains, valley floors and, critically, to areas and practices which may alter the stability of the basin's sediment storages. We make poor judgements about land management, especially erosion control, without proper surveys - as recorded in environments as different as the Himalayas (Ives and Pitt, 1988) and the Colorado (Graf, 1992).

Very little progress has been made towards this end. Sutherland and Bryan (1988) have pointed to the value of surveys of the intermediate colluvium store in a Kenyan basin so as to properly understand the links between the favoured plot and basin scales for sediment investigations.

However, we are (even in Kenya) content with fairly arbitrary allocations of buffer strips to aid channel protection. Geomorphological maps for tectonically-active zones such as the Himalayas are gaining engineering credibility; what is needed is an approach to the land use of a developed basin (agricultural, urban and industrial) based upon hazards of the 'normal' sediment system. One thing is certain: proper protective zones for rivers will not be parallel lines of 'x' metres but will reflect rock, soil, aquifer and drainage properties of landforms.

At present we load all our ignorance of the system into a concept of 'delivery ratio' from source of sediment to project site (Walling, 1983); how can we possibly guide soil protection policies in headwaters, for example, when the routing characteristics and morphological effects of the (reduced) sediment yield are unknown beyond a spuriously accurate look-up table? As an illustration that there are alternatives, Table 1 offers a basic impact matrix linking major development aims to the problems posed by the sediment system; clearly this approach is infinitely expandable if and when the demand is made to geomorphologists for a handbook of sustainable management principles.

Table 1 Impact matrix of basin development taking a longer-term geomorphological viewpoint

Development Problem	Geomorphological Status of Basin	Outcome/risk (*)
DAM CONSTRUCTION (sedimentation)	a. soil erosion active	sedimentation esp. small/montane basin **
	b. erosion Holocene or historical	sedimentation from valley-floor or colluvial stores ***
	c. soil erosion controlled	(see below)
RIVER REGULATION (downstream channel stability/habitat)	a. dam above active alluvial zone or silt flushing	channel change by direct impact or by tributary inputs *
	b. dam with flood control/HEP	habitat effects via bed structure and flooding regime **
	c. dam near coast	coastal erosion *
LAND-USE PLANS - erosion control - forestry - irrigation - urbanisation (sustainability)	a. land on slopes with active tectonics	failure unless small-scale, adopting local strategies ***
	b. land on active or semi-active alluvium/colluvium	changed hydrology releases sediment from storage ***
	c. floodplains	longer-term function conserved, or loss of habitat/spontaneous regulation

World water development faces many problems but there are signs of reform, notably in the economics of development funding; sadly, the recent World Bank paper on water resources management (World Bank, 1993) refers very little to the costs of watershed degradation, in spite of the fact that economists such as Winpenny (1991) are able to include the effects of abuse of the basin sediment system as examples of applying financial costing systems to environmental misuse.

REFERENCES

Bogen, J., Walling, D.E. & Day, T.J. 1992. Erosion and Sediment Monitoring Programmes in River Basins, *IAHS Publication No.210*, Wallingford, U.K.

Church, M. & Slaymaker, H.O. 1989. Disequilibrium of Holocene sediment yield in glaciated British Columbia. *Nature, 337*, 452-454.

Dedkov, A.P. & Moszherin, V.I. 1992. Erosion and sediment yield in mountain areas of the world. In: Walling, Davies and Hasholt (eds) *Erosion, Debris Flows and Environment in Mountain Regions*, IAHS Publication No. 209, Wallingford, U.K., 29-36.

Downs, P.W., Gregory, K.J. & Brookes, A. 1991. How integrated is river basin management? *Environmental Management, 15(3)*, 299-309.

Ergenzinger, P. 1992. A conceptual geomorphological model for the development of a Mediterranean river basin under neotectonic stress (Buonamico basin, Calabria, Italy). In: Walling, Davies and Hasholt (eds) *Erosion, Debris Flows and Environment in Mountain Regions*, IAHS Publication No.209, Wallingford, U.K., 51-60.

Falkenmark, M. 1986. Fresh water: time for a modified approach. *Ambio, 15(4)*, 192-200.

Ferguson, R.I. 1981. Channel form and changes. In: Lewin (ed) *British Rivers*, Allen & Unwin, London, 90-125.

Graf, W.L. 1985. *The Colorado River, Instability and Basin Management*, Assoc. of American Geographers, Washington DC.

Graf, W.L. 1992. Science, public policy and Western American rivers. *Trans. Inst. of British Geographers*, NS17, 5-19.

Ives, J. & Pitt, D.C. 1988. *Deforestation: Social Dynamics in Watersheds and Mountain Ecosystems*, Routledge, London.

Larrone, J.B. & Reid, I. 1993. Very high rates of bedload sediment transport by ephemeral desert rivers. *Nature, 366*, 148-150.

L'vovich, M.I., White, G.F., Belyaev, A.V., Kindler, J., Koronkevic, N.I., Lee, T.R. & Voropaev, G.V. 1990. Use and transformation of terrestrial water systems. In: Turner, Clark, Kates, Richards, Matthews & Meyer (eds) *The Earth as Transformed by Human Action*, Cambridge University Press, 235-252.

Madej, M.A. 1984. *Recent Changes in Channel-stored Sediment, Redwood Creek, California*. Technical Report, Redwood National Park, California.

Marchand, M. & Toornstra, F.H. 1986. *Ecological Guidelines for River Basin Development.*, Centrum voor Milienkunde, Rijksuniversiteit, Leiden, The Netherlands.

Mitchell, B. 1990. Integrated water management. In: Mitchell (ed.) *Integrated Water Management*, Belhaven, London, 1-21.

Newson, M.D. 1992a. *Land, Water and Development. River Basins and their Sustainable Management*. Routledge, London.

Newson, M.D. 1992b. Geomorphic thresholds in gravel-bed rivers — refinement for an era of environmental change. In: Billi, Hey, Thorne & Tacconi (eds) *Dynamics of Gravel-bed Rivers*, John Wiley, Chichester, 3-20.

Passmore, D.G., Macklin, M.G., Brewer, P.A., Lewin, J., Rumsby, B.T. & Newson, M.D. 1993.

Variability of late Holocene braiding in Britain. In: Best & Bristow (eds) *Braided Rivers*, Geological Society, London, Spec. Publ. 75, 205-229

Petts, G.E. 1987. Timescales in ecological change. In: Craig & Kemper (eds) *Regulated Streams: Advances in Ecology*, Plenum, New York, 257-266.

Rosgen, D.L. (in press). A classification of natural rivers. *Catena*.

Schumm, S.A. 1977. *The Fluvial System*. Wiley, New York.

Sear, D.A. & Newson, M.D. (in press). Sediment-related river maintenance: the role of fluvial geomorphology. In: Thorne (ed) *Geomorphology at Work*, John Wiley, Chichester.

Sutherland, R.A. & Bryan, R.B. 1988. Estimation of colluvial reservoir life from sediment budgeting, Katiorin experimental basin, Kenya. In: Bordas & Walling (eds), Sediment Budgets. *IAHS Publ. No. 174*, Wallingford, UK, 549-560.

Swanson, F.J., Janda, R.J., Dunne, T. & Swanson, D.N. *1982. Sediment Budgets and Routing in Forested Drainage Basins*. General Technical Report PNW-141, USDA, Forest Service, Portland, Oregon.

Thorne, C.R., Russell, A.P.G. & Alam, M.K. 1993. Planform pattern and channel evolution of the Brahmaputra River, Bangladesh. In: Best & Bristow (eds) *Braided Rivers*, Geological Society, London, Spec.Publ. 75, 257-276.

Thornes, J.B. (ed) 1990. *Vegetation and Erosion: Process and Environments*, John Wiley, Chichester.

Trimble, S.W. 1974. *Man-induced Soil Erosion on the Southern Piedmont 1700-1970*. Soil Conserv. Soci. of America, Akeny, Iowa, USA.

Walling, D.E. 1983. The sediment delivery problem. *J. Hydrol.* **65**, 209-237.

Winpenny, J.T. 1991. *Values for the Environment. A Guide to Economic Appraisal*. HMSO, London.

World Bank 1993. *Water Resources Management*, World Bank Policy Paper, Washington D.C.

2 River basin development to sustain megacities

ROGER L BROWN
Binnie & Partners, Redhill, Surrey, UK

INTRODUCTION

Considerable concern is developing internationally over the very rapid expansion of many large cities towards "megacity" status with populations of 10 million or more. A recent project funded by the Asian Development Bank (ADB) involved a wide-ranging sector review of eight megacities in their region, entitled " Managing Water Resources to Meet Megacity Needs".

Guidelines were drawn up for water supply and sewage disposal based on lessons from these megacities but also using experience from other large urban developments outside the region. The objectives are to aid provision of services in an appropriate, efficient and controlled manner.

This paper presents some of the aspects and ideas, related to basin development, which were considered in drawing up the guidelines with illustrations from selected cities.

IMPACT OF MEGACITIES

In areas of water shortage, megacities create extreme pressure on their local natural resources, requiring innovative approaches to achieve sustainable solutions. Their levels of water demand, and of waste water generated, can be similar in size to local irrigation needs, especially where urban growth is over agricultural land.

The traditional engineering approaches of impounding reservoirs and importing water from adjacent catchments are often opposed strongly on a variety of grounds. The high initial cost and adverse cash flow, especially over early years when the facility can be under-utilised, makes funding difficult and raises the issue of "affordability".

Water-sharing mechanisms, "prioritisation" of schemes and agreement to major investments can be achieved only after considerable debate and negotiation, based on sound data with analyses that support a strong case.

TYPES OF SITUATION

The appropriate approaches and constraints to development of river basins which include large cities depends partly on where they are located: e.g. inland, near the mouth of the river or on the coast. Inland megacities may be short of surface water because of limited upstream catchments while disposal of waste water needs careful consideration because of downstream users. Examples include Delhi (India), Beijing (China), and Kathmandu (Nepal).

At the time that they were originally established as ports, coastal megacities had adequate surface water or groundwater which has become deficient through increase in demand, pollution or over exploitation. Waste water disposal has been to the sea, usually via natural water courses running through the urban area. Examples include Karachi (Pakistan), Dhaka (Bangladesh), Manila (Philippines), and Lima (Peru).

An intermediate group consists of cities located near the mouth of major rivers (also originally ports), where catchments are large and pollution is less critical because of the size of rivers and the low impact on downstream users. However, many compete with numerous upstream users. Examples include Bangkok (Thailand), Seoul (South Korea), Buenos Aires (Argentina), and London (UK).

With the exception of Buenos Aires, all these cities require substantial infrastructure upstream in order to meet current and future water demands including, in many cases, inter-basin transfers. Many cities relied on groundwater in the past but for various reasons most water is now obtained from surface sources. In some water-short areas, groundwater may be used conjunctively with surface water in the future.

CONSTRAINTS ON INTEGRATED DEVELOPMENT

The concept of integrated basin development is recognised as highly desirable but is very far from being applied universally. Administering a river basin through a single organisation may appear to have advantages over establishing various bodies. However, even where one organisation has overall responsibility for a basin, the same compromises have to be reached internally which may be easier, but does not have the advantages of transparency resulting from more open debate through public scrutiny.

The most extreme debate on water rights and development projects occurs on international rivers, such as those flowing eastward from the Himalayas out of Nepal, through India and into Bangladesh. Allocation of the westward flowing rivers between India and Pakistan was resolved by the Indus Basin Treaty in 1960, but only after a long-standing water dispute which brought the two countries almost to the brink of war.

The same problems of allocation procedures occur where water rights are held by local states or provinces, as in India and Pakistan. For example, Delhi has no water rights, being a Union Territory and not a State, and so has to negotiate some new allocations to meet increasing needs.

Conflict arises between water demands for large urban areas, agricultural and horticultural irrigation, major industry (particularly power stations for cooling water), navigation, and the environment. Adjustments in allocations can be achieved during critical dry periods by such measures as drought orders as in England and Wales, and trading in water rights, as in California where farmers are prepared to transfer their rights to those who can afford to buy them, such as urban dwellers. These mechanisms achieve prioritisation through legislation and market forces respectively.

Traditionally, access to water has been a "right" and still tends to be considered as such in many cultures. Politicians are often reluctant to authorise adequate tariff levels to sustain appropriate levels of service, including provisions for the future.

INLAND CITIES

Delhi is located in the centre of the Indo-Gangetic plain, which experiences a summer monsoon climate. The city mainly occupies the right bank of the Yamuna river, which is bordered on the east by the Ganga river and to the north and west by the Sutlej river. The area is highly developed for irrigated agriculture with numerous major canals fed direct from barrages on the rivers, and also from major impounding reservoirs at Bhakra and Nangal on the Sutlej.

In India all water rights belong to the individual States but Delhi, being a Union Territory, has no water rights. In 1992 the total water supply was 26.3 $m^3 s^{-1}$, with 3.2 $m^3 s^{-1}$ coming from groundwater, 5.7 $m^3 s^{-1}$ from Uttar Pradesh out of the Upper Ganga canal to the east through a 28 km long, 2.8 m diameter reinforced concrete pipe, cast *in situ* (to minimise losses), and 17.4 $m^3 s^{-1}$ from Haryana State out of the Bhakra reservoir 325 km to the north. Transfer of surface water from the reservoir is by a series of non-dedicated canals which also serve for irrigation outside the monsoon season.

Negotiations have been under way to increase the abstraction from Bhakra reservoir. One element of the negotiation is to consider the city's sewage effluent as a water resource. An increased abstraction upstream of the city, at the expense of irrigation, would be compensated for by discharging treated sewage effluent directly into irrigation canals on the downstream side. Existing sewage works sell treated effluent to local farmers, sewage sludge is used as a fertiliser and methane is piped to local housing estates at about half the price of town gas.

Beijing is not situated on any sizeable river, nor is it close to the major rivers of China. The whole Beijing-Tianjin region suffers from severe water deficiency compared with rapidly increasing demands from expanding industry, agriculture and domestic users. In 1992 the city used 16.7 $m^3 s^{-1}$, of which 10.5 $m^3 s^{-1}$ was groundwater and 6.2 $m^3 s^{-1}$ was surface water from the Miyun and Guanting reservoirs, which are 80 km to the north and west respectively. These sources are deteriorating because of pollution and increased abstraction further upstream.

Beijing experiences severe water shortage, especially during drought periods, and groundwater levels are falling. Domestic water supply has priority over industry. Many communities, large organisations and industry have their own private water supply systems.

Sewage is discharged to designated rivers. Fifteen sewage treatment works are planned and some are already under construction. Water re-use in industry has increased from about 45% in the 1980s to up to 80% now for some large industries, as a result of various measures including raised charges for pollution and general shortage of water.

Various new water sources are being considered, including a gravity canal 1200 km long from the Yellow river to import water northwards to the whole region for all purposes.

Kathmandu is not a megacity in the sense defined earlier, but nevertheless faces a severe water shortage because of rapid development and few options for new sources. The city is surrounded by steep hills and is built on an old dried-out lake bed infilled with a few hundred metres of impermeable fine-grained deposits overlying sand. Large rivers draining from the Himalayas have cut down on the west (Trisuli river) and east (Melamchi/Sun Kosi rivers) sides by a few hundred metres and their tributaries have eroded to the north and south, leaving Kathmandu as an isolated pinnacle.

Originally the city relied on surface water from spring sources, to which the Bagmati river was added in 1966. Since the 1970s groundwater has been developed for public supply, mainly from northern aquifers. However, supply to consumers has been irregular and has lagged well behind demand, so that industry, embassies and hotels have sunk their own wells to deep aquifers under the city. At present, groundwater and surface water each provide about half the public supply.

COASTAL CITIES

Karachi is located on the Arabian sea to the west of the mouth of the River Indus. The total water supply of 18.5 $m^3 s^{-1}$ comes mainly from the Indus (13.5 $m^3 s^{-1}$), by 170 km long canals and pipes, but with 4.7 $m^3 s^{-1}$ from a small impounding reservoir on the Hub river and the rest from diminishing groundwater.

The catchment and flows of the Indus are enormous, but so are the requirements of water for irrigation, recession agriculture in Sindh province and the ecology of the estuary. Disputes over conflicting demands by different users and between provinces have lasted for decades but a sharing mechanism has finally been agreed and daily allocations are being implemented to optimise use of water. In spite of the large adjacent source of water, Karachi has severe shortages, partly because of inadequate transmission conduits, but also, in the medium term, through lack of water allocation.

A major steel works is a substantial water user, but salt water cooling is being introduced there now.

Dhaka relies almost entirely on groundwater extracted from the aquifer underlying the city, with each well serving the immediately surrounding area. However, increasing urban density has increased the demand per unit of ground area and reduced infiltration and recharge, so groundwater levels are falling.

The nearest rivers are the Buriganga to the west and the Lakhya to the east. Both are wide rivers which are used extensively for river transport, as a water source by industry on the banks and for discharge of waste water and effluent. Flows are seasonal so the rivers stagnate.

About 25 km to the east of Dhaka is the Meghna river which has a very large continuous flow and is not polluted. However, a conduit from this river would have to cross weak waterlogged ground subject to flooding.

Manila has many similarities to Jakarta as natural water courses flowing through the city carry the city's waste and are severely polluted.

The present trend is to develop multipurpose surface water schemes by staged diversions from eastward flowing rivers towards Manila to the west. Currently, the public supply of 26.1 $m^3 s^{-1}$ is almost entirely surface water but because of the inadequate public service many residential "villages" and industry have installed their own private schemes abstracting 9.7 $m^3 s^{-1}$ of groundwater, which is now falling rapidly and becoming saline.

In many poor areas, authorised vendors sell water legally bought from official standposts, while in others, unauthorised vendors sell "stolen" water. In order to reduce agricultural use, and thereby to release water for domestic consumption, crop substitution was tried, but without success because of poor markets for the new products.

Lima has suffered from inadequate water supply for decades, partly because it very seldom rains, so that natural recharge into the alluvium on which the city is built is virtually nil.

Surface water drawn from the river Rimac is treated at La Atarjea, which has a capacity of 15 $m^3 s^{-1}$, compared with average low flows of 12 $m^3 s^{-1}$ and minimum flows of 7.5 $m^3 s^{-1}$. Groundwater extraction capacity from the aquifer under the city is about 10 $m^3 s^{-1}$ but water levels are falling rapidly because of over abstraction. Leakage from the water distribution system is a major source of recharge.

Transferring water from the Atlantic to the Pacific catchments has been under discussion for 25 years.

RIVER MOUTH CITIES

Bangkok at the mouth of the Choa Phraya is one of the last users on this large and well-developed catchment, with strong competition from agriculture and hydro-electricity, as well as environmental needs.

The city used to rely extensively on local groundwater. However, as abstraction increased with the growth of the urban demand, much of the underlying aquifer became saline. Even worse, the soft deposits consolidated as a result of groundwater lowering, at rates up to 100 mm per annum. As the city is only a few metres above mean sea level, the risk of flooding and the cost of flood protection became unacceptable and groundwater was abandoned for the public supply and severely restricted for private use. The rate of settlement has now slowed down.

The numerous water courses (klongs) within the city have become so polluted from overflow of on-site sanitation and rubbish disposal that they are a major health hazard, even for traditional transport routes.

New major developments are required to install their own small sewage treatment works before discharge to the klongs. Piped sewerage is starting through a system of interceptor sewers followed by treatment.

The main environmental concern on the Choa Phraya is to prevent the sea water entering the city. A minimum flow of 70 $m^3 s^{-1}$ has to be maintained which, in effect, limits the extraction permitted for the city. However, the waste water discharged by the city through public and private systems is now a substantial proportion of this compensation flow so a case is being argued for considering it as part of the total flow.

Seoul is situated on the Han river, which drains most of the north of South Korea and part of North Korea. Storage has been built upstream but sites are limited. Abstraction for the city is 59 $m^3 s^{-1}$ and for the metropolitan region it is an additional 37.3 $m^3 s^{-1}$.

The river has been transformed into an amenity through the city so that minimum flows and quality are controlled. All sewage is now discharged downstream of the city after treatment. Reuse is being considered for industry, agriculture, toilet flushing and street cleaning.

Buenos Aires has the luxury of being on the bank of the enormous River Plate. The opposite bank is visible only on clear days. Raw water is taken from the river opposite the city itself, through intake structures about 1.5 km off the bank. Groundwater is deep and contaminated, so usage has dropped to about ten per cent of the total supply of 41 $m^3 s^{-1}$.

Rivers passing through Buenos Aires carry waste water and rubbish, which they discharge into the River Plate, so bathing is prohibited off its beaches. This serious contamination is restricted to the shore and does not extend to the intake.

WATER MANAGEMENT

The assessment of the conditions in Megacities highlighted a number of key points and factors which have general application.

Water cycle
Considering the full water cycle through urban areas shows aspects which affect overall strategy (political), planning (legal, engineering and economic) and organisation (institutional). These aspects are all involved in effective integrated basin development. Not only must schemes be technically sound, but they must also be sustainable financially and controlled efficiently.

Many cities are suffering from falling levels of service, because of insufficient funds to maintain the infrastructure properly and to expand to meet growth in demand. Quite naturally decision makers and politicians are reluctant to authorise increases in tariffs if the service is deteriorating, so once this starts, cash becomes even tighter.

The sector also has to compete for scarce capital funds. Inadequate money leads to short-term expedients, which are seldom part of sound basin development, and to deterioration of distribution systems, causing excessive losses and wastage of water resources.

A simple diagram such as Figure 1 can be understood by the many disciplines involved. Stages are separated into resources, water supply and waste water disposal, with a return to resources. Each interface offers opportunities for offering a service, at a price. Water supply can be to individual properties or to groups (supply of water in bulk). Waste water can be disposed of on-site, collected by piped sewers or reused. A part of the sewage disposed of by underground methods of on-site sanitation may return to the water resource, but treated sewage and reuse can be quantified through measurement.

Points of particular importance are:

(1)   Interfaces at which a **sale or charge** is appropriate, e.g. for raw water, and collection of sewage. Raw water is often taken without charge or for a nominal fee. In many cases a substantial investment is made in storage reservoirs, which is part of the "cost" of that raw water, e.g. Beijing. These costs may be shared with others in multipurpose schemes serving water supply, irrigation and hydroelectricity. Generally, the operating needs of each user differ leading to potential conflict and the need to compromise, as in Manila where irrigation and public water supply compete for water and those responsible for hydro electricity try to limit water releases in order to maintain generating head. Higher charges lead to the ability to pay for measures and practices which conserve water.

(2)   Opportunities to fix **tariffs** to reflect the "value" of goods. The market values of water services are most clearly illustrated in urban poor areas of large cities. Water is sold to tanker operators in Karachi at a price which enables them to on-sell to households. It must be low enough to discourage tanker owners from taking water illegally. Public toilets and washing facilities are constructed in Delhi by the authorities and charity organisations, which are then operated privately with a scale of maximum charges. In Dhaka service contracts are being let through competitive bidding for operation of public toilets. In Manila, water vending is the recognised method of delivering water to poor areas, at prices well in excess of the tariff for piped water.

The actual price paid for delivered water by poorer consumers shows that there is no real ceiling to "affordability" but rather that communities become accustomed to a certain level of expenditure relative to income and other costs and therefore oppose increases. This is particularly so where charges have been highly subsidised for years, as in Eastern Europe.

(3)   **Bulk purchase** by large users such as industry, blocks of flats, housing associations etc., who can, if they wish, arrange for their own internal charging, water conservation, reuse and waste water treatment and disposal. In Beijing, new

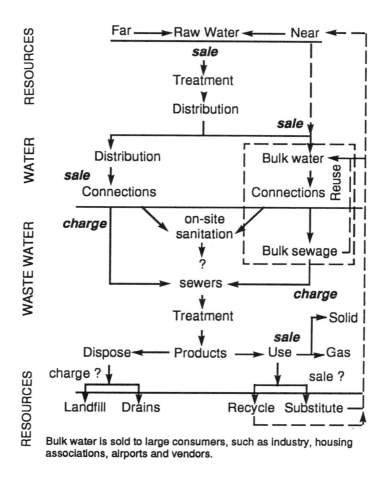

Figure 1 Outline of urban water use

large blocks of flats, 20 to 25 storeys high, purchase water in bulk and charge each occupier against their metered consumption. Karachi supplies water to large industrial groups and areas, which simplifies billing, but these powerful bulk consumers can negotiate from a position of political strength. A low price tends to reduce their willingness to reuse water and to pay for conservation.

   (4)   **Value of sewage** treatment products, and impact on the water resource if no treatment is given. Raw water substitution is being considered in Delhi and Bangkok, using treated sewage as irrigation and compensation flows respectively. Reuse and recycling is increasing in Beijing because of restrictions on pollution and the shortage of water. In Bangkok, new public housing developments are required to treat sewage before discharge to the already polluted canals, which are becoming unusable even for transport. In Delhi, sewage sludge is sold as fertiliser and in Bangkok is removed

from small sewage works by local gardeners. Similarly, gas is sold in Delhi to communities near sewage works. Golf courses in Karachi are irrigated with treated effluent.

The prices charged for products from sewage works are not sufficient to cover all direct costs of collection and treatment but the value of the benefits of better basin management should be added.

(5) **Wastage** of water resource if sewage is treated by on-site sanitation rather than collected, treated and "used". Sewers carrying sewage to treatment works in Kathmandu are intercepted by local farmers as a source of water and nutrients. One of the benefits of the interceptor sewers being installed in Bangkok is to quantify return flows. In poorer areas of Karachi shallow sewers have been installed which discharge to local rivers, which have no natural flow for part of the year. The sewage discharged to these water courses is "lost" as well as causing severe pollution. The extra cost of collection and treatment locally would provide water for environmental improvements.

Legal

In many water-short countries the importance of institutional and legal issues is clearly demonstrated including the need for:

- **policy** towards water issues, which establishes the approach to what can be a matter of survival, including the value given to the environment, water sharing particularly during dry periods, and charging and subsidies;
- **laws** to implement policies, including a system of "water rights", which need to be modified to meet changing conditions, but which tend to be numerous and related to specific sectors or organisations which are granted powers and responsibilities;
- **organisations** and mechanisms at central and, in some cases at local, government level to undertake all aspects of implementation of water projects. Political boundaries seldom match hydrological catchments and large river basins may include a number of bodies who control water allocation;
- **involvement levels** for water issues:
  — local, to identify needs of different users and to minimise interference;
  — basin or regional, to develop rational use of resources, including negotiation between owners of water rights;
  — national, to establish standards of water quality for abstraction and effluent discharge, preferably based on designated use of each water course with applicable standards e.g. recreation, and bathing beaches;
  — international, for cross border sharing and agreement on operations especially during very dry and wet periods.

Generally, laws and organisational responsibilities have grown up over many years and are not necessarily still appropriate or best able to deal with future conditions. Three systems of law can be identified:

- **customary** water laws, covering established users, riparian rights for those

adjacent to water, and municipal preference giving priority to human consumption;
- **traditional** water laws, often on specific water usage with the intention of controlling competing interests (Bangladesh, India, Indonesia, Pakistan and Thailand);
- **modern** water laws, providing for water allocation, distribution and regulation but also for management including planning at basin level, optimum utilisation through shared schemes, and integration of quantity and quality control in recognition of the return flows from any water use (China and Philippines).

The doctrines behind legal systems differ in policy and in their impact on the efficiency of water use in the following ways:
- **preference allocation,** e.g. domestic first, then agriculture, hydropower, industry, environment (Beijing);
- **prior appropriation**, "first in time, first in right";
- **riparian rights**, which do not cover volume or number of users, and so are not controllable;
- **economic planned allocation**, where the state decides on best uses (Pakistan);
- **free market allocation**, allowing transfer of water rights, which are traded as a commodity (California).

Public and private sectors

Most urban water supply and sanitation utilities are operated in the public sector. Notable exceptions are England and Wales where assets have been divested to the private sector, and concession contracts such as Buenos Aires where the private sector operates, maintains and expands the water supply and sanitation systems. As water supply and sewerage are natural monopolies, the private sector must be regulated to an appropriate degree by an independent organisation to ensure that consumers receive the service that they wish and are paying for.

In order that its forward planning, implementation and operation can be most effective the private sector demands better defined rights, under normal and extreme hydrological conditions, than might be accepted by a publicly owned utility. In addition, the regulator assesses if water availability and reliability are reasonable.

Transfer from the public to the private sector of any water using activity, such as urban water supply, urban sewage disposal, agriculture and electricity generation, and even recreation, is likely to cause close scrutiny of water allocation and control mechanisms.

FUTURE TRENDS

In order to obtain integrated basin development and management certain new aspects may need to be addressed or tightened.

(1)   Water rights policy and law should recognise the need to include management of the resource.

(2)   Basins should be considered as a whole, with each major user optimising his own use through abstraction rates and quantity and quality returned to the hydrological system.

(3)   Emergency planning should establish rights, priorities and actions to be taken during shortages, in order to allocate water on the most effective basis. This may mean sharing of scarce resources on a rational and controllable basis with disproportional restrictions to penalise least critical users.

(4)   The concept of economic pricing of water raises important issues, but relative values are subjective and may depend on availability of water compared with total demand.

3 Integrated and sustainable development of the water resources of the Niger basin

LEKAN OYEBANDE
University of Lagos, Lagos, Nigeria

INTRODUCTION

The concept of integrated river basin development involves the solution of development problems which takes into account in a coordinated manner the interests of all sectors of the economy, branches of water management and social groups. Future development perspectives and requirements of the economy are also accorded adequate consideration (Degren 1976).

Over the years, as integrated basin development became more or less universally accepted, there has been a progression from single-purpose to multi-purpose water projects. In recent years, the environmental criterion has become one of the important objectives of integrated basin development, as a major improvement over the narrow cost-benefit approach. Together with efficiency of demand management, environmental friendliness constitutes the centre-piece of sustainable development of water resources in a given river or lake basin.

The Niger is the largest river system in West Africa. It drains a total area of nearly 2 million km². Its main course runs from the head waters in the Fouta Djallon Highlands in Guinea for 4 200 km and drains nine countries: Guinea, Côte d'Ivoire, Mali, Niger, Burkina Faso, Benin, Nigeria, Chad and Cameroon. The river system has two deltas: the interior delta of lakes and swamps, some 20 000 km² in area, and the ecologically more diverse Atlantic delta.

Available socio-economic and environmental indicators paint a picture of underdevelopment, degradation and poverty of the majority of the inhabitants of the basin. Per capita income ranges from US $150 to $970 with a median value of $310, while national debt (as per cent of the national income per capita) exceeds 70% in five countries. Annual deforestation rates average 1.7%, but exceed 5% in one of the countries in the headwaters of the Niger river system (Côte d'Ivoire).

HYDROLOGY OF THE NIGER SYSTEM

The river traverses two humid catchment areas which are separated by a wide expanse of semi-arid environment (Figure 1). The average rainfall is over 2,200 mm in the headwaters of Niandan and Milo rivers in Guinea and in the Niger Delta in Nigeria. Consequently the river exhibits different regime types and anomalies at certain sections which reflect the climatic and physiographic characteristics of the component sub-basins.

In particular, two distinct floods occur annually in the river. The first is the "black flood" which originates from the high rainfall area in the headwaters (Figure 2). This flood arrives at the Kainji (Nigeria) in November and lasts until March at Jebba after attaining a peak rate of about 2000 m^3 s^{-1} in February (Oyebande *et al.*, 1980).

The second flood which becomes prominent only downstream of Sabongari soon after the river enters Nigeria, is the "white flood" usually heavily laden with silt and other suspended particles. The flood derives its flow from the local tributaries and reaches Kainji in August in the pre-Kainji Dam River Niger and attaining peak rates of 4000-6000 m^3 s^{-1} in September-October at Jebba.

In the broad alluvial plains of the interior Niger Delta (Figure 1) vast areas are flooded and the distributaries, lakes and swamps have become Mali's lifeline in the

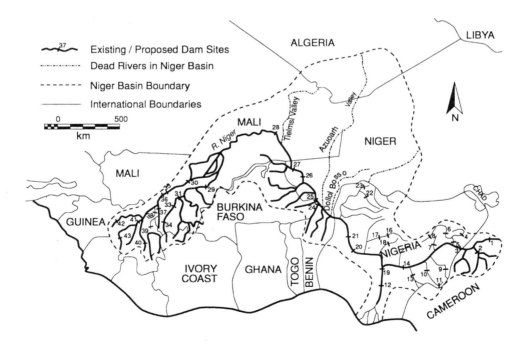

Figure 1 The Niger Basin: existing and proposed dam projects

Table 1 Key to Figure 1: list of dam sites

1.	Dao Kumi	P		25.	Park 'W'	P
2.	Lagdo	E		26.	Kandaji	P
3.	Dasin Hausa	P		27.	Labezanga	P
4.	Hawal	P		28.	Tossaye	P
5.	Kiri	E		29.	San	P
6.	Dadin Kowa	E		30.	Markala	E
7.	Shemankar	E		31.	Baoule IV	?
8.	Dindima	P		32.	Banifing	?
9.	Taraba	P		33.	Baoule III	?
10.	Manya	P		34.	Baoule II	?
11.	Gembu	P		35.	Kenie	?
12.	Katsina Ala (1)	P		36.	Sotuba	E
13.	Katsina Ala (2)	P		37.	Selingue	E
14.	Makurdi	P		38.	Bafandougou	P
15.	Gurara	P		39.	Mandiana	P
16.	Shiroro	E		40.	Kerouane	P
17.	Zungeru	P		41.	Baro	P
18.	Onitsha	P		42.	Dabola	P
19.	Lokoja	P		43.	Faranah	P
20.	Jebba	E		44.	Bagoe II	P
21.	Kainji	E		45.	Fomi	P
22.	Bakolori	E		46.	Bambari	P
23.	Goronyo	E				
24.	Mekrou	P				

P — Planned (34 Nos.)
E — Existing (12 Nos.)

Sahel zone (Barth 1990). During high water period, the inundated surface area of the delta reaches more than 20 000 km², and the huge storage in it attenuates the Niger's flow for 4-5 months, so that while peak water levels occur in September in Koulikoro, they are delayed till mid-November in Mopti and December in Diré. High evaporation rates of over 2000 mm and high infiltration and seepage rates cause a loss of more than 45% in the black flood in the delta. Much silt is also trapped in it. Indeed, the monthly mean flow attains a peak of 5280 m³ s⁻¹ at Koulikoro (upstream of the Delta), but drops to 2290 m³ s⁻¹ at Diré (delta zone) and only 1800 m³ s⁻¹ at Niamey (downstream of the delta).

VARIATION OF RIVER FLOWS

Over the basin, rainfall is concentrated in the headwaters and near the outlet of the Niger system. The seasonal distribution of the rainfall and its variability from year to year are as important as the long-term average. The seasonal pattern is reasonably similar over most of the area, with the maximum around August and a dry season centred on January-February. However, in southern Nigeria as well as in Garoua (Cameroon), July or September records the maximum mean fall.

The balance between rainfall and evaporation determines the river flow regime. The first rains after the long dry season serve to replenish the soil moisture deficit and it is the surplus that gives rise to runoff and soil erosion as well as groundwater recharge. The river flows are sensitive to periods of low rainfall as well as to areas of low rainfall (Sutcliffe *et al.*, 1990). The seasonal variations of river flows are illustrated in Table 2 with the Niger at Koulikoro and the hydrographs in Figure 2.

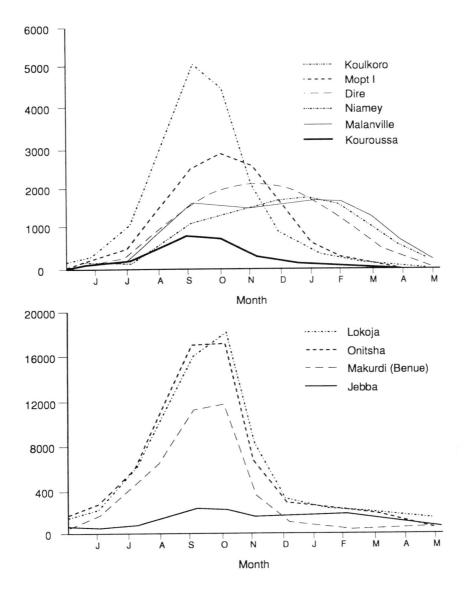

Figure 2 Monthly mean hydrographs for the Niger system

The range of the seven-year moving average flows for the Niger and other West African rivers such as the Senegal and the Chari is very wide. For the Niger it varies from about $25 \times 10^9 \, m^3$ to some $65 \times 10^9 \, m^3$ (a ratio of 1:2.5) during 1907 to 1985 as a consequence of the recent Sahelian drought (since 1968) which affects the sub-region more severely than the previous ones of 1901-1915 and 1940-1945 (Table 2). Table 2 further shows that the contribution of the Upper Niger during 1968-1985 was only 77% of the volume for the period of 1903 up to 1967. The year 1984 recorded the lowest river flows, contributing a mere 28% of the flow for the period 1907-1967, while the five driest years on record are 1982-1986.

Table 2 Upper Niger water resources & their variations: *from* Sutcliffe & Lazenby (1990)

(a)   Annual contributions $(10^9 \, m^3)$

Period	Niger at Koulikoro	Bani at Douna Douna
Record up to 1967	48.7	22.1
up to 1985	46.2	17.3
1968-1985	37.7	8.3
1984	20.1	2.2
1984 as % of record up to 1967	28	10

(b)   Monthly means flows $(m^3 \, s^{-1})$ of Niger at Koulikoro (1907-79)

J	F	M	A	M	J	J	A	S	O	N	D	Year
392	196	96	64	92	342	1198	3122	5763	4484	2079	856	1506

The pronounced seasonal distribution of river flows as well as the variation in annual flows make the use of storage reservoirs a necessity. Reasonable levels of control can be obtained from reservoirs on selected tributaries. However, as was rightly noted by Sutcliffe *et al.* (1990), the variability of annual flows is greater on the drier and less productive tributaries and presents a problem for a multipurpose dam project. If, for instance, a reservoir is sited on the tributary with the highest average runoff in order to ensure adequate inflows even in dry years, then during wet years when large floods occur, a smaller proportion of the total river flow will enter the reservoir than in average or dry years. This implies that significant flood control may not be possible when it is most needed, except perhaps if the uncertainties of climate variability are incorporated in terms of some over-design of the dam structure.

The persistence of the recent Sahelian drought has taken its toll on water availability and severely undermined the reliability of some of the West African rivers, particularly the Niger, as the major sources of surface water in West Africa's Sudano-Sahelian zone (SSZ). The total yield of the Upper and Middle Niger into the Kainji Dam reservoir in Nigeria, for instance, decreased steadily from 51×10^9 m^3 in 1969/70 through the all time low of 24.3×10^9 m^3 at the height of the drought in 1973 to a new level much below 35×10^9 m^3 in 1989/90. The downward trend has seriously undermined the envisaged potential of the hydro-electric project. Indeed, for several weeks in May and June of 1993 the computed inflow was virtually nil, so that the power station operators were drawing down the reservoir. The power station which still generates 33% of the country's electric power was almost shut down, an event which led to the government's approval of a workshop on *Climate change and power production* scheduled for the first quarter of 1994.

The seasonal fluctuations of unregulated flows are even more worrying. In particular, the low levels of the River Niger at Niamey have caused great concern in the Niger Republic over the last 15 years. During the drought in 1974, the river had a flow of less than 1 m^3 s^{-1} at Niamey. The capital's water supply was in jeopardy. Ten years later (1984) the low water mark fell again to 3 m^3 s^{-1} causing disturbance in Niamey's water supply and in the irrigation projects along the river. However, the worst was yet to come in 1985 when the river's flow fell practically to zero on 15th June and remained so for over a fortnight posing a severe threat to the welfare of the city's population (Beidou, 1987).

Fortunately the HydroNiger Project's forecasting system was able to predict the drastic reduction of 1985 which convinced the government of Niger of the necessity of building a temporary dam, thus averting the worst impact of the water shortage.

ENVIRONMENTAL HAZARDS AND PROBLEMS

Improper land use with associated deforestation and soil loss, particularly in the upland areas, reduces available water during the dry season because little water infiltrates during the rain storms. The result is an increase in the variability of flow with higher flood peaks and sediment loads during the rainy season. The Fouta Djallon Massif, known as the "water tower" of West Africa, is one such area. It is a hydrological complex of nine catchments or watersheds. The headwaters of the River Niger in Guinea account for 30 per cent of the total area of the complex. The land and water resources of the highlands have deteriorated rapidly owing to the intensifying problems of droughts and deforestation, accelerated soil erosion, sediment transport and desertification. The hazards have reached crisis proportions and are threatening the resource base of the complex; and the whole basin — particularly the Upper Niger watershed — requires protection.

As in most of the world's largest shared basins, current water use is generating demands that are fast approaching the easily available supplies. Unfortunately, the most pronounced deficits in the basin are occurring in regions very ripe for high-intensity conflicts, with high flow variability as discussed earlier. There is increasing pressure on the Niger as a source of water for irrigation, hydro-electric power generation, drinking and industrial water supplies as well as for fisheries and recreation. Some of the adverse environmental impacts of water development projects have been very serious and sometimes disastrous. Of particular importance is the reduction of areas of wetland and hence disruption of recession (floodplain) farming which engages millions of farmers in Mali and Nigeria. Unfortunately too, there are cases of faulty design and operation of storage dams which have coincided with the period of severe droughts (1980-1984) thus magnifying the drought intensity.

Climate variability and change, particularly the severe and persistent drought(s) has afflicted the basin's Sahel countries since 1967 with severe losses of life, livestock and crops and created environmental refugees. Following the 1984 drought, the Touregs of Burkina Faso having lost their livestock were forced to settle down to farming on the flood plains and practise recession farming. Their poor farming practices caused much degradation to the river bank zone while they remained refugees.

THE PAST AND CURRENT APPROACH TO WATER RESOURCES DEVELOPMENT

Water resources development in most of the Niger basin has proceeded for a long time in an uncoordinated manner and without a serious attempt to evolve an ecologically sound plan to orientate the management of the vital resources. It was not surprising that water shortage for all uses became so acute as to constitute an increasing constraint to the economic growth and development of the sub-region. A number of pressing needs have made the developing economies occupying the Niger basin aware of the need to develop their water resources on sound environmental principles. Among these are the expansion of agricultural, domestic and industrial activities as well as the ever-increasing population which grows annually at the rate of 2.5 to 3.3%. Natural disasters have further aggravated the poor living conditions of the inhabitants, as the Sahelian droughts of 1970-76 and 1984-86 inflicted severe losses of crops, animals and human lives and intensified the process of desertification.

Incidentally, the above needs, crises and ecological deficits may further intensify during the 1990s, especially with the daunting threat of climate change. It is increasingly appreciated that the unfavourable conditions can be contained if the river system management is undertaken as a planned scientific, legal and socio-economic activity for optimum coordination of the basin-wide resources to meet the needs of the society in the riparian states. This need for cooperation by riparian states has been long appreciated. In fact, river systems have tended to form the basis of general economic cooperation which transcends water resources development. Several treaties and agreements were concluded in West Africa between 1963 and 1972 and more are expected to be signed while existing ones are upgraded in the 1990s to meet the need for sustainable and conflict-free development.

As far back as 1969 an interdisciplinary mission was set up by UNDP to examine ways and means of supporting the economic development of the riparian countries of the Niger basin in the light of available natural resources in the area. One of the conclusions of the mission was that a basic pre-requisite was to achieve a full and detailed understanding of the water resources available in the basin and their behaviour as a function of hydro-climatological differences between the various portions of the basin. It was argued that such knowledge constitutes basic information required for planning the economic development of the riparian countries (UNDP/ WMO 1976). The mission also emphasized the need to develop existing meteorological and hydrological networks throughout the basin and stressed the need to establish a comprehensive hydrological forecasting system for the whole basin to achieve better control of the surface water.

In order to mitigate the effects of droughts and in view of the findings of the 1969 mission as well as to ensure harmonious use and management of the common water resources of the Niger basin, the nine riparian countries jointly evolved the Niger Basin Authority (NBA). The NBA was charged with the following activities:

● collection of regular and reliable hydrological data;
● control of water resources projects and hydraulic structures on the river system;
● real-time hydrological forecasting for development planning and operational purposes.

The request to establish a hydrological forecasting system, made to UNDP in 1974, was granted by UNESCO/WMO, and the Hydro-Niger Telemetric project came into being. The system uses data collection platforms (DCP) equipped with a timer and teleprinter. The air pressure bubble method is used to sense water level while the tipping bucket gauge is used to record rainfall data from 67 stations basin wide. The data are transmitted via the TIROS-NOAA satellites to the International Prediction Centre (IPC) at Niamey and also to the National Forecasting Centre (NFC) located in each of the member states.

The Hydro-Niger Project is currently afflicted by funding and management problems as the original donors have withdrawn their support while the parent institution, the N.B.A. has been dissolved. It is hoped that when the World Hydrological Observing System (WHYCOS) of the WMO and World Bank comes on stream it will help sustain and possibly upgrade much of the current activities and products of the Project.

To date, 12 large dams exist while 34 are at different stages of planning, design and construction (Figure 1, Table 1). Faulty design and/or operation coupled with lack of basin- wide coordination has caused flood hazard or water shortage to downstream communities and countries. For instance, the operation of the Selingue dam (Mali) abandoned one of the multi-objectives, that of releasing water to sustain the inland delta ecosystems and to meet the need of traditional rice cultivation. The result was a disaster for the fishing and rice industries and for over 1 million inhabitants of Mali in 1984. Also in Nigeria, the implementation of multi-purpose projects is far from perfect. Adequate flood control is missing in most large dams, and in some cases where hydropower plants are added, the power capacity is designed on the basis of the incidental use of downstream water release, without the necessary optimization study for hydropower generation.

Some dams are designed with larger active reservoir capacity than the average inflow. One such dam in the basin is at Goronyo on the Sokoto River in Nigeria whose inflow is 80% of the active capacity. The result is that often the wetland downstream is starved of water for its farming and fishing activities. This has led to violent protests by the deprived downstream users.

As noted earlier decreasing water inflow into Kainji Dam in Nigeria is attributed to drought events. However, water projects in Mali, Burkina Faso and Benin are also being held responsible for the depletion. The message for the 1990s is clear: in the absence of an effective basin-wide coordination, Nigeria will have less water for its water projects in the basin. On the other hand, uncontrolled water releases from Lagdo dam on the Upper Benue in the Cameroon have caused much flood damage both in Nigeria and Cameroon. A committee had to be set up to open bilateral talks to find ways of controlling the hazard while the EEC is also helping Cameroon to integrate its water resources development through adoption of a multi-objective approach.

STRATEGIES FOR INTEGRATED AND SUSTAINABLE DEVELOPMENT

A key aspect of any strategy is a diagnostic study to review the state of knowledge in the basin, including identification of institutional and technical capabilities, a review of water quality and quantity problems, and an inventory of existing and proposed water development projects. The study should include the review of available information on the delta-ecosystems and should identify immediate and future programmes. It should also identify suitable hydrometric and water quality sites and associated laboratories and data processing centres. Other aspects are a survey of land use patterns, the identification and inventory of major sources of pollution, and investigation of long-term supportive capacity of different ecosystems and of investment opportunities.

Another important aspect concerns the preparation of comprehensive basin-wide operational action plans for suitable development and management of the basin's water and other resources, using an appropriate combination of the increasing array of water resources management strategies and tactics: multi-objective planning, risk analysis, decision-support systems and simulation models. The latter can be used for defining a plan for optimization, distribution and joint control of available water resources among the riparian countries with priority for agricultural, hydro-electric power generation potentials and for navigational improvements.

Other aspects include rehabilitation, upgrading and operation of hydrometeoro-logical and water monitoring networks and laboratories, including a support hydrological information system, forecasting, formulation of sustainable land-use policies, as well as innovative erosion and pollution control measures. However, the strategies listed are feasible only if the riparian countries are determined to make necessary sacrifice and with substantial technical and financial support from the external support agencies (ESAs).

CONCLUSION

The very seasonal and variable nature of flow of the Niger river system make storage projects vital instruments to conserve water for various uses. Such projects are of particular importance during periods of prolonged drought and with the threat of climate change. The riparian countries however need to be guided by some of the many international declarations which emphasise principles such as
- protection of the environment and safeguarding health through integrated management of water resources;
- institutional reforms which promote an integrated approach with effective linkages at all levels, including the education and participation of women;
- sound and transparent financial management practices, improved management of assets, and greater application of appropriate technology; and
- adoption of community management services backed by measures to build the capacities of local institutions to implement and sustain water and sanitation programmes (New Delhi Statement, 1990).

Adequate machinery should be established to prevent and resolve international water disputes. Some means of preventing disputes have been discussed elsewhere (Oyebande, 1992). Treaties are needed for resolution of disputes. In the final analysis, reconciliation of the issue of the waters of the Niger requires a comprehensive water-sharing agreement among some or all of the riparian countries.

REFERENCES

Barth, H.K. 1990. Environmental and agricultural implications of dam construction in the Niger valley of Mali. In: Stout & Demissie (eds), *The State of the Art of Hydrology and Hydrogeology in the Arid and Semi-Arid Areas of Africa*, UNESCO & International Water Resources Association (IWRA), Urbana, 858-868.

Beidou, B. 1987. The role of operational and applied hydrology on socio-economic development of the population of the Niger. In: *Water Lectures*, World Meteorological Organization, Geneva, 63-71.

Degren, I. 1976. Integrated development of river basins — overview and perspectives. In: *River Basin Development — Planning and Policies*, Budapest, 3-20.

New Delhi Statement, 1990. *The New Delhi Statement*, New Delhi, India (Sept.14), 1-6.

Oyebande, L. Sagua, V.O & Ekpenyong, J.L. 1980. Effects of the Kainji dam on the hydrological regime, water balance and water quality of the River Niger. *IAHS Publ. No. 130*, 215-220.

Oyebande, L. & Balogun I. 1992. The Niger river system: need for environmentally sound and integrated management. In: *Nile 2000 — Protection and Development of the Nile and Other Major Rivers*, Cairo, 7-12-1-7-12-16.

Sutcliffe,J.V. & Lazenby,J.B.C. 1990. Hydrology and river control on the Niger and Senegal. In : Stout & Demissie(eds.)*The State of the Arts of Hydrology and Hydrogeology in the Arid and Semi-Arid Areas of Africa*. UNESCO/IWRA, Urbana, 846-856.

UNDP/WMO. 1976. Provisional Report on Project 'Hydrological Forecasting System for the Middle and Lower Basins of the Niger River'. United Nations Development Programme / World Meteorological Organization, Geneva.

4 The role of artificial flooding in the integrated development of river basins in Africa

MIKE ACREMAN

IUCN - The World Conservation, Gland, Switzerland

(On secondment from the Institute of Hydrology, Wallingford, UK)

INTRODUCTION

Floods are truly the 'Jekyll and Hyde' of environmental processes. Recent floods in Bangladesh and the USA have claimed thousands of lives and destroyed millions of dollars worth of property and infrastructure. Consequently, the reaction of many people throughout the world, including those responsible for the integrated development of river basins, is to prevent inundation and engineering works such as embankments and walls, and is a common response to this perceived evil threat. In contrast to this negative view, flood waters are the life blood of many environments and rural communities. As a result of periodic inundation, the floodplains of the major rivers of Africa including the Zaire, Senegal, Niger, Nile and Zambezi (Figure 1) support wetland ecosystems of exceptional productivity, particularly in comparison with the surrounding arid and semi-arid rangelands where the dry season is very long. For centuries, these floodplains have played a central role in the rural economy of the region providing fertile agricultural land which supports a large human population.

The flood waters provide a breeding ground for large numbers of fish and bring essential moisture and nutrients to the soil. Water which soaks through the floodplain recharges the underground reservoirs which supply water to wells beyond the floodplain. As the flood waters recede arable crops are grown, but some soil moisture persists into the dry season and provides grazing for migrant herds. The floodplains also yield valuable supplies of fish, timber, medicines and other products and provide essential habitats for wildlife, especially migratory birds (Dugan, 1990).

In recent years drought, increasing populations of people and livestock and rising poverty have combined to put increasing pressure on the floodplains and have led to over-exploitation of their resources. In the face of such pressure, the key to development has often been seen as the implementation of major river engineering schemes, such as dams, for hydro-electric power generation and for intensive cereal

Figure 1 Major rivers and wetlands of sub-Saharan Africa

cultivation. Similar schemes have been very successful in other parts of the world. Dams can provide an all year supply of water in a highly seasonal climate and intensive rice plantations in south east Asia give sustainably high yields. However, few schemes in Africa have ever realised their full potential and many are facing serious technical, economic, administrative and social problems. Salinisation is particularly prevalent as many schemes were built without adequate drainage. A further problem is that there is insufficient manpower to maintain the irrigation plots. Added to this the reduction in floodplain inundation caused by the retention of flood water behind the dams, has had disastrous effects on the traditional rural economy. Thus, in some cases, these schemes have diminished rather than improved the living standards and economy of the region as a whole.

There is clearly a need for the integrated development of river basins in Africa to combine the best of both intensive and extensive floodplain farming systems, traditional and modern techniques. Given that such dams already exist, the best option is to develop an operational management plan for the dams to release water at certain times of the year to produce an artificial flood, to inundate the floodplain and thus rehabilitate the indigenous farming system, whilst retaining sufficient reserves for power generation and irrigation. Using examples from across Africa, this paper presents a review of the multiple benefits of flooding, economic comparisons of extensive floodplain agriculture and intensive irrigation, why floodplain users must be involved in the decision making process and how artificial floods can play a central role in the integrated development of African river basins.

THE BENEFITS OF FLOOD WATERS TO THE RURAL ECONOMY AND ENVIRONMENT

The inundation of the floodplain provides a natural irrigation system. On many African floodplains, such as the Inner Niger delta in Mali, floating rice (*Oryza glaberrima*) is seeded just prior to annual flood. As the flood waters recede the rice is harvested and another crop is grown on the exposed moist soil, such as sorghum or cowpeas, which are often interplanted with vegetables. On the drier margins drought resistant crops, such as millet, are grown. As some varieties of sorghum are flood-tolerant, a succession of crops can be grown right up until the following year's flood. More often, as the dry season progresses, migrant herds of cattle move in to graze on the crop stubble or on grass where soil moisture is still sufficient. The herds are welcomed by the farmers as manure helps fertilise the soil and animal products, such as milk and yoghurt can be traded for cereals and vegetables. Horowitz (1991) has calculated that a full commitment to the modern alternative of intensive irrigation is economically irrational because the traditional activities of recession farming, herding and fishing have higher returns per unit labour and per unit capital investment. One of the potential benefits of dams is that they can release water during dry periods which can be used for irrigation. Indeed in the Hadejia-Jama'are floodplain constant flows from reservoirs combined with the use of small pumps has allowed land to be cultivated during the dry season which would normally be available for cattle grazing. However, this led to pitched battles in the late 1980s between farmers and herders and several people were killed.

Annual flooding also stimulates the breeding of riverine fish as the flood waters release a rich source of food and the floodplain vegetation provides protection for spawning fish, eggs and fry. Many species of edible fish breed exclusively on inundated floodplains and it has been estimated that over 100 000 tons per annum are caught from the inner delta of the Niger alone. Sophisticated techniques have been developed to catch fish as they provide an important source of protein. On some floodplains, small trenches are dug, occasionally over one kilometre long, to connect the river to depressions. As the water recedes the fish move along these canals and are caught in traps. The construction of the Bakolori and Kainji dams in north-western Nigeria led to a 50-70% decrease in income from floodplain fisheries as flood flows were reduced.

Flood waters which reach the sea are not wasted but support extensive inshore fisheries. For example, the shrimp fishery on the Sofala bank at the mouth of the Zambezi provides Mozambique with an important source of foreign income worth some US$ 50-60 million per year. Gammelsrod (1992) has shown that shrimp abundance is directly related to wet season freshwater runoff and earnings could be increased by US$ 10 million per year by correctly releasing flood waters from the Cabora Bassa dam which are not currently utilised.

Recharging of groundwater has long been recognised as an important function of wetlands and Hollis et al. (1993) concluded that recharge in the Hadejia and Jama'are river basins of northern Nigeria occurs primarily during flood flows, since the floodplain provides a large surface area and the river bed is often impermeable. Recently constructed dams have reduced the area inundated, but insufficient monitoring of groundwater levels is being undertaken to assess the impact, at a time when the World Bank is promoting pumping from shallow aquifers beneath the floodplain, which may not be a sustainable form of river basin development.

The Okavango delta is one of the world's outstanding wildlife areas with diverse plant communities and numerous species of macro- and micro-invertebrates, herbivores and birds which owe their existence to annual flooding. The delta is home for over 15 species of antelope including the shy sitatunga and large herds of lechwe (Dugan, 1993). Likewise the nearby floodplains of the Zambezi river basin including, for example the Kafue and Luena flats, support an outstanding diversity of wetland organisms, including over 4500 species of higher plants, particularly ferns, grasses and orchids and more than 400 species of birds. The aquatic environment is equally diverse with 120 species of fish (Howard, 1993). Floodplains in Sahelian Africa are just as important for wildlife. Annual inundation of the Hadejia-Nguru wetlands has made them an internationally important site for birds, with over 265 species either resident or visiting the area.

Floodplains also provide a range of other products including fuelwood, reeds for thatching and mat-making, medicines and fruits which are a key component of the income of local villages. They have an important cultural significance and the annual flood has been an essential element in the lives of many African people for centuries.

UN-INTEGRATED DEVELOPMENT OF A RIVER BASIN - THE CASE OF THE SENEGAL RIVER

As part of the development of the Senegal River basin, the *Organisation pour la Mise en Valeur du fleuve Senegal* (OMVS) was set up, with representation from the three main riparian states of Mali, Mauritania and Senegal. Two dams were constructed, one near the river mouth (Diama) and a second in the headwaters

(Manantali). Diama dam inhibits salt-water intrusion into the river to allow its use for irrigation and for regulating water levels to facilitate transport. Manantali dam was built to generate hydro-electric power and to regulate flows in the river. The prospect of irrigation made the floodplain an attractive investment and on the Mauritanian side, the Moorish elite took over land from black communities in the valley at a time when drought and minimal releases increased pressure on the remaining floodplain. This led to conflict between farmers and herders which fuelled long standing inter-ethnic friction. In the ensuing violence, 200 Wolof and Peul were killed in Nouakchott and in retaliation 35 Moors were killed in Dakar. The Senegalese government placed more than 15 000 Mauritanians under military 'protection' and eventually some 180 000 people were moved across the international border.

Prior to construction which began in 1986, natural inundation of the floodplain in the middle Senegal valley supported up to 250 000 hectares of flood recession agriculture, forests, which provide fuelwood and construction timber, and wildlife habitat. Because of the recognised delay between dam construction and development of the irrigation plots and the installation of turbines, OMVS agreed that man-made floods should be released for a transitional period of 10 years (Scudder, 1992). However, the releases made have been small and have inundated only around 50 000 hectares, even though the turbines have not yet been installed. OMVS claim that releasing a flood would result in 25% less electricity being generated and is therefore not possible. In a cost-benefit analysis, Horowitz & Salem-Murdock (1990) calculated that the best economic option is to use the Manantali dam both to release an artificial flood and to generate some hydro-power. Any decision is politically sensitive since electricity benefits the urban elite, commerce and industry (there being little rural electrification) whilst floods benefit rural poor.

THE VALUE OF A FLOOD-BASED ECONOMY — THE CASE OF THE HADAJIA-NGURU WETLANDS, NIGERIA

In northeast Nigeria, where the Hadejia and Jama'are rivers combine within the Komodugu-Yobe basin, an extensive floodplain of around 2000 km² used to be inundated annually. Since 1971, a series of dams has been constructed on the main tributaries and during recent droughts the area inundated has reduced, with only 300 km² flooded in 1984 (Hollis et al, 1993). The dams are used primarily to provide water for cereal irrigation and their construction was initiated partly by a ban on imported wheat in Nigeria which made irrigation profitable overnight. In 20 years the Nigerian Government spent US$ 3 billion in irrigation development, though by 1991 only 70 000 hectares had been farmed making an investment of US$ 43 000 per hectare (Adams, 1992).

It is clear that the yields from intensive irrigation schemes are higher per hectare than from floodplain agriculture, although the high operational costs of the schemes reduce substantially the relative benefits. However, because the economy is limited by water resources, it is more appropriate to express the benefits of various development options in terms of water use. Barbier *et all* (1991) undertook an economic analysis of the Kano River project, a major irrigation scheme in the headwaters of the Hadejia river. They showed that the net economic benefits of the floodplain were at least US$ 32 per 1000 m³ of water (at 1989 exchange rates), whereas the returns from the crops grown on the Kano river project were only US$ 0.15 per 1000 m³ and when the operational costs are included, this drops to only US$ 0.0026 per 1000 m³! (see Table 1). Furthermore, this analysis did not include the other benefits of flooding, such as groundwater recharge or flows downstream to Lake Chad.

At a meeting in Kuru in 1993, representatives from the responsible authorities including state water boards, River Basin Development Authorities, members of research institutes and government departments met to discuss the water resources of the Komodugu-Yobe basin. They agreed unanimously that artificial flooding should have a central role in the integrated development of the river basin. One of the main recommendations was that *"flooding in the wetlands made possible by artificial releases from dams in the wet season should be maintained to make possible the production of rice, dry season agriculture, fuelwood, timber, fish, wildlife, as well as biodiversity and groundwater recharge"* (HNWCP/NIPSS, 1993). In addition it was agreed that *"existing facilities at Tiga dam that could be used for artificial flood releases should be tested, the adequacy of outlets for releases from Challowa Gorge* (dam) *should be assessed and detailed reassessment of Kafin Zaki dam* (being planned) *should incorporate the capacity for adequate flood releases."*

COMMUNITY BASED HYDROLOGICAL MANAGEMENT - THE CASE OF THE PONGOLO RIVER

In the late 1960s the Pongolapoort dam was constructed on the Pongolo River in north-east South Africa near its borders with Swaziland and Mozambique. The

Table 1 Comparison of productivity (in $US) of the Hadejia-Nguru wetlands compared with an alternative use of water in the Kano River Project

	Production per hectare	Production per 1000 m³ water
Kano River Project		
crops	115	0.15
project (including operating costs	2.5	0.0026
Hadejia-Nguru wetlands		
agriculture, fishing, fuelwood	58	50

reservoir was filled in 1970 with a view to irrigating 40 000 hectares of agricultural land for white settlers, with no provision for hydropower generation. No assessments were undertaken of impacts of the impoundment, neither on the floodplain where 70 000 Tembe-Thonga people were dependent on recession agriculture, fishing and other wetland resources, nor on the biodiversity of the Ndumu game reserve. In the event no settlers came to use the irrigation scheme. The dam changed the whole flooding regime of the river which led to crop failure on a massive scale and represents a classic case of uni-sectoral rather than integrated river basin management.

In 1978 a workshop was held on the Pongolo floodplain to review the future of irrigation and how to minimise the negative impacts on the floodplain. This led to a plan for controlled releases to rehabilitate the indigenous agricultural system and the wildlife. However, initial releases of water from the dam were made at the wrong time of the year and crops were either washed away or rotted (Poultney, 1992). In 1987 the Department of Water Affairs and the tribal authorities agreed to experiment with community participation. As a result, water committees were established, representing five user groups: fishermen, livestock keepers, women and two kinds of health workers (new primary health care workers and traditional herbalists and diviners). They were given the mandate to decide when flood waters should be released. These committees were very successful at implementing people's views and have led to management of the river basin to the benefit of the floodplain users. Indeed they have been so successful that attempts to disband them have been made by the KwaZulu government as they were seen as a threat to power. This is a unique example of where floodplain users are participating directly in the decision making process and influencing development and management of the river basin.

RESTORATION OF A DEGRADED FLOODPLAIN - THE EXAMPLE OF THE LOGONE RIVER, CAMEROUN

The River Logone drains from the Adamoua plateau in central Cameroun. Before it joins the River Chari, flood water from the river inundates annually a large floodplain, originally around 6000 km². This wetland has a high biodiversity with large herds of giraffe, elephant, lions and various ungulates (including topi, antelope, reedbuck, gazelle, kob). Part of the floodplain has been designated as the Waza National Park which attracts around 6000 tourists per year.

In the flood season, the entire floodplain becomes a vast fish nursery. Up to the 1960s, fishing was the primary economic activity amongst the local Kotoko people who could earn US$ 2000 in four months. The Fulani name for the floodplain is yaérés, which means "dry season pasture", and annually some 300 000 cattle and 10 000 sheep and goats grazed on the floodplains (Schrader, 1986). Pastures became accessible when surrounding savanna grasses withered and their protein content was depleted. The carrying capacity has been estimated at 1-2 cattle per hectare, compared with 0.2 for surrounding savannah (Broer & Tejiogho, 1988). Floating rice is the main arable crop, since it has low labour demands and fits well into the fishing calendar. Yields were not high, but enough land was available to ensure self sufficiency in rice.

Since 1979, the area inundated has reduced, partly by climatic factors, but primarily due to construction of a barrage across the floodplain which created Lake Maga to supply water to the Semry II irrigation project. Flooding is now insufficient in large areas to grow any floating rice and fish yields have fallen by 90%. The Semry rice schemes which cover around 5000 hectares are not making full use of water stored in Lake Maga and the potential to release water to rehabilitate the floodplain, whilst retaining enough to maintain rice production, has been identified. Drijver and van Wetten (1992) describe improvements which have resulted from minor reflooding in the north of the floodplain. Here, the nutritional quality of grasses has improved, the fisheries have been revitalised and herders and migratory birds have been attracted back.

A model of the floodplain has been developed to assess the effect on floodplain inundation of various options for releasing water from Lake Maga and through the embankments. In addition, a socio-economic model has been used to determine the benefits of each option in terms of improved fisheries, agriculture and herding (Wesselink & Drijver, 1993). The net present value over 30 years for one option has been estimated as up to 1300 million CFA (c. £3 million) assuming a discount rate of 8%, though much depends on the available flood discharges and the response of local people and their environment (in terms of vegetation, fish etc.) to renewed flooding. A potential negative effect is a large influx of people who would over-exploit and thus degrade the rehabilitated resources of the floodplain. Several options have additional, indirect benefits, such as the provision of a spillway which would lower the risk of failure of Maga dam that currently has no such safety feature and was almost overtopped in 1988.

IMPLICATIONS FOR INTEGRATED DEVELOPMENT OF RIVER BASINS IN AFRICA

Although there is clearly a need to protect some residential and industrial property from flooding, a dynamic flooding pattern is essential for the development and sustainability of African floodplain systems for the short term economic importance of fisheries, agriculture and pastoralism and the longer term importance of soil fertility and biodiversity. River flows have other benefits, such as providing water for domestic and commercial use, as well as irrigation and for generating hydro-electric power. The problem is that river basin development has been undertaken in a uni-sectoral manner with insufficient collaboration between the various water using communities including energy, water supply, agriculture, fisheries (both coastal and inland), forestry and environment. All too often the natural resources and functions of the floodplain have been the main casualties, even though their economic value can be shown to be greater than the modern development schemes which have replaced them.

Annual floods have sustained rural economies for many centuries and Scudder (1980) was the first to suggest controlled flooding as a development strategy so that the loss of floods does not have a zero opportunity cost. The aim of reflooding is not to restore former conditions, this would not be possible - even if desirable - due to infrastructure changes and socio-economic adjustment to the new regime, but to have fully integrated river basin development which includes coastal fisheries and combines the best of traditional practices and modern techniques. In many cases infrastructure, such as small embankments can be of great benefit, particularly if integrated into local water management practices which are governed by customs and rules. As demonstrated by the Pongolo river floodplain, releasing artificial floods is not straight-forward and requires hydrological expertise, a detailed understanding of the land use and ecology of the flood-plain, and river and coastal fisheries, and close collaboration with floodplain users. Nevertheless, if correctly managed, artificial floods can play a central role in the integrated development of river basins.

ACKNOWLEDGEMENTS

Information for the paper was provided by numerous colleagues in Africa and comments on a draft of the text were provided by Professor Ted Scudder of the California Institute of Technology and Patrick Dugan, Regional Director of IUCN and former Wetlands Programme Coordinator.

REFERENCES

Adams, W.M. 1992. *Wasting the rain: rivers, people and planning in Africa*. Earthscan, London.

Barbier, E.B. Adams, W.M. & Kimmage, K. 1991. *Economic Valuation of Wetland Benefits: the Hadejia-Jama'are Floodplain, Nigeria*. London Environmental Economics Centre Paper DP 91-02. International Institute for Environment and Development, London.

Broer, W & Teijiogho. 1988. *La capacité de charge de yaérés au Nord du Cameroun*. Centre for Environmental Science, University of Leiden, Netherlands.

Drijver, C.A. and van Wetten, J.C.J. (Ed.) 1992. *Sahel wetlands 2020*. Centre for Environmental Science, Leiden, The Netherlands.

Dugan, P.J. (Ed.) 1990. *Wetland Conservation: A Review of Current Issues and Required Action*. IUCN, Gland, Switzerland.

Dugan, P.J. (Ed.) 1993. *Wetlands in danger*. Mitchell Beazley, London.

Gammelsrod, T. 1992 Improving shrimp production by Zambia river regulation. *Ambio*, **21**, 145-147.

Hadejia-Nguru Wetlands Conservation Project and the National Institute for Policy and Strategic Studies. 1993. Proc. workshop on the management of the water resources of the Komadugu-Yobe basin. National Institute Press, Kuru, Nigeria.

Hollis, G.E., Adams, W.M. & Aminu-Kano, M. (Eds) 1993. *The Hadejia-Nguru Wetlands: environment economy and sustainable development of a Sahelian floodplain wetland*, IUCN Gland Switzerland and Cambridge, UK.

Horowitz, M. 1991. Victims upstream and down. *J. Refugee Studies*. **4**, 164-181.

Horowitz, M. & Salam-Murdock 1990. *Senegal River Basin Monitoring Activity Synthesis*. Institute for Development Anthropology, Binghamton, New York.

Howard, G.W. 1993. *Ecology and biodiversity in the conservation of wetlands in Zambia*. IUCN, Nairobi, Kenya.

Poultney, C. 1992. Water committees take action. *ILEIA Newsletter*, 1, 92, 18-20

Schrader, T. H. 1986. Les yaérés au Nord du Cameroun: pâturages de saison sèches; aspects socio-économique du développement pastoral dans la plaine d'inondation du Logone. Ecole de Faune/ IRZ Garoua / Centre for Environmental Science, University of Leiden, The Netherlands.

Scudder, T. 1980. River basin development and local initiative in African Savanna environment. In: Harris, D.R. (Ed.) *Human ecology in savanna environments*. Academic Press, London.

Scudder, T. 1992. The need and justification for maintaining trans-boundary flood regimes: the African case. *Natural Resources Journal, School of Law, University of New Mexico*.

Wesselink, J.W. & Drijver, C.A. 1993. *Waza Logone flood restoration study. Identification of options for re-flooding*. Report to IUCN, Delft Hydraulics/University of Leiden, The Netherlands.

5 Multi-cell groundwater balance model to support sustainable development of aquifers in narrow alpine valleys

J. FÜRST
Universität für Bodenkultur, Vienna, Austria

INTRODUCTION

The term "Sustainable Development" has become a buzzword since it became obvious that — even on a global scale — the current degradation of the environment and resources will soon lead to widespread problems. On a rather conceptual and global level, the goals of sustainable development policies are specified as:

- environmental integrity;
- economic efficiency; and
- equity, including present and future generations (Young, 1992).

A large part of the literature on sustainable development focuses on government policies that influence the economies to maintain environmental integrity (e.g., Schmidheiny, 1993). Ecological theory contributes by focusing on ecosystem functioning.

For water resources development on the basin scale, the main challenge for "Sustainability" is implied by the necessity to extend the time horizon. Multiple criterion decision making (MCDM), risk analysis and conflict resolution techniques are appropriate tools for this purpose (Nachtnebel, 1988). However, the specific elements for identifying sustainable water resources projects have not yet been introduced into engineering practice.

It is beyond the scope of this paper to give a thorough discussion of suitable criteria and the methodology for their measurement. Rather, a modelling concept is developed for aquifers in narrow alpine valleys, that provides the basic information required to derive a variety of possible criteria. Even without a formal application of advanced decision techniques this should help regional authorities to follow the principles of sustainable development.

Although usually formulated for a global scale, the goals for sustainable development are equally applicable to small scale (basin scale) water resources management:

- Changing the groundwater system can cause severe environmental problems, e.g., a drawdown of the groundwater table can endanger the existence of riverine forests.

- Availability of well protected water of good quality, without the need for purification or long distance transport, is of course an economically valuable resource.
- Equity is an important aspect as well: first, groundwater extraction is usually based on long term water rights, i.e. it will anticipate the rights of two or three generations; second, transport of water from mainly agricultural areas with low per-capita income, to urban areas with higher incomes, might cause social irritation.

A useful tool to support groundwater management under criteria of sustainability must therefore model the groundwater system with reasonable spatial resolution, should clearly distinguish between components of the groundwater balance that follow different hydrological processes, enable long term predictions, and account for the uncertainties due to model parameters and the stochastic nature of inputs.

This paper presents a multi-cell balance model that calculates the groundwater balance for several subsections (cells) of the aquifer. The groundwater system is modelled by a cascade of reservoirs and yields time series of the components of the groundwater balance: storage, change of storage, rate of exchange between river and groundwater, groundwater recharge by infiltration, discharges, inflow through boundaries, flux between cells.

This ability to indicate components of different origin, quality and temporal behaviour directly supports the objective of sustainable groundwater development. Groundwater extraction rates as well as the location of wells can easily be assessed to match those components that are of the desired quality and also in the long term yield the required quantities without negative impacts on the groundwater regime.

The simplicity and moderate data requirements of the model allow the performance of extensive Monte Carlo simulations of various management scenarios for long periods, taking into account the stochastic nature of the uncontrollable system inputs of precipitation and fluctuations in water level. A case study of an Austrian aquifer will demonstrate the use of the model.

CHARACTERISTICS OF THE CONSIDERED AQUIFERS

Aquifers in narrow alpine valleys (Figure 1) offer considerable groundwater resources and are increasingly utilized for regional water supply. However there is a deficit in the available methodology for modelling and managing these aquifers. Well developed techniques based on finite difference (FD) or finite element (FE) solutions of the differential equation for groundwater flow, which are often applied to large alluvial groundwater systems (Bachmat *et al.*, 1985; Istok, 1989), are not appropriate for alpine valleys for several reasons:

- The ratio of area to perimeter is small. The influences of boundaries predominate; a model that only solves the groundwater flow equation as a boundary problem would not be properly applied. The processes of lateral inflow, from the tributaries and their catchments need to be included.

NATURE

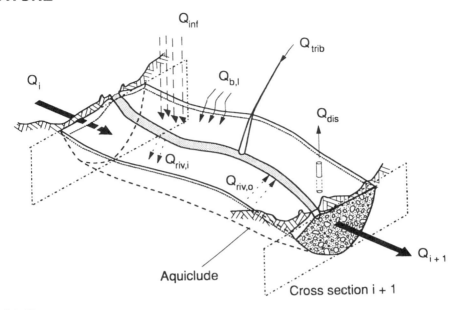

MODEL

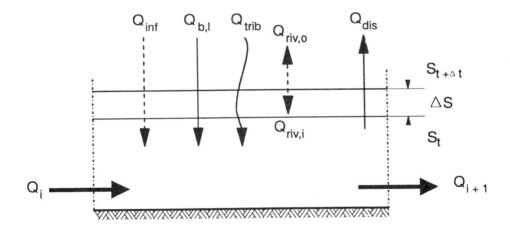

Figure 1 Schematic representation of a groundwater system and a "cell " of the model

- The aquifers are often rather deep as compared to the width. Two-dimensional groundwater models could violate the Dupuit-Forchheimer assumptions.
- The spatial distribution of the hydrogeological parameters tends to be very inhomogeneous, and a detailed exploration of these parameters is not affordable.
- The magnitude and importance of the aquifers do not economically allow

sufficient exploration of the hydrogeologic parameters and groundwater observation networks needed for running and calibrating finite difference or finite element models.
- The problem under consideration often does not require the detailed results provided by FD or FE models.

MULTI-CELL GROUNDWATER BALANCE MODEL

For the multi-cell groundwater balance model the aquifer is divided into several (5 - 10) subsections or "cells" by cross sections that are placed according to the required spatial resolution of the model and with respect to available data. The lower part of Figure 1 schematically shows such a cell and its representation as a reservoir. Inflows and outflows within a cell are only distinguished by their type and origin.

The inputs to the system are the components of the groundwater balance for each cell. Fluxes across the balancing cross sections as well as the infiltration from and exfiltration to the river are a function of the stored groundwater volume in the adjacent cells. Other components of the groundwater balance are lateral subsurface inflow, infiltration from tributaries, groundwater recharge from precipitation and artificial discharges. These inputs have controllable and uncontrollable components.

The uncontrollable components of the input are determined by precipitation, meteorological data and stream flow. Precipitation and evaporation directly affect groundwater recharge and lateral subsurface inflow. Stream flow is an input for the interaction of river and groundwater. The fluxes through the balance cross sections and the infiltration from tributaries are also considered uncontrollable.

Controllable components of the inputs are described by the decision variables: discharge rate of wells, decoupling of groundwater and river, impermeable areas, cutoff of lateral inflow. The state of the system at time t is given by the stored groundwater volume in each cell. The state transition function is described by a simple balance equation, which gives the storage at time $t + t$ as the sum of the storage at time t and the change of storage S. The change of storage in a cell within a time step t is expressed as a function of the storage S_t at the beginning of the time step and the inputs I_t.

The output of the system is the available groundwater volume in each cell. For convenience, the contributions of the individual components of the groundwater balance are given separately. The output function expresses the objectives of management. Various management objectives can arise for the type of aquifers considered. Therefore no output function is explicitly incorporated in the model. Instead the model is designed for flexible scenario analysis that transparently supplies a variety of information that can be used as criteria for decision analysis.

Submodels for the components of the groundwater balance
Each component of the groundwater balance is described by a separate submodel. These submodels were selected to be as simple as possible, to depend on physically

meaningful parameters, and to provide an easy option to represent management alternatives. The simple models described below were applicable in two Austrian case studies. Modifications might be necessary where the hydrogeological situation is different.

Flux through cross section is described by Darcy's law, where a representative hydraulic conductivity in the cross section and an average hydraulic gradient between the adjacent cells is assumed. This component is not subject to management alternatives.

Groundwater and river exchange water in both directions. The flux depends on the gradient of the water table and the hydraulic transmissivity of aquifer and river bed. Within a reach of the river between two cross sections there may be parts with infiltration as well as exfiltration. The exfiltration of groundwater to the river is described by the simple analytical solution of the model "flow to a partially penetrating ditch" (Bear, 1979). The representative groundwater table in cell i is derived from the storage $S_{i,t}$. The transmissivity of the aquifer is considered to be valid for the computation of the exfiltration.

The infiltration from the river into the groundwater is reduced by partial clogging of the river bed by fine sediments, which is described by a clogging factor to correct the river bed permeability.

The exchange between river and groundwater is severely inhibited by cutoff walls for runoff hydropower stations. In a management scenario, the effects of hydropower stations are simply represented by the ratio of impounded and free flowing length of the river between two balancing cross sections.

Infiltration from tributaries takes place where the bed of the tributary is above the aquifer. The infiltration per unit length is assumed to be proportional to the depth of water. For a few reference water levels, the infiltration was derived from precise discharge measurements in the tributaries.

Discharge from wells is at first sight simply a decision variable without need for further modelling. However, if a well is located close to the river, the cone of drawdown locally induces increased infiltration of river water to the well. Therefore the well discharge cannot be fully accounted for in the groundwater balance. The analytical solution for the situation of a "well in an infinite, parallel strip of land with constant head boundaries" is used for computation (Bear, 1979).

Lateral inflow is the subsurface outflow from the bordering catchment and is described by a linear reservoir. The input is given by the climatic soil water balance (Renger *et al.*, 1974) which is based on soil parameters, precipitation, surface runoff and evapotranspiration. The recession constant of this reservoir can be estimated from the recession of hydrographs of local tributaries.

This component is affected by landuse in the catchment area as well as by construction works, especially construction of roads, that cut off lateral inflow at the boundaries of the valley. The model can account for this by the ratio of open and cutoff length of the boundary.

Groundwater recharge from precipitation in the valley floor is represented by a simple model, based on the climatic soil water balance and a linear reservoir describing the outflow into the groundwater. Soil type, thickness, porosity and field capacity are roughly estimated from soil maps (Müller *et al.*, 1982). Extension of residential areas, construction of roads and highways reduce the area for infiltration, which can be assumed to be proportional to groundwater recharge. Again, a simple coefficient introduces this type of management alternative into the model.

CASE STUDY

The model has been applied to a section of the river Drau in the province of Carinthia, Austria (Figure 2). It has a length of 14.4 km, a width between 1 and 2 km and a total area of 21 km². The catchment area for this section is 185 km². The thickness of the aquifer varies between 10 and 120 m.

The aquifer is expected to provide resources for regional domestic water supply, but there are also plans to develop the river Drau for hydropower production. The main

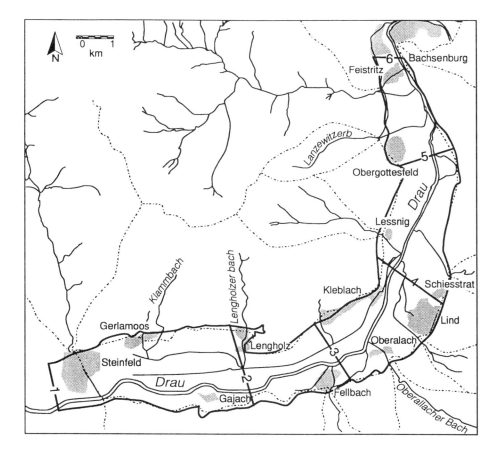

Figure 2 Map of the study area with balance cross sections

objectives of the case study therefore were to determine:
- the capacity of the aquifer for long term extraction of drinking water
- the location of possible wells with minimum regional impact
- possible conflicts between use for water supply and hydropower development.

A multi-cell groundwater balance model, representing the aquifer by five cells (Figure 2), was applied to derive the time-dependent groundwater balance and to provide a tool for the simulation of management scenarios. To establish the required database, a four-year observation programme provided weekly groundwater levels at 18 wells and geophysical measurements and boreholes explored the hydrogeological characteristics especially at the balancing cross sections. Additionally, daily precipitation time series and surface water hydrographs were available from the basic hydrometric observation network.

Groundwater balance
A calibration of the model was possible for the groundwater observation period of four years, using the hydrograph of observed available groundwater quantity in each cell. As an example, Figure 3 displays the hydrographs of computed and observed groundwater volume for the year 1988 with a time step of seven days. Figure 4 summarizes the groundwater balance for the same year in a flow chart of the annual means of the individual groundwater balance components. Note that cell 1 and 2 are dominated by inflow from the river, while cell 4 and 5 only have outflow to the river. Also, there is a large flux through cross sections 2 and 3, while flux in cross sections 4 and 5 is very low and negligible in cross section 6. The most important inputs in cell 4 and 5 are from the tributaries and from lateral inflow.

Uncertainties of groundwater balance due to uncertainties of model parameters
The limited knowledge of the aquifer and the use of only representative values of the aquifer parameters in the submodels induce uncertainties of the model parameters which propagate to the groundwater balance. A sensitivity analysis indicated that representative hydraulic conductivities in the cross sections and the clogging factor of the river bed, which determines the interaction between river and aquifer, have a strong influence on the groundwater balance.

A Monte Carlo simulation generated sets of hydraulic conductivities and clogging factors that were used to compute the groundwater balance. The corresponding distributions of the affected components of the groundwater balance are characterized by box plots in Figure 5. The inner quartiles of the distribution of the annual means, as well as of minima and maxima are very narrow for the groundwater volume and the exchange between aquifer and Drau. Only the fluxes through the cross sections, which are proportional to the hydraulic conductivities, reproduce the shape of the distribution of the conductivities. The extreme exfiltration and infiltration rates are physically infeasible, i.e., the range of the simulated conductivities was probably too pessimistic. The results of the model are thus quite reliable.

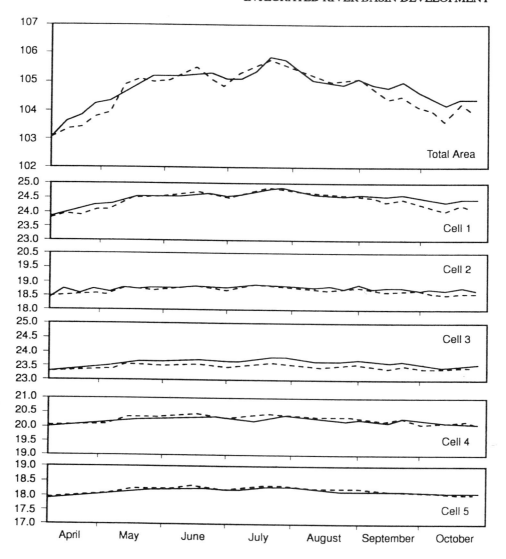

Figure 3 Hydrographs of computed and observed groundwater volume for 1988

Uncertainties due to stochastic inputs

The time frame for both water supply and hydropower development is in the range of 100 years. To achieve sustainable management, it is thus necessary to consider also the uncertainties due to the stochastic nature of the precipitation and stream flow inputs.

Available precipitation and streamflow data of 15 years were used to estimate the natural fluctuations in the groundwater balance. The resulting distributions of the weekly computed components of the groundwater balance are presented similarly in Figure 5. As an example, for the contribution of the tributaries and lateral inflow in

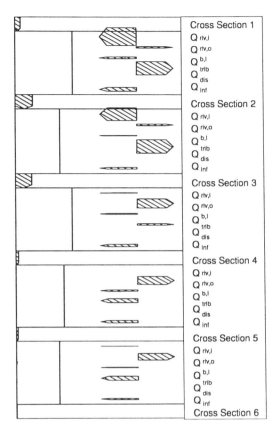

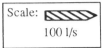

Figure 4 Flow chart of annual mean groundwater balance (the direction of the arrows
indicates inflow or outflow; the width is proportional to the magnitude of the flow)

cell 4, an average annual mean of 81 l s^{-1}, a minimum annual mean of 53 and a
maximum annual mean of 114 l s^{-1} was computed.

Management of aquifer
The results above already give hints for the location and possible discharge rate of
a well and the following conclusions for the management were easily proved through
scenario simulations.

Cells 4 and 5 have the advantage that there is only negligible inflow from their
upstream cross sections and very low cross section outflow. An extracting well in
these two cells would therefore have no negative impacts outside the cell. This
increases regional equity as possible future opportunities in the remaining areas are
not affected.

We can also expect that a discharge rate of approx. 50 l s^{-1} in these two cells is
covered by inputs from lateral inflow and infiltration from the tributaries. These
inputs come from mainly forested catchments with negligible changes of land use

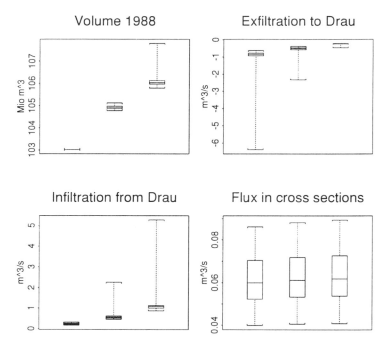

Figure 5   Boxplots of simulated groundwater balance components. (The distribution of the
annual minima, means and maxima is displayed for each component.)

expected in the future. Thus also in the long term there is low risk for the high quality
of these inflows. The discharge would only reduce the outflow to the river Drau.
Having a mean discharge of 75 m^3 s^{-1} this reduction is negligible for the Drau.

Would hydropower development be compatible with water supply? The construction
of hydropower plants requires cutoff walls to prevent seepage from the impounded
river to the aquifer. In our case, this would severely affect cells 1 and 2. However, in
cells 4 and 5, there is no significant inflow from the Drau, so a cutoff wall would not
change the situation for the aquifer. The outflow of groundwater to the river can be
handled by a drainage system. The important aspect of sustainability is that both
options are compatible in cells 4 and 5 and to opt in favour of either one would not
restrict opportunities for future generations.

CONCLUSIONS

To provide information specific to sustainable development of groundwater resources,
the models used in decision support need to extend the spatial domain of the problem,
allow simulation of long time periods, provide flexible mechanisms for scenario
simulation and have options to express the uncertainties.

For aquifers in narrow alpine valleys the multi-cell groundwater balance model
provides the necessary spatial resolution for the aquifer and also represents the

processes of groundwater recharge from the surrounding catchment area. The computational efficiency resulting from its simplicity makes it suitable for the investigation of uncertainties through Monte Carlo simulations.

Very simple yet flexible controls for the representation of a variety of decision alternatives allow the performance of extensive scenario simulations and "what if" analyses. The data required for the model and the presentation of the results make the analysis of the aquifer transparent and simplify the decision-making process.

The case study demonstrated how two specific issues of sustainable development could be addressed, namely uncertainty and regional and intergenerational equity.

ACKNOWLEDGMENTS

The case studies discussed in this paper were funded by the provincial governments of Styria and Carinthia.

REFERENCES

Bachmat, Y., B. Andrews, D. Holtz & S. Sebastian 1985. Groundwater Management: The Use of Numerical Models. Water Resources Monograph 5, AGU, USA.

Bear, J. 1979. Hydraulics of groundwater. McGraw-Hill.

Istok, J.D. 1989. Groundwater Modelling by the Finite Element Method. Water Resources Monograph 13, AGU, USA.

Müller, W. 1982. Bodenbeurteilung und Bodenmelioration vor dem Hintergrund moderner physikochemischer und bodenkundlicher Erkenntnisse. *Mitteilungen der Österreichischen Bodenkundlichen Gesellschaft,* **24**, Wien.

Nachtnebel, H.P. 1988. Wasserwirtschaftliche Planung bei mehrfacher Zielsetzung. *Wiener Mitteilungen* **78**, Wien.

Renger, M., Strebel, O. & Giesel, W. 1974. Beurteilung bodenkundlicher, kulturtechnischer und hydrologischer Fragen mit Hilfe von klimatischer Wasserbilanz und bodenphysikalischen Kennwerten. *Zeitschrift für Kulturtechnik und Flurbereinigung* 15.

Schmidheiny, S. 1992. *Changing Course*. MIT Press, USA.

Young, M.D. 1992. Sustainable investment and resource use: equity, environmental integrity and economic efficiency. *Man and Biosphere Series 9*. UNESCO, Paris.

6 Stochastic analysis of compensatory water releases in a semi-arid environment

**BARBARA DE BRUINE, GUIDO VAN LANGENHOVE and
AUNE HUTUTALE**
Ministry of Agriculture, Water and Rural Development, Republic of Namibia

INTRODUCTION

Figure 1 shows the important features of the Oanob River, which originates to the south-west of Windhoek, the capital of Namibia, and then flows in a south-easterly direction past the southern outskirts of the town of Rehoboth.

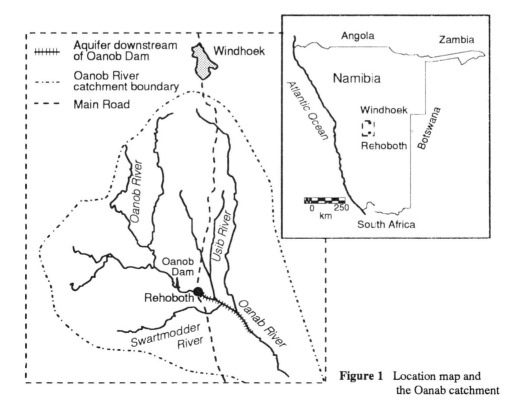

Figure 1 Location map and the Oanab catchment

At the site of the Oanob Dam, which was completed in 1990 and which has a full storage capacity of 34.5 Mm³, the drainage basin has an area of 2730 km². The alluvial aquifer downstream of the dam consists of two compartments, which have to rely on river flow for recharge.

Before the construction of the Oanob Dam commenced, fears were expressed about the possible adverse effects its presence might have on the downstream river flow and recharge, and consequently on the riverine vegetation, in particular the *Acacia* tree populations.

An attempt was thus made to model the situation that would exist after completion of the dam in order to assess the volume of water that would be available for compensatory releases from the dam, and the effect thereof on the groundwater table in the aquifers.

HYDROLOGICAL CHARACTERISTICS

Namibia is a semi-arid to arid country: the evaporation rate exceeds precipitation throughout the whole year for the largest part of the country. The few rainfall events, which are confined to the short rainy season from December to April and which mostly result from convective thunder storms, are erratic in frequency and magnitude. The hard and impermeable geological surfaces, the thin loose top soils, the hilly and mountainous terrain and the absence of a dense vegetation mean that river flow only occurs as a direct and short-lasting runoff response to these rain storms. All internal rivers in Namibia are ephemeral and may not flow at all during the driest years.

As in many other developing countries throughout the world, there is an absence of long stream flow records for most catchments, and in most cases flow information is available only for the last twenty years. Rainfall observations, however, often date back more than fifty years. For the purpose of hydrological analysis, the observed flow sequence can be extended with synthetic values generated by a hydrological catchment response model fitted for the shorter period with concurrent rainfall and flow records. A flow record for 52 seasons could be generated in this manner for the Oanob River. Figure 2 shows the seasonal flow volumes and Figure 3 the average monthly distribution, illustrating the high variability — or rather the unreliability — and the seasonal nature of flow in the Oanob River. The main features of the seasonal flow volumes are listed in Table 1.

Table 1 Characteristics of seasonal flow volumes for the Oanob Dam site

Mean (Mm³/a)	14.9
Minimum (Mm³ for 52 seasons)	nil
Lower quartile (Mm³/a)	0.8
Median (Mm³/a)	5.4
Logarithmic mean (Mm³/a)	8.0
Upper quartile (Mm³/a)	21.0

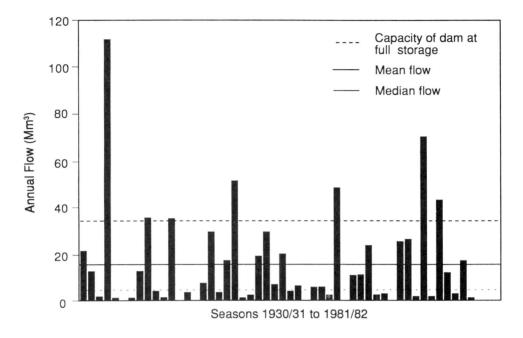

Figure 2   Mean annual flow at Oanob dam

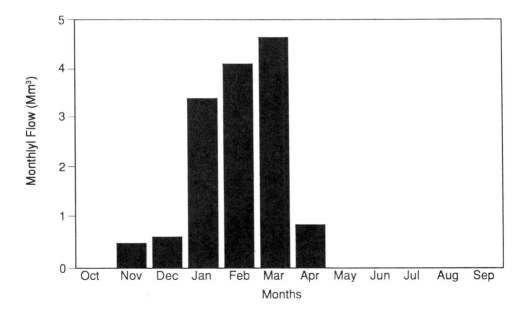

Figure 3   Mean monthly flows at Oanob dam

Maximum (Mm³ for 52 seasons)	111.7
Standard deviation (Mm³/a)	21.0
Coefficient of variation	1.48
Skewness	2.62
Number of zero values	8 (15.4 %)
Number of values < average	35 (67.3 %)
Number of values > average	17 (32.7 %)
Number of values ≫ dam capacity	4 (7.7 %)
Averages (Mm³/a) for driest periods:	
2 seasons	0.0
3 seasons	0.1
4 seasons	2.7
5 seasons	3.7
6 seasons	3.7
7 seasons	6.1
8 seasons	6.1

STOCHASTIC METHODOLOGY

To assess the feasibility of releasing water for compensatory purposes, whilst at the same time maintaining an acceptable degree of reliability for domestic water supply, stochastic simulation techniques have to be applied. The available historic sequence of stream flow events represents only one possible scenario for the future, in particular for critical dry periods extending over more than one season, as is illustrated in Table 1 for the averages for the driest periods with lengths from two to eight years. In order to determine the statistical reliability of the system other sequences should be generated.

The rainfall and runoff characteristics in the interior of Namibia are such that all flow is immediate runoff as a result of the few convective rainfall storms occurring in the short rainy season from December to April. Analysis of the records has shown that serial correlation between monthly or seasonal flows can be neglected. The important parameter to consider for a representative flow record is the sequence and magnitude of the seasonal flow volumes.

A commonly used method in such cases is based on the derivation of an analytical statistical function, often of normal or lognormal distribution. The high positive skewness of Namibian flow records points to the use of extreme value distributions, but difficulties arise from the incidence of zero values in the flow records.

A more simplified approach was used to overcome this problem which involved selecting at random whole seasons with their monthly flows from the available records thus avoiding the necessity to derive some method of disaggregation. This method allows a flow series of any length to be generated, even if the single values are restricted to those present in the record used.

DISCUSSION OF STOCHASTICS

Because of the limitation to the specific flow volumes present in the available record (in this case 52), the system has in principle major disadvantages; however, these are less — even not at all — relevant in the context of the hydrological conditions prevailing in Namibia:

(i)   Extreme monthly or seasonal flow values, higher or lower than those of the record, cannot be generated. In the prevailing semi-arid to arid conditions, however, the lowest values in the flow records are zero, and no lower values are feasible, while the highest vales result in overflow at any realistic reservoir, and higher values would not make any practical difference for yield analysis.

(ii)   Due to the short duration of the rainy season, it is not necessary to make a refined model of the monthly flow distribution.

(iii)   It is also not necessary to give attention to month-to-month or season-to-season serial correlations, because all flow is as a direct response to rainfall events, which are of a random nature.

The real power of the method is that droughts, stretching over two or more seasons, which determine the yield-reliability characteristics, of varying length and severity will be present in the longer record, which will therefore be adequate for long-term statistical analysis. This is illustrated in Appendix 1, where a comparison is made between random selection from a limited set of discrete values and random generation of values using a continuous analytical distribution for the case of the flow record discussed.

In 1991 a project, funded by the German Gesellschaft für Technische Zusammenarbeit (GTZ), was initiated to compile a Master Water Plan for the Central Area of Namibia. One aspect of the first phase of the project involved the reassessment of the hydrological flow data and it was found that the technique of stochastic flow generation described above produced realistic and acceptable results for the rivers in the Central Area, as described by McKenzie et al. (1993).

CRITERIA FOR RELIABILITY FOR SUPPLY

Three competing consumers had to be considered, in order of priority:

'primary' users — domestic use

'secondary' users — irrigation

'tertiary' users — compensatory releases

Absolute safety for supply is not feasible under the unreliable flow conditions prevailing in Namibia. It is operational policy to stipulate a 95% reliability for primary users, which corresponds to a 1-in-20 year failure risk, and a 80% reliability for secondary users, which is the 1-in-5 year risk. Any surplus water, which is available because the demand will initially be less than the ultimate supply capacity of the dam, would then be available for compensatory releases. The investigation was targeted at finding — for any combination of primary and secondary demand to be met at the set reliability levels — the possible magnitude and frequency of additional releases.

TRIAL-AND-ERROR SIMULATION

The simulation was carried out using a monthly water balance of the reservoir, combining inflow, evaporation, if applicable, spill, and supply, including possible releases. Whereas a customary analysis of a multi-purpose dam could investigate month-to-month operating rules based on storages and inflow patterns or predictions, this approach was not adhered to. Operating rules for meeting the three demands were determined, based only on the storage in April, at the end of the rainy season, and then not altered for the following twelve months. The rationale behind this simple decision rule is:

- Rules based on storages in the dam could be complicated as they should differentiate according to the position in the season.
- It is unrealistic to expect further inflow between April and January of the next year, and the status of the storage in the dam for supply to the primary users is highly predictable.
- Once a decision for supply to the secondary user is made, it cannot be altered for the duration of the growing seasons, starting in April and October, because speculation on inflow is not acceptable.
- To simulate natural floods, compensatory releases should be carried out as one short exercise in April, and need not be reconsidered again every month.

The mechanics of the calculations are then to repeat the same simulation process for a period of 5 000 seasons, giving different combinations of values to five variables: the domestic demand, the irrigation demand, the compensatory release, the April threshold storage for irrigation supply, and the April threshold storage for releases. The computer program then calculates the statistical reliabilities for the three users. As the objective is to find specifically 95 % and 80 % reliability levels for the two first users, interpolation between the demands that straddle these percentages is carried out to obtain the final results.

The approach is straightforward, but the necessity to repeat the simulation calculations for a range of combinations of five variables necessitated approximately 50 000 computer runs, making it, in terms of computing time, a very expensive operation, requiring approximately 120 hours, at the time the calculations were done (in 1989). The advances in computer technology, however, have greatly reduced this disadvantage of the method.

RESULTS

For ease of interpretation, the results were given in the form of "control curves", which give, for every target release, what the threshold April storage in the dam for releases should be for various combinations of primary and secondary demands. An example is given in Figure 4.

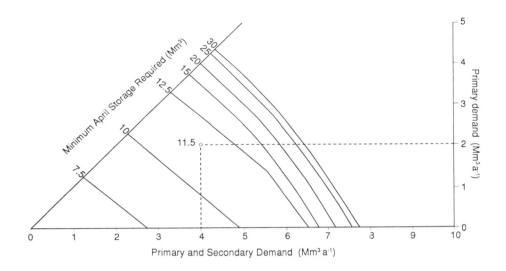

Figure 4 Proposed control curves for Oanob dam (target for releases = 3 m³

EFFECT OF RELEASES ON AQUIFERS

The water balance and behaviour of the aquifers could be modelled in a similar manner by using simplified recharge, throughflow, evaporation and evapotranspiration characteristics. The reaction of the aquifers to the implementation of the Oanob Dam and the possible effect of releases has been investigated by comparing the situations without dam, with dam but with no releases, and with dam and with releases, for the demands realistically expected during the next 25 years. The releases will result in a partial recovery by at least 50 % of the natural state of the aquifer.

ON-GOING INVESTIGATIONS AND PROPOSED IMPLEMENTATION

Sufficient inflow was received in the Oanob Dam during the 1992/93 rainy season to warrant compensatory releases of water for ecological purposes. At the beginning of September 1993, a time of the year when the river bed is dry, but during which active growth occurs in the *Acacia* trees, approximately 2.5 Mm³ water was released from the dam. The release pattern was alternated between periods of flow of approximately 6 m³ s⁻¹, for pre-wetting the river bed, and 12 m³ s⁻¹ for extending the influence of the artificial flow further downstream.

Preliminary results obtained highlighted the problems associated with accurately monitoring changes in the aquifer. It has been proposed, however, that the exercise be repeated, using a similar release pattern, but attempting to make the releases in a wet period toward the end of the rainy season to assess the influence of the condition of the river bed.

CONCLUSIONS

The hydrological patterns of stream flow in the ephemeral rivers in Namibia are such that some widely used methods of stochastic flow analysis are not very suitable. However, it has been found that, despite the fact that the technique is rather simple, generation of a flow sequence by random selection from the stream flow record can provide adequate flow sequences for use in stochastical hydrological analyses. In the case investigated it was used to generate control curves whereby decisions could be made concerning compensatory ecological releases from a reservoir.

This method could have wider applications in other areas with similar hydrological characteristics, but the suitability thereof should be evaluated for that particular region.

ACKNOWLEDGEMENTS

The encouragement by the Department of Water Affairs in Namibia to submit this paper is acknowledged, as well as the financial support of the GTZ in making it possible to attend the conference.

REFERENCES

McKenzie,R.S., Pegram,G.G.S. & Van Langenhove,G. 1993. Stochastic modelling of streamflow sequences in arid areas. *Proc. Sixth South African National Hydrology Symposium.*

APPENDIX 1: COMPARISON OF RANDOM GENERATION OF DISCRETE AND CONTINUOUS SETS OF VALUES

The question can be raised whether a long sequence of flow values generated by random selection from a limited set of discrete values will be as representative as a similar sequence generated from a continuous and smooth statistical distribution fitted to these points.

To avoid fitting problems caused by the zeroes in the record, and also by the high skewness, no attempt was made to select and fit an analytical statistical distribution to the Oanob seasonal flow volumes. Instead a piece-wise linear function through all data points was adopted for the cumulative probability distribution.

The two methods to generate data were:

(i) Discrete sampling, by generating an integer number between 1 and 52 and then selecting the flow volume for the corresponding season.

(ii) Continuous sampling by generating a fractional number between 0.5 and 52.5 and then interpolating (or extrapolating, if applicable) in the ranked record of flow volumes.

For both cases, 10 000 sequences of 52 values were obtained. The critical droughts are compared in Table A, which also shows the corresponding values for the flow record itself.

Table A: Comparison of critical drought periods

Length of driest periods (seasons)	Averages for record (Mm^3/a)	Averages for discrete sampling (Mm^3/a)	Averages for continuous sampling (Mm^3/a)
1	0.0	0.00	0.00
2	0.0	0.11	0.11
3	0.05	0.57	0.58
4	2.68	1.26	1.27
5	3.66	2.09	2.08
6	3.74	2.92	2.91
7	6.08	3.71	3.72
8	6.14	4.45	4.46

The results confirm that the difference between discrete and continuous sampling is negligible, and that there is little justification to apply the latter, more complex, method.

Figure A shows that the stochastic generation of flow records results in a more logical relation between droughts of different lengths, which is evidence of the statistical smoothing gained by application of the method. It is also to be noted that the historic record would be too optimistic for dry periods of longer than three years.

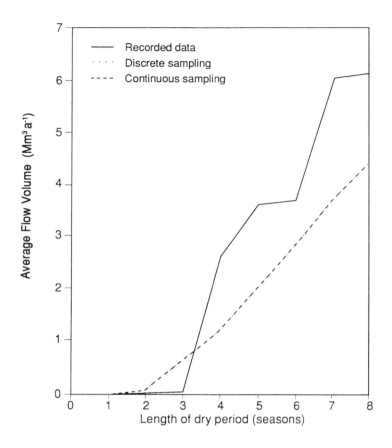

Figure A Graphic representation of drought periods

7 Stochastic modelling as an aid in the planning of reservoir systems

S VAN BILJON and J A VAN ROOYEN
Department of Water Affairs and Forestry, Pretoria, South Africa

INTRODUCTION

South African river flows are characterised by considerable variability and like the rest of the African continent we are subject to long periods of deficient runoff. This requires extreme care in the planning and phasing of new water resources schemes and in the operation of existing schemes. Analysis of streamflow time series by means of critical period techniques proved to be unsatisfactory due to single valued estimates of yield which must be revised after or during extreme droughts.

For a long time the 1925-1933 drought in South Africa provided the basis for planning. Extremely dry conditions were experienced during the late 1970s and early 1980s, in most parts of the country exceeding the 1925-33 drought in severity. These extreme conditions were followed by another very dry spell beginning in 1990, from which most reservoir systems had not recovered by early 1994. Managing water resources amidst climatic extremes and changing land use, in addition to ever increasing demands from a growing population and strategic industries demanding high reliabilities of supply, requires better analytical techniques for integrated reservoir systems than are afforded by the single valued deterministic and critical period methods.

Due to the wealth of minerals and coal in the Vaal river supply region the Pretoria Witwatersrand Vereeniging complex grew rapidly; and so did the demand for water. The water supply problems in the Vaal River System are extremely complex and the Department initiated a study in 1985 to ensure the optimal utilisation of the system. Results obtained in the study have been published as a series of reports under the name "Vaal River System Analysis". A number of water resources analysis techniques were introduced, including a multivariate stochastic model. This model underwent stringent verification and was recommended for use in the study. At present it forms part of a standard technique for all water resources analyses in the Department of Water Affairs and Forestry.

STOCHASTIC MODEL

One of the requirements of good planning is the investigation of alternatives. A single historical trace provides excellent information about past events, but the probability is remote that specific high or low flow sequences will retain the observed intensities and durations. Stochastic modelling provides a means of generating synthetic sequences which does not contain new information but retains the overall statistical properties (moments) of the original time series. It provides insight into possible future flow scenarios with different flow intensities and durations. Observed time series have to be naturalised to correct for changing land use practices, thereby obtaining stationarity. Since each synthetic trace is equally likely to occur in future, all traces are analysed to provide an estimate of likely outcomes. "Box and whisker" plots provide a handy means to display the likely variance in results.

The reliability of a reservoir for over year storage depends on a number of factors such as the annual mean, variance, seasonal variability, serial correlation and probability distribution of annual flows. Accepting that river flows may be described as consisting of deterministic and stochastic processes, a multivariate ARMA(1,1) model (Salas *et al.*, 1980; Pegram, 1986) was selected for modelling the annual flows. The marginal distributions of naturalised streamflows are estimated and then converted to normal by means of a suitable transformation. The technique is due to Johnson as coded by Hill *et al.*, (1986). If the Johnson transform chosen is lognormal, a lognormal distribution is fitted to the annual data by the method of maximum likelihood.

After verification an upward bias of the first two moments was found (the standard deviation more than the mean) when the 3-parameter lognormal distribution was fitted by means of maximum likelihood. About a third of the flow series analysed for the Vaal river systems analysis were described by the Sb (bounded) distribution as fitted by the Hill algorithm. On close examination, after 10 000 simulations for each gauge, it was found that the Sb/Hill distribution resulted in maximum flows not being greater than about 20% of the original. Elderton and Johnson (1969) state that the estimation of the moments of the Sb distribution is insensitive to the positioning of the bounds, especially the higher bound for highly skewed data. The Sb distribution was modified by setting the lower bound equal to zero and the upper bound to b = 2*max-mean. This was called the Sb3 distribution to distinguish it from the Sb distribution.

A comparison of the first four moments of the sample to resampling from a 10 000 simulation obtained by Pegram (1986) is presented in Table 1 for five different sites.

Since zero monthly flows are common in rivers over the drier parts of the country it was decided to disaggregate the annual flows to monthly flows by comparison to the observed monthly distribution. For the Vaal River study the generated (annual) values at ten key gauging stations were compared to the nearest observed (annual) values of the original time series. The generated sequences were disaggregated according to the flow distribution of a specific calender year for all the synthetic sequences. This ensures retention of the monthly cross-correlation between stations. The model preserves the annual serial correlation while the method of disaggregation preserves the monthly serial correlation.

Two disadvantages stem from this method. The serial dependence of flows for the first month of a year is not retained with respect to the flows of the last month of the previous year. This effect can be minimised by the choice of hydrological year. The second disadvantage concerns the upper bound of observed flows biasing the distribution of higher generated flows. The effect on storage reliability is, however, negligible, since high flow values are up to four times higher than the mean annual runoff with the result that most reservoirs fill during simulation runs.

Table 1 Comparison of re-sampled statistics to the sample

Site	Dist.	Mean	St. dev.	Skew	Kurt.
Allemanskraal	Sample	111.7	123.4	2.42	9.40
	Sb3	108.6	105.4	2.24	9.89
	LN3	112.0	138.4	4.92	47.82
Grootdraai	Sample	456.8	346.4	1.31	4.29
	Sb3	459.3	347.8	1.32	4.91
	LN3	461.4	389.9	2.54	14.71
Klipbank	Sample	192.3	181.0	1.72	5.46
	Sb3	188.6	169.6	1.80	6.09
	LN3	198.6	225.1	3.84	27.21
Vaal	Sample	1436.5	1098.3	1.43	4.52
	Sb3	1451.9	1040.9	1.43	5.31
	LN3	1431.3	1260.2	3.16	20.55
Westoe	Sample	56.1	33.4	0.55	2.45
	Sb3	56.1	33.5	0.56	2.46
	LN3	56.2	35.1	0.90	3.35

VAAL RIVER SYSTEMS ANALYSIS

Water demand
The large water demand in the Vaal river supply region developed mainly as a result of the wealth of minerals and coal found in the area. Water is supplied to the following important users:

● the Pretoria Witwatersrand Vereeniging (PWV) complex which is the industrial and economic heartland of South Africa;

● the gold fields of the Eastern Transvaal, Western Transvaal and the Orange Free State;

● Eskom power stations situated in the Eastern Transvaal and Orange Free State coal fields;

● petrochemical industries;

● iron and manganese mines in the Northern Cape;

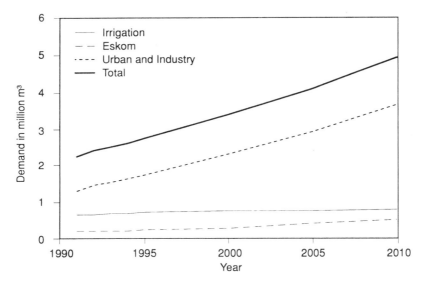

Figure 1 Most probable Vaal system water demand

- the diamond fields and Kimberley;
- various smaller towns along the main stream of the Vaal River;
- irrigation areas along the Vaal River; and
- Vaalharts, the largest irrigation scheme in the country.

These users are located not only in the Vaal River catchment area but also in adjacent catchment areas of the Olifants, Crocodile and Harts Rivers and even as far afield as the Gamagara River catchment area.

In Figure 1 the total demand projection for the Vaal River System, based on the most probable demand projection reported in Technical Report 134, is given broken down into the major user groups. The main drive behind the growth in water demand is from the urban sector which includes the domestic and industrial users. The irrigation demand stays basically constant as it is not economical to use expensive water transfer schemes to supply water for irrigation in the Vaal River basin, especially when that water could be used for irrigation in the donor catchment.

System for water supply
Unfortunately the Vaal river supply area is very sparsely endowed with water. The runoff from the Vaal River catchment at the Vaal Dam amounts to only 4. 3% of the total surface runoff for the country. The large water resources are located in the upper reaches of the Orange River catchment, the Transkei and Natal.

To further complicate water supply, the runoff in the Vaal River is highly variable, with years of very high runoff followed by long periods with flows sometimes very far below average. This is shown in Figure 2.

To overcome the problems of supplying the continually rising water demand of the very important Vaal River supply region, a complex water supply system has been

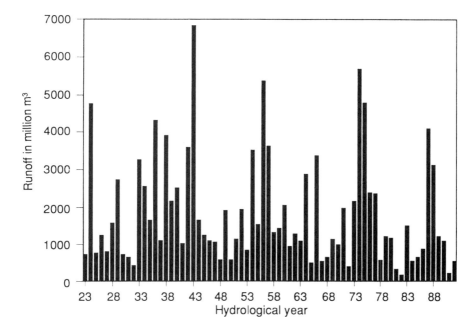

Figure 2 Annual runoff of the Vaal Basin

developed over the past 50 years. The system consists of a large number of dams, canals, pumping stations and pipelines, situated not only in the Vaal River catchment area, but also in neighbouring catchments from where water is imported or where water is distributed. The Vaal River System was divided into seven different subsystems, each of which was individually analysed to ensure the best utilisation of each subsystem. The different subsystems were then combined into one large system and rules were established by which the subsystems should support one another.

This integrated model of the Vaal River System makes it possible to manage the system properly as it can simultaneously balance the demand, supply and quality. The model makes provision for the following:
- the linking up of all the subsystems forming the Vaal River System;
- an increasing water demand;
- a system configuration that changes over time, for example it can simulate that the Katse Dam will begin storing water in October 1995 while the water will only be released into the Vaal River by January 1997;
- water restriction rules;
- return flows; and
- the salinisation aspect of water quality.

For the optimal planning of water supply schemes, the different water users' claims have been categorized into priorities with a risk level and restriction level linked to each category. This is shown in Table 2.

Table 2 User sectors, priorities, risk and restriction levels

User sector	Demand of each sector divided into priorities		
	Low	*Medium*	*High*
1. Strategic	0	0	100%
Domestic (60%)	40%	30%	30%
Industrial (40%)	0	30%	70%
2. Urban (weighted)	24%	30%	46%
3. Irrigation	50%	30%	20%
Risk level	5%	1%	0. 5%
Restriction level	0 to 1	1 to 2	2 to 3

When restrictions are imposed, use in the low priority class is restricted first, followed by medium and finally high priority.

A Level 1 restriction means that the low priority use of water is completely restricted. Urban use is then restricted by 24% (mainly garden irrigation) and agricultural irrigation by 50%, while no restrictions are imposed on industrial or strategic use. The risk levels assigned to the different user priorities mean that water should be supplied so that:

● the risk that the low priority class will experience water restrictions in any year should be equal to or less than 5% ;
● the risk for the medium priority class should be equal to or less than 1% ; and
● the risk for the high priority class should be equal to or less than 0. 5%.

The acceptable risk levels are exceeded when, for the water curtailment figures, any of the following conditions apply:

● more than 5% of the sequences are restricted (this is indicated by the "whisker" of the "box and whisker" plots showing above the 0-line);
● more than 1% of the sequences give restrictions above Level 1 (this is indicated by the 1% line rising above the Level 1 line); or
● more than 0. 5% of the sequences give restrictions above Level 2 (this is indicated by the 0. 5% line rising above the Level 2 line).

The integrated model is used among other things to determine when new schemes should be phased in; to investigate new management strategies for water quality problems; and to test annual operating procedures.

PHASING OF NEW SCHEMES

With Phase 1A of the Lesotho Highlands Water Project (LHWP) now already well

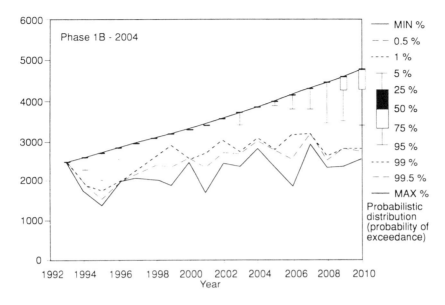

Figure 3   Total integrated Vaal system: demand and supply

under way, analyses were done to determine when Phase 1B, as well as the augmentation scheme following on that, should be implemented.

The analyses were based on 201 stochastic flow series, commencing with the actual dam levels as at the beginning of May 1993. For these analyses it was assumed that the Katse Dam would begin storing water in October 1995 and that water would flow to the Vaal River by January 1997. One annual decision date, on 1 May each year, is used to make the appropriate water allocation decisions. The results show the status of the system on 1 May of each year.

For the first analysis it was assumed that Phase 1B would be implemented as currently scheduled. Mohale Dam would start filling during October 2002 and the tunnel would be completed by January 2004, for water to flow to Katse Dam and via the transfer and delivery tunnels to the Vaal River.

Figure 3 shows the most probable demand curve and the risk of shortages experienced for the first analysis. This figure shows that there is always some risk that shortages may develop due to droughts, and where water restrictions will then have to be implemented. A general indication of serious problems can be determined from this graph, by looking at the situations where there is a risk of more than 5% that shortages may develop. The same information, in a slightly different but much more refined way, is presented in Figure 4.

Figure 4 shows the risk of water restrictions for the years up to 2010 and relates to the user priorities and water restriction criteria given in Table 2.

The risks for restrictions of the low priority users during the period 1994 to 1996 are higher than 5%, and also higher than 1% for the medium priority users. This exceeds the risk level set for both categories of users and is due to the low starting state

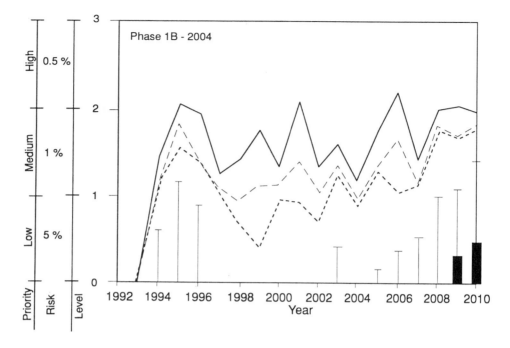

Figure 4 Total integrated Vaal system: curtailment

of the system in May 1993 as a result of the drought. It is not possible to supply water from any new scheme to improve the situation. In practice, the management of this situation is reviewed by conducting a short term systems analysis at the beginning of May each year.

The implementation of Phase 1A of the LHWP in 1997, has a dramatic effect on the risk of water restrictions. The risk drops immediately to below 5% for the low priority users and below 1% for the medium priority users. It stays like this until 2003, a year before Phase 1B is scheduled, when the restriction criteria are again exceeded. Phase 1B reduces the risk for one year but it becomes unacceptable again in 2005 and the risks increase drastically thereafter as no projects were scheduled in the model after Phase 1B.

It is significant to note that there is very seldom any risk that the users in the high priority class will be restricted. This means that although very severe restrictions may have to be implemented, water will be available for basic human need and for essential industrial use.

Figure 5 shows the result if Phase 1B is commissioned one year earlier, with the filling of Mohale Dam to start in October 2001 and the delivery to Katse Dam in January 2003. The result of this is that the risk of water restrictions in 2003 drops to acceptable levels and shows that it is preferable to implement Phase 1B one year earlier than scheduled. The date for commissioning of Phase 1B can not be determined unilaterally by the Department since a treaty with Lesotho on the whole

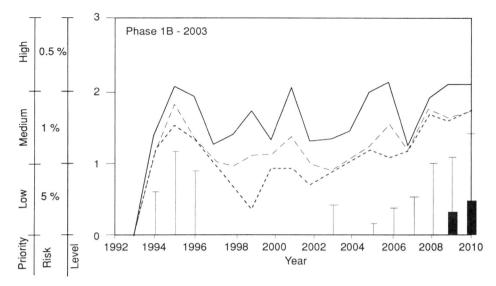

Figure 5 Total integrated Vaal system: curtailment for advancing Phase 1B

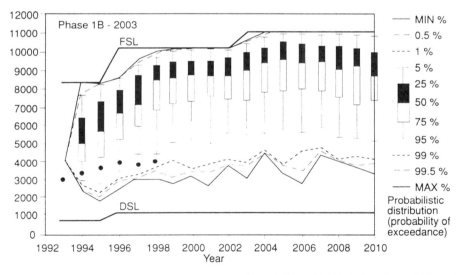

Figure 6 Total integrated Vaal System: storage

of Phase 1 already exists. The possibility of the earlier implementation of Phase 1B will be investigated.

The analysis also shows that the risk of water restrictions is exceeded in 2005. A new project will have to be phased in by then to supplement the resources of the Vaal River.

Figure 6 shows the projections of the storage state of the system at the beginning of May each year. The full supply level (FSL) of the system rises when the Katse Dam and later the Mohale Dam are phased in. Also shown on the graphs as large dots are the levels below which water restrictions will have to be implemented. These levels are a function of the rising demand and the FSL of the system.

CONCLUSION

It has been found that stochastic modelling provides an insight into possible future hydrological scenarios whereby risk based decisions for planning water supply schemes can be taken. It is argued that the stochastic technique is superior to the single valued estimates of reservoir systems yield, especially when flow variabilities are high and water supply systems are multi-objective and highly complex.

ACKNOWLEDGEMENT

The permission of the Director-General of Water Affairs and Forestry to present this paper is gratefully acknowledged.

REFERENCES

Basson, M. S. , Triebel, C. , & Van Rooyen, J. A. 1988. Analysis of a multi-basin water resource system: a case study of the Vaal River system. *IWRA VIth World Congress on Water Resources*, Ottawa, Canada.

Basson, M. S. & Van Rooyen, J. A. 1989. The integrated planning and management of water resources systems. *IVth South African National Hydrological Symposium*, Pretoria.

Elderton, W. P. & Johnson, N. L. 1969. *Systems of Frequency Curves*. Cambridge Press.

Department of Water Affairs. 1988. Water Demands in the Vaal River Supply Area, forecast to Year 2025. *Technical Report* 134.

Department of Water Affairs. 1986. *Management of the Water Resources of the Republic of South Africa*.

Hill, I. D. , Hill, R. & Holder, R. L. 1986. Fitting Johnson curves by moments. In: Griffiths, P. & Hill, I. D. (eds) *Applied Statistics Algorithms*. Ellis Horwood Series, Chichester.

McKenzie, R. S. & Allen, R. B. 1990. Modern water resource assessment techniques for the Vaal System. *Proc. Inst. Civ. Eng.* , Part 1, 995-1014.

Pegram, G. G. S. 1986. Stochastic Modelling of Streamflow. Department of Water Affairs. *Report* PC000/00/5186.

Salas, J. D. , Delleur, J. W. , Yevjevich, V. & Lane, W. L. 1980. *Applied Modelling of Hydrologic Time Series*. Water Resources Publications, Littleton, Colorado.

Van Rooyen J. A. 1993. Water supply to the Vaal River supply area. In: *Proc. Vaal River Liaison Forum*; Dept. of Water Affairs and Forestry.

PART II

INTEGRATED QUANTITY AND QUALITY OBJECTIVES

8 Integrated hydrological modelling with geographic information systems for water resources management

J RIETJENS, J GAILHARD, PH GOSSE, K MALATRE, C SABATON and F TRAVADE
Electricité de France, Chatou, France

INTRODUCTION

The need for water resource management which considers water resources as part of a country's patrimony and which would take into account the needs of all users is vital in France today, where 75% of all water resources are managed by Electricité de France. The law on water of 3 January 1992, expresses this development by officializing balanced management as the priority water management approach. However, while the new law orients those concerned towards procedures for consultation, the tools for a concrete application of such procedures are lacking.

Establishing locally "balanced" management of a water resource is not without difficulty in that the relationships between the users are often conflicting, notably in times of shortage. This procedure implies the choice of priorities, choices established in the following ways:

- the definition of quality objectives for the aquatic environment;
- the definition of a rule for water-sharing between users;
- consultation between those locally concerned.

In such cases, recourse to data-based tools becomes a necessity. Today's available technologies, as much in the Data-Base (DB) areas as in simulation tools or in Geographic Information Systems (GIS), enable us to design computer systems adapted to integrated management of a river basin (Fourcade *et al.*, 1993).

In this context, the Research and Development Division at EDF has developed a software tool to aid integrated water resource management: the AGIRE program (Rietjens, 1993). The AGIRE prototype was tested on two rivers in south-west France, the Gave de Pau and the Echez, in the context of a convention involving the EDF Research and Development Division, the Adour-Garonne River Agency, the French Ministry of the Environment and the Interdepartmental Institution for Hydraulic Planning in the Adour River Basin.

AGIRE is a communication tool enabling the user to test different approaches to water resource management:

- according to climatic hazards;
- according to various scenarios for use development (agriculture, drinking water, industry);
- according to environmental restrictions (minimum authorized flow, for example);
- according to the priorities agreed upon by the users;

and to visualize the effects on these uses or on the ecosystem.

This software tool, in the form of an interactive program operating on work stations, is based upon the following:

- a spatial and/or temporal data base
- a cartographical data base
- various simulation models (simulation tools)
- a user interface developed from the Geographic Information System ARC INFO (visualization tool).

THE SPATIAL AND TEMPORAL DATA BASE

The data base contains two types of variables:

Spatial statistics:

These are data that change little or not at all over time. We distinguish between *physical data* such as the surface of the river basin, the length or the slope of a segment of river; and *data related to human activities* such as community limits, useful agricultural surfaces, the location of release points in water purification plants or in industries.

Temporal data, requiring regular up-date. Again, we distinguish between *temporal data related to "natural" events*. These are meteorological data, the discharges and concentrations measured *in situ; temporal data in the models, characterizing each of the uses* under consideration. These are as follows:

- for agricultural use, the data from the General Agricultural Census (irrigated surfaces, types of crops, amount of livestock, etc.) as well as other variables given by different agricultural organizations, such as annual fertilizer sales, by department;
- for domestic use, population data, as well as data concerning purification plants (capacity, yield, etc.);
- for industrial usage, the data related to water consumption (quantity, origin), to treatment plants, to the work rhythms of companies.

The setting up of such a data-base underlines the problem of the availability and reliability of information. The task is not without difficulty in that a large number of different organizations in France share the management and control of water resources. Numerous organizations or administrations are active in this field, each possessing their own data bases. The greatest difficulty, then, lies in determining the most reliable sources. Furthermore, when the administration of a department is involved, obtaining facts depends on the kind of policy put into practice locally.

We must specify that this problem conditions the reliability of the simulation tool, in which the simplifying hypotheses cannot always be chosen advisedly because of inappropriate data.

THE CARTOGRAPHICAL DATA BASE

The complex study site is broken down into "covers" (or themes), each of which contains a type of information—geographical (position of the object in space); topological (relative position of different objects); and descriptive (values of the characteristic variables from the preceding data base, or calculation variables related to this cover).

We distinguish the covers as follows:
- regional hydrographical network,
- cities,
- departments,
- hydrographical sectors,
- local hydrographical network,
- river segments,
- sills and dams,
- industries,
- purification plants.

We use dynamic segmentation to describe the river, enabling us to attribute to it a notion of measurement (kilometric point or KP). Once this measurement is defined, every event in the river can be described by its KP. Points of in-take and release, sills and dams are then defined as so many distinct events, while the hydraulic and quality variables, calculated segment by segment, are identified as linear events.

As the numerical bases of the geographical data necessary to the tool were not available on the French market when the prototype AGIRE was completed, all the necessary maps for viewing the study zone Gave de Pau, Echez (hydrographic network, hydrographic sectors, cities, dams, factories, etc.) have been digitized and formatted manually.

SIMULATION TOOL

The simulation tool is made up of a group of numerical programs developed in FORTRAN that are linked together by modules. The AGIRE program today includes the following:
- a hydrological model
- a model estimating water needs for agriculture, industry and drinking water
- a hydraulic model (Gosse, 1990)
- a thermal model

- a calculation model for nitrate influx due to agricultural practices in a river basin (Geng, 1988)
- a model estimating release quality for industrial usage and for drinking water
- a model of river nitrogen
- a model of river bacteriology
- a model of fish-breeding habitat quality for Fario trout (Sabaton and Miquel, 1993)
- a model estimating satisfaction for agriculture and industrial usage.

Its overall operation can be described as follows (see Figure 1):

Starting from a given meteorological scenario, a first model reconstitutes, over one or several annual cycles, the so-called "pseudo-natural" flow of the water course, which would be the flow of the river not undergoing, in the modelled section, any intake or discharge linked with human activities. These calculations are made on each basic segment (or section, on an average length of one kilometre) of the river. At the same time, and on the basis of the same meteorological scenario, as well as on a given water-use scenario, we calculate the "desired" water needs for each of the so-called "active" uses of the water resource, which is to say the agricultural, industrial and

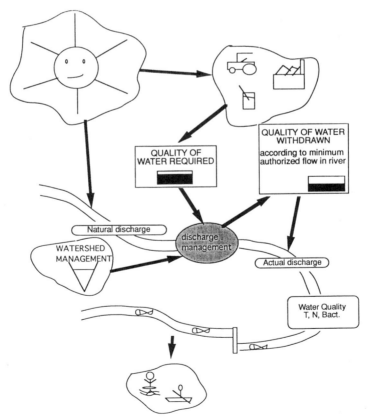

Figure 1 Model chart

drinking water needs. On this level it is equally possible to make up a management scenario involving low-water support.

A "volume management" module then enables comparison of the water available in the stream and the needs of the different users. The users may be totally or partially satisfied, depending on one or several authorized minimum flow rates in the river, (a salubrity flow level for which one chooses a value at any point.) Two types of responses are given at this stage: the volumes actually withdrawn and released by the different users and which, compared with the demand, show the degree of user satisfaction, and the "real" volumes in the river, taking into account the level of human activity.

It is then possible to calculate:
- the hydraulic values for the water course (flow rate and depth averaged over given segments);
- the variables chosen to describe water quality, taking the quality of releases into account;
- the quality of the river for fish-breeding.

Lastly, given the results of the preceding calculations, it is possible to assess the satisfaction of so-called "passive" uses of the water resource (neither withdrawing nor releasing water), or in other words, leisure activities such as swimming, kayaking, etc.

In the AGIRE prototype, the time step is equal to one month. The choice of the time-scale results from worries over limiting the calculation time in order to satisfy user-friendliness limits imposed on the AGIRE tool. This restriction, compared to the existing tool, also involves simplifying the depiction of certain calculation variables. However, the progress of new calculators should enable us in the short term to refine the time-scale.

VISUALIZATION TOOL

The visualization tool makes the AGIRE program more user-friendly. This interface enables the user to easily "drive" the simulation tool (Gailhard, Rietjens, 1994) by:
- constructing simulation scenarios (choice of meteorological hazards, choice of usage and of management scenarios);
- defining the graphic output (maps, graphs, geographical extension, time step);
- using different functions (choice of scales, zoom),
- gaining access to pertinent information on water use by simply activating the mouse.

The screen is divided into four work zones (Figure 2):
- a menu scroll bar allowing access to the different functions of the tool;
- a screen for the cartographical depiction of the study zone;
- a screen for graphically depicting time;
- a screen for graphically depicting space.

The visualization of calculation variables linked to the river is done for given time

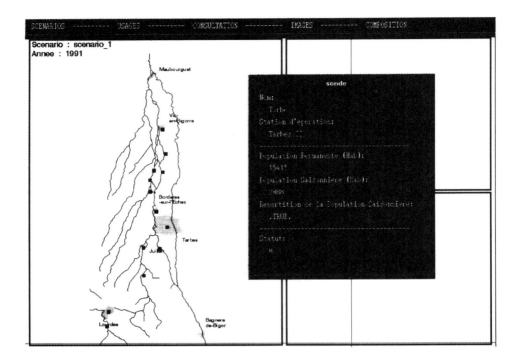

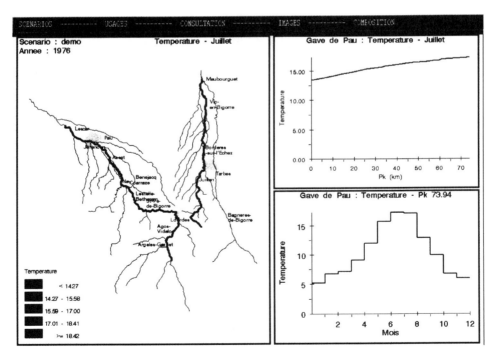

Figure 2 Examples of AGIRE's screen display

steps. At the same time the following appears on the screen:
- the map of the study zone depicting the variable calculated with the help of a colour scale associated (or not) with line thickness;
- a graph (X,Y) depicting the evolution of the variable selected on the basis of the kilometric point of the river;
- a graph (X,T) depicting the evolution of the variable selected at the outlet over an annual period. To trace this graph to a certain point in the river, the user only has to "click" the point on to the neighbouring map.

To obtain information on any of the uses, the user simply "clicks" on to the appropriate icon.

CONCLUSION

Perfecting the AGIRE software has emphasized the interfacing of the different types of systems in use (models, GIS). Besides the different disciplines such as hydrology, chemistry, and modelling, this tool has called for considerable programming expertise. Passing from the "scientific" tool to the "communication" tool and the lack of appropriate data has led us to simplify the depiction of certain calculation variables.

The results of the calculations must be treated with caution,, owing to the limitations on the reliability of the simulation models used.

Several recent developments are now under way, notably with
- the introduction of hydroelectric use by coupling with the PARSIFAL model developed by the General Technical Division of EDF, the software for which enables the optimization of hydro-electric releases within minimum flow limits;
- the introduction of an economic value for the "economically" quantifiable uses (hydroelectricity and irrigation being dealt with in priority);
- the introduction of a dynamics model for trout populations;
- the introduction of a management model for migrators;
- the improvement of the software's portability, the main difficulty in using such a tool on any river basin being of course the availability of data.

There are many possible paths towards development, and these depend on the questions that might arise locally. Thanks to its "modularity," this tool should provide an efficient way to help managers to define their local long-term planning strategies and to open discussion between the different parties.

REFERENCES

Fourcade, F., Quentin, F., Rietjens,J., Delcambre, J. and Manoha, B. 1993. Combining optimized management and preservation of the environment: one example hydroelectricity and irrigation. IEA, Vattenfall, OECD Int. Conf. on Hydropower, Energy and the Environment, Stockholm, Sweden, 14-16 June 1993 (in press).
Rietjens, J. 1993. Etude Gave de Pau-Echez: mise au point d'un outil d'aide à la gestion intégrée de la ressource en eau: le logiciel AGIRE. Colloque H20 sur la gestion de l'eau, Grenoble, 12 May

1993 (in press).

Rietjens, J. 1993. Note de présentation du logiciel AGIRE: Outil d'Aide à la Gestion Intégrée de la Ressource en Eau. Rapport EDF/DER/HE31/93.22., Electricité de France, Paris, 11 pp.

Sabaton, C. and Miquel, J. 1993. La méthode des micro-haabitats: un outil d'aide au choix d'un débit réservé à l'aval des ouvrages hydroélectriques. Expérience d'Electricité de France. *Hydroéco. Appl. Tome* **5**, 1, 127-163.

9 Optimum reservoir operation for water quality control

K D W NANDALAL
University of Peradeniya, Sri Lanka
JANOS J BOGARDI
Wageningen Agricultural University, The Netherlands

INTRODUCTION

Reservoirs are built to store water to improve the temporal availability of water. Operating these reservoirs optimally is very important in water resources planning and management. A large number of simulation and optimisation models has been developed over the past several decades for the optimal operation of reservoirs (Wurbs, 1993). In most of these models only the quantity of water has been considered. But there are occasions when both the quantity and quality of water available from the reservoirs are important.

There are a few studies of optimum reservoir operation in which water quality in the reservoir has been considered (Foruria *et al*, 1985; Crawley & Dandy, 1989; Dandy & Crawley, 1992). In these cases, water quality in the reservoir was modelled assuming complete mixing of water within the reservoir.

This paper presents a deterministic optimisation model based on the Incremental Dynamic Programming (IDP) technique to derive optimal operational policies for a reservoir. The reservoir is to be operated to improve the quality of water available while satisfying the quantity requirement.

OPTIMISATION MODEL

The optimisation model has been formulated based on the IDP technique. Nandalal (1986) described in detail the formulation of an optimum reservoir operation model based on IDP technique for a two serially linked reservoir system. The general scheme of the IDP procedure described in that model was adopted in the formulation of the model in this study.

The reservoir operates on a monthly basis. The rates of inflow, outflow and spill for the reservoir are assumed to be constant during each month. The objective is to improve the quality of water both within the reservoir and of that discharged from the reservoir. In this study only salinity will be used to characterise the water quality status. The model assumes complete mixing in the reservoir while determining the concentration of salt in the outflowing water, even though the salinity distribution in a reservoir is not uniform due to the process of mixing and stratification occurring in it. The forward algorithm of dynamic programming is used in the optimisation procedure.

The objective function used in the model is to minimise the weighted summation of the squared deviation of the release salinity and the reservoir salinity from the respective target levels, over the total period considered. The downstream quantity demand is treated as a constraint that must be always satisfied.

$$\text{O.F.} = \text{Minimise} \sum_{i=1}^{n} \left[\frac{W_1}{C_{trg}^2}(C_{rel,i} - C_{trg})^2 + \frac{W_2}{\hat{C}_{trg}^2}(C_{res,i+1} - \hat{C}_{trg})^2 \right] \qquad [1]$$

where
$C_{trg}, \hat{C}_{trg}$ = target release salinity, target reservoir salinity (ppm);
$C_{rel,i}$ = average salinity of release during period i (ppm);
$C_{res,i+1}$ = average salinity of reservoir at the end of period i (ppm);
W_1, W_2 = weighting factors;
i = time period $(1,2,...,n)$, and
n = number of periods.

The reservoir storage and release are assumed to be the state variable and decision variable, respectively. The minimisation is subjected to the following constraints.

STORAGE VOLUME CONSTRAINT

The storage volume in the reservoir at the beginning of the first period and at the end of the last period are fixed. For all the other periods the storage capacity of the reservoir belongs to the set of admissible storage volume.

$$S_{min} \leq S_1 \leq S_{max} ; \quad i = 1, 2, ... ,n \qquad [2]$$

where
S_1 = storage volume at the beginning of period i in MCM;
S_{max} = maximum storage of reservoir in MCM, and
S_{min} = minimum storage of reservoir in MCM.

RELEASE CONSTRAINT

The maximum allowable release from the reservoir is limited to the allowable release through the outlet. The minimum release is specified by the downstream irrigation demand, which is an implicit objective to be satisfied in the operation of the reservoir.

$$R_{min,i} \leq R_1 \leq R_{max} \; ; \quad i = 1,2, \dots ,n \tag{3}$$

where

R_1 = release during period i in MCM;

R_{max} = maximum allowable release through outlet in MCM, and

$R_{min,i}$ = irrigation demand during period i in MCM.

CONSERVATION OF SALT

The constraint that represents the conservation of salt in the reservoir is

$$S_{i+1} C_{res,i+1} = S_i C_{res,i} + I_i C_{infl,i} - R_i C_{rel,i} - O_i C_{o,i} \tag{4}$$

where

$C_{infl,\, i}$ = average salinity of inflow during period i in ppm;

$C_{o,i}$ = average salinity of spill during period i in ppm;

I_i = inflow during period i in MCM, and

O_i = spill during period i in MCM.

Other variables are as defined above. The evaporation terms do not enter the salt balance, since it is assumed that no salt is contained in the evaporating liquid.

The following equations are used to assess the salinity in the reservoir at the end of period i. The derivation of these equations is described in detail by Dandy and Crawley (1992).

If the reservoir volume is changing during period i,

$$C_{res,i+1} = \frac{1}{(Q_i+b)} \left[I_i C_{infl,i} - [I_i C_{infl,i} - C_{res,i}(Q_i+b)] \left(\frac{S_{i+1}}{S_i} \right)^{-(Q_i+b)/b} \right] \tag{5}$$

If the reservoir volume is constant during period i,

$$C_{res,i+1} = \frac{1}{Q_i} \left[I_i C_{infl,i} - [I_i C_{infl,i} - C_{res,i} Q_i] \exp(-\frac{Q_i \Delta t}{S_i}) \right] \tag{6}$$

The average salinity of spill during period i,

$$C_{o,i} = \frac{I_i C_{infl,i}}{Q_i} + \frac{S_i}{Q_i^2 \Delta t} [I C_{infl,i} - C_{res,i} Q_i] \left\{ \exp(-\frac{Q_i \Delta t}{S_i}) - 1 \right\} \tag{7}$$

where

Q = total outflow (total of release and spill) during period i, and

b = rate of change of reservoir storage during period i, i.e. $(S_{i+1} - S_i)/\Delta t$.

STATE TRANSFORMATION EQUATION

Based on the principle of continuity for the reservoir,

$$S_{i+1} = S_i + I_i - R_i - E_i - O_i \qquad [8]$$

where
E_i = evaporation during period i (MCM); other variables are as defined above.

RECURSIVE EQUATION

The DP recursive equation is formulated as

$$F^*_{i+1} (S_{i+1}) = \underset{R_i}{\text{Min}} \left\{ S_i + F^*_i (S_i) \right\} \qquad [9]$$

where
$F^*_{i+1} (S_{i+1})$ is the minimum accumulated value of the objective function from
stage 0 to stage i+1, when the state at stage i+1 is S_{i+1} and

$$A_i = \left[\frac{W_1}{C_{trg}^2}(C_{rel,i} - C_{trg})^2 + \frac{W_2}{\hat{C}_{trg}^2}(C_{res,i+1} - \hat{C}_{trg})^2 \right] \qquad [10]$$

SIMULATION MODEL

A simulation model was formulated to evaluate the results obtained from the IDP optimisation. Equations 4 to 8 are used in the regulation of the reservoir in this simulation model. Furthermore, the simulation procedure considers the constraints for reservoir storages and releases as given in Equations 2 and 3 respectively. This model furnishes end-of-month reservoir salinities and monthly average release salinities for a pre-specified release pattern.

APPLICATION

The Jarreh Reservoir Project on the Shapur river in Southern Iran was selected in order to study the applicability of the models developed. The climate of this basin, which has a drainage area of 4000 km², is classified as arid: the average annual rainfall is less than 20 per cent of the total annual potential evaporation. Except in occasional wet years, most precipitation is confined to the winter months; the dry season lasts from April to October.

The total annual rainfall varies from 600 mm in the upper part of the basin to less than 200 mm at the downstream end. Rainfall occurs mainly during the six months from November to April with a peak in mid-winter. The mean annual potential evaporation is high, about 2000 mm. The average annual streamflow volume of the Shapur river is about 530 million m³.

The Shapur river originated from karstic springs flowing down over saline and erodible formations and is being increasingly contaminated by saline sources of natural origin. The river flows are very saline, especially during the summer when they are needed for irrigation. The average salinity of the river water at the area to be irrigated is about 2130-2440 ppm in winter and rises to about 3700-4000 ppm in the summer.

Besides the shortage of water, this high salinity level has been found to impede the agricultural development in the Shapur plains. A storage dam has been planned to be built at Jarreh to improve the conditions for developing the agricultural resources in the downstream plains. The model presented here demonstrates that the same reservoir could be managed in such a manner as to improve the quality of water supplied (i.e. to decrease high peak salinities).

The Jarreh reservoir has a total storage capacity of 470 MCM and a dead storage capacity of 150 MCM. In this study the maximum release in a month was limited to 150 MCM. Observed data for 15 years are available for this reservoir.

RESULTS AND ANALYSIS

The objective function used in the model has two parts:
(a) to minimise the deviation of release salinity from a target, and
(b) to minimise the deviation of reservoir salinity from a target. (see Equation 1)

Initially, the impact these two components have on the final aim of reducing the release salinity was examined. This was carried out by giving different weighting factors to the two components. The weightings given and the results obtained are compared in Table 1; Figure 1 shows the monthly average release salinities. The model was run for 15 years in this study.

According to Table 1 and Figure 1, the differences observed between the three different objective functions mentioned above are almost negligible. This shows that the improvements obtained in quality of release water, by considering the quality of release or quality of reservoir or both, in the objective function, are almost similar. In this model improvements in the quality of release water are attempted by manipulating releases only. Finally, the second alternative that gives equal weighting to the two components was selected for use in the study.

The optimum operation obtained from the IDP model was compared with a reservoir operation simulation. In this simulation model complete mixing is assumed to be occurring in the reservoir throughout the year. The primary operation criterion adopted in the simulation model is to make the mandatory releases (downstream demands) only. However, if this criterion is strictly followed it is

Table 1 Comparison of different objective functions

Weightings		Total release (MCM)	Total spill (MCM)	O.F. value
$W_1 = 1.0$	$W_2 = 0.0$	7920	155	230.5
$W_1 = 0.5$	$W_2 = 0.5$	7987	90	228.2
$W_1 = 0.0$	$W_2 = 1.0$	8079	--	224.8

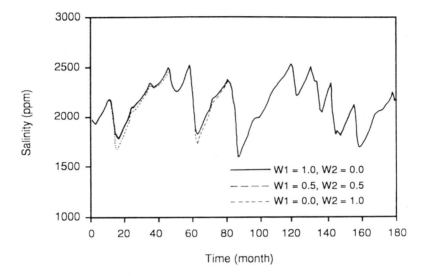

Figure 1 Average monthly release salinities — comparison of alternative objective functions

with perfect knowledge of both inflows and outflows.

The average monthly releases obtained from the IDP model and the demands are given in Table 3. The additional releases represent the amount of water released beyond the compulsory downstream demand. This volume of water is used for flushing or scouring the reservoir. Table 3 shows that the flushing (or cleansing) of the reservoir occurs mainly in the winter period during which the inflows are high.

The validity of the assumption of complete mixing was then examined. The reservoir stratification model "DYRESM"(Imberger *et al.*, 1978) was used for the comparison. It is a one-dimensional (in the vertical) numerical model for the prediction of temperature and salinity profiles in lakes of small to medium size. Shiati (1991) showed the applicability of the model DYRESM to Jarreh reservoir. The reservoir operation was simulated initially by employing the simulation model which assumes complete mixing in the reservoir. The operation rule described above was used at this step. The reservoir operation was then simulated using the

Table 2 Comparison of IDP optimum operation with simulation

Operation	Total release (MCM)	Total spill (MCM)	O.F. value
IDP optimum operation	7987	90	228.2
Simulation (releasing demands only)	7325	697	255.9

unavoidable that the reservoir storage reaches the maximum volume before the end of the period in certain months. If this happens then the excess volume of water has to spill. In such instances the above policy to release only the demand is over-ruled and the excess volume of water is released through the outlet subjected to the maximum allowable release. If it exceeds the maximum limit, the additional volume spills. This operation simulation will be called "simulation with releasing demands only" in the paper.

The results obtained from the IDP optimum operation were compared with that obtained from the aforesaid simulation. The results are presented in Table 2. The end-of-month reservoir salinities and monthly average release salinities are compared in Figures 2 and 3 respectively.

Table 2 indicates that the IDP optimum operation results in an improved operation pattern. If only the outflow could be manipulated then this is the best release pattern to give the minimum release salinities. Because the optimisation model was run

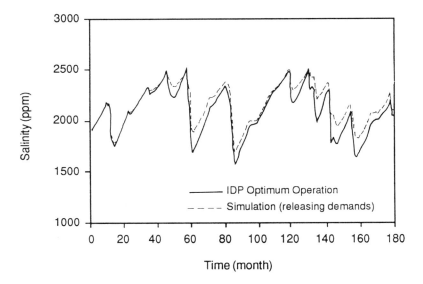

Figure 2 Reservoir salinity — comparison of IDP optimum with simulation

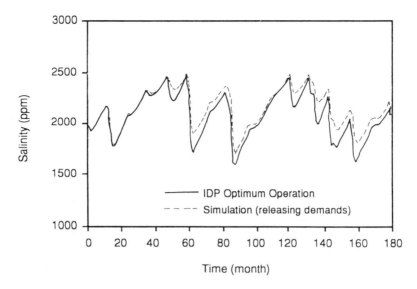

Figure 3 Average monthly release salinity — comparison of IDP optimum with simulation

Table 3 Releases from the IDP optimisation

Month	Average monthly release (MCM)	Demand (MCM)	Average of additional releases (MCM)
January	64.45	17.50	46.92
February	50.32	23.00	27.32
March	61.55	34.00	27.55
April	40.36	26.50	13.86
May	22.29	19.50	0.86
June	22.86	22.00	0.21
July	26.71	26.50	0.20
August	30.20	30.00	8.28
September	35.28	27.00	8.28
October	31.54	22.00	9.54
November	57.68	10.50	47.18
December	89.26	12.00	77.26

model which takes stratification into consideration. The same time series of releases was used in both simulations to make the comparison consistent. The reservoir salinity distributions obtained from the models were used for the comparison.

Figure 4 reveals that the reservoir salinities obtained from the simulation with DYRESM were less than that obtained from the fully mixed reservoir. The salinity

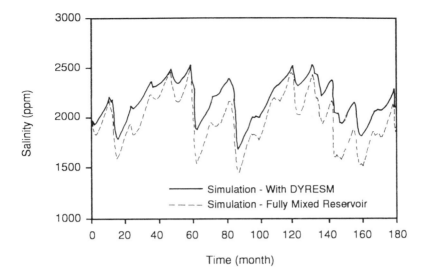

Figure 4 Reservoir salinities — effect of stratification

in the reservoir decreases rapidly in the winter months as, for example, in the months 14, 15 (Feb, Mar 1976) or 61, 62 (Jan, Feb 1980) etc. These drops in salinity level are substantial in the simulations with DYRESM compared to those in the fully mixed reservoir. These occasional larger drops bring the average salinity level in the stratified reservoir down to always less than that in the fully mixed reservoir. The inflows are very large during these months. Additionally, these larger inflows are cold and have a low salinity concentration. These cold inflows of lower salinity concentration travel to the bottom of the reservoir in the case of the stratified reservoir while making the water in the top layers spill over. The water spilled from the reservoir from top layers are at a relatively higher salinity level and when this high salinity water is spilled from the reservoir the average salinity level in the reservoir drops.

In the fully mixed reservoir inflow of less salinity concentration mixes fully with the reservoir and the salinity level of the water spilled is finally less when compared to the previous real situation. Therefore, in the fully mixed reservoir the final reservoir salinity is higher than that of the stratified reservoir.

The impact that the results obtained from the IDP optimisation model for a fully mixed reservoir has on a stratified reservoir was studied next. For this the reservoir operation was simulated employing the stratification model DYRESM with the releases obtained from IDP optimisation model. The resulting reservoir and release salinities were compared with those obtained from the simulation with DYRESM (releasing demands only) discussed earlier. In both simulations water was withdrawn from the stratified reservoir through the bottom outlet.

The reservoir salinities and release salinities are shown to have improved in

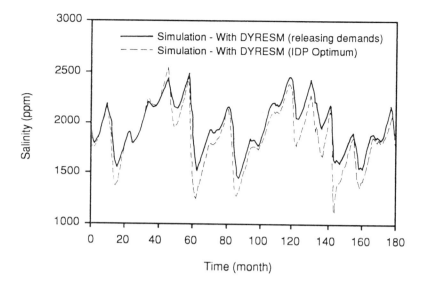

Figure 5 Reservoir salinities — effect of IDP optimisation on a stratified reservoir

Figures 5 and 6 when the operation pattern obtained from IDP optimisation (with the assumption of fully mixing in the reservoir) is used. This indicates the possibility of deriving better operation policies for a reservoir even with the simplification of fully mixing is occurring in the reservoir. This enables one to avoid the large amount of computational effort required when reservoir stratification is considered in the operating policy derivation.

CONCLUSIONS

The operation pattern obtained from the model based on the IDP technique was proven to improve the quality of water supplied. It is also shown that the model could successfully be used for the operation of a stratified reservoir. The optimum operation pattern is to flush the reservoir during the winter when the inflows are considerably high.

The optimisation model could be applied as an aid in deriving optimum operation policies for a reservoir when the quality of water to be supplied is of as much interest as that of satisfying the quantitative demand. The following approach could be used: several sets of streamflow data could be generated initially, to be followed by a deterministic optimisation, using the developed IDP model, for each generated data set. The resulting optimum strategies could then be used in the derivation of operation rules employing a least squares regression analysis.

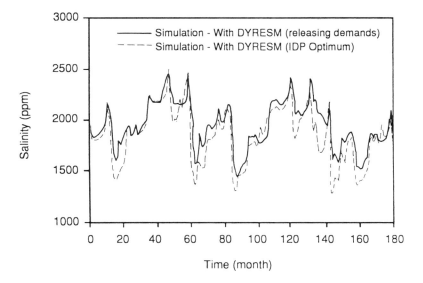

Figure 6 Monthly average release salinities — effect of IDP optimisation on stratified reservoir

REFERENCES

Crawley, P. & Dandy, G. 1989. Optimum reservoir operating policies including salinity considerations. *Proc. Hydrology and Water Resources Symposium*, Christchurch, New Zealand, 289-293.

Dandy, G. & Crawley, P. 1992. Optimum Operation of a Multiple Reservoir System Including Salinity Effects. *Wat. Resour. Res.* **28**, 979 - 990.

Foruria J., Sivakumar, M. & Volker, R.E. 1985. A simulation model for management of water quality in a water supply reservoir. *Proc. 21st IAHR Congress* **3**, Australia, 305-309.

Imberger, J., Patterson, J., Hebbert, B. & Loh, I. 1978. Dynamics of reservoirs of medium size. *J. Hydraul. Div. ASCE* **104** (HY5), 725-743.

Nandalal, K.D.W. 1986. Operation policies for two multipurpose reservoirs of the Mahaweli Development Scheme in Sri Lanka. M.Eng Thesis No. WA-86-9, AIT, Thailand.

Shiati, K. 1991. Salinity management in river basins: modelling and management of the salt-affected Jarreh Reservoir (Iran). Doctoral dissertation, Wageningen Agricultural University, The Netherlands.

Wurbs, R.A. 1993. Reservoir-system simulation and optimisation models. *J. Wat. Resour. Plan. Manage.* ASCE **119**, 455-472.

10 Proposed management plan for the Yesilirmak river basin

N ALPASLAN, N B HARMANCIOGLU and E SANER
Dokuz Eylul University, Turkey

INTRODUCTION

The primary objectives of water resources management in Turkey have for more than sixty years been water supply, irrigation, energy production and flood control. These objectives have taken priority above all other goals in natural resources development plans to meet the demands of the growing population, agriculture and industry. Accordingly, the majority of large river basins have been engineered into systems of numerous hydraulic structures that impound water and regulate runoff for these basic purposes.

Within the last decade, the governmental institution for Planning and National Policy Development has stressed the need for enhancement of freshwater productivity and management of fishery resources of inland waters. Investigations on the current status of freshwater fisheries in Turkey revealed that inland fish production increased from 7000 to 49 000 t a^{-1} within the 25 years between 1963 and 1988 (Ministry of Environment, 1991). Studies also disclosed the presence of 192 species of freshwater fish living in riverine systems. However, as of 1988, the proportion of inland fishing in the total fishery production of the country is stated to be as low as 7.2%.

There are basically two factors underlying this rather poor development. First, freshwater fish production has not been properly addressed in natural resource development plans because other objectives (such as irrigation, water supply, energy production, and flood control) were given the first priority. Second, the recently emerging issue of water pollution has reduced the fishery value of water courses. Intensive industrialisation and agricultural practice have significantly degraded the quality of rivers which, in turn, led to decreases in freshwater productivity. On the other hand, no extensive study has been carried out to define the ultimate goals of fishery management and to establish a scientific basis for multi-faceted guidelines to be followed in achieving these goals (TKB, 1992).

Because of the apparent need to manage the freshwater fishery, the Ministry of Agriculture (TKB) and the Scientific and Technical Research Council of Turkey (TUBITAK) started a research programme in 1991 — with contributions from Istanbul Technical University and later from Dokuz Eylul University — to collect, analyse and synthesise the scientific information needed to conserve and manage the fish resources of the country. The Yesilirmak river basin along the Black Sea coast was selected as the pilot study because of its potential for development of fishery production. This potential is indicated by the quality of river runoff and by the possibility of river regulation via a series of existing and planned reservoirs. The management plan proposed for the basin is essentially an integrated scheme that covers scientific, technical, institutional, social, economic and legal aspects of decision-making to define the optimum fishery yield. Thus, the pilot study is a multi-faceted large scale project to be realised through a number of stages, each of which is undertaken as a separate sub-project.

The study presented outlines the basic aspects of the research program and describes the objectives and stages of the Yesilirmak management plan together with the approaches proposed for each stage. Current investigations are devoted to modelling of the inputs to the riverine system, where the spatial and temporal variation of water quantity and water quality are simulated by two different modelling programs: SALMON-Q, devised by HR Wallingford, and QUAL2E-UNCAS, from the United States Environmental Protection Agency (EPA). The paper presented discusses the application of these models within the pilot study, pointing out to their merits and shortcomings in the case of the Yesilirmak basin. Finally, further stages of the proposed plan are summarised to include sub-projects on biological survey, development of a basin-wide data bank and production of a management model via GIS (Geographic Information Systems) and expert system technologies.

The Study Area

The Yesilirmak river basin in the Black Sea region has a drainage area of 36 114 km² and an annual average flow of 5.80 km³. The basin permits a storage of 3464 hm³ of water with the establishment of 57 dam reservoirs which serve energy production, irrigation and flood control purposes. In addition, the existing reservoirs along the river, together with those planned, constitute significant potential for the development of fishery production. At its full development stage, the systems planned along the Yesilirmak and its tributaries will irrigate 1 326 046 ha of land, provide flood control over a total area of 73 000 ha, and will supply domestic water to the extent of 144 hm³ a^{-1} (DSI, 1991). With these characteristics, the Yesilirmak ranks as the third largest basin in Turkey.

The Yesilirmak and its tributaries receive pollutant discharges from domestic and industrial sources. A wide range of pollutant constituents are discharged into the river by different types of industry including leather, chemicals, sugar, cigarettes, dairy, and food industries. Since no significant treatment is applied to domestic and industrial wastewaters, the water quality in the river is gradually being degraded. Non-point pollution from intensive agricultural activities contributes significantly to this degradation. The annual loads of water constituents are estimated to be 1532 µmho cm^{-1} of electrical conductivity (EC), 5.67×10^6 kg of total dissolved solids, 13.8×10^8 kg of Cl, 3.96×10^6 kg of NH_3-N, 3.08×10^5 kg of NO_2-N,

4.5 × 10^8 kg of NO_3-N, 18.4 × 10^8 kg of SO_4, and 2.31 × 10^8 kg of Mg (Ministry of Environment, 1991).

Two agencies, the State Hydraulic Works Authority (DSI) and the TKB, run the major water quality monitoring stations in the basin, where the longest period of observation is approximately 15 years. The Electrical Works Authority (EIE) is basically involved in streamflow gauging, although water quality is also sampled at some of the these gauges, particularly in areas of significant pollution.

Currently DSI collects data at seven sampling points and TKB at 49, the majority of them located downstream of significant industrial discharges. The data collected so far have not yet been extensively evaluated to disclose the spatial and temporal distribution of the water quality constituents.

Objectives of Management

The pilot study initiated by TKB and TUBITAK foresees the following basic objectives for the management of the Yesilirmak basin:

- determination of the basin-wide potential for fishery production and identification of the optimum fishery yield (t a^{-1});
- delineation of potential sites and appropriate fish species for economic freshwater production;
- development, conservation and proper management of fishery resources;
- development of a regulatory system to control and maintain a high quality natural environment that will not have adverse effects upon fishery production;
- development of the scientific and technical basis to realise these four objectives.

Essentially, the stated goals of the study necessitate a management model to define appropriate policies to maximise fishery production in the basin. Such a model is also expected to provide realistic answers to the following specific questions:

- What is the potential (t a^{-1}) of the Yesilirmak basin with respect to fishery production?
- Which regions of the basin constitute potential sites for economic production?
- Which species can be harvested in these different selected regions?
- How can a regulatory system be established to control and protect the quality of the environment for purposes of maintaining favourable conditions required by fishery production?
- Which polluting sources must treat their wastewaters before discharging into the riverine system?
- What are the scientific, technical, administrative, financial and legal requirements in defining a national water quality monitoring system?
- What kinds of databases are required for application of GIS (Geographic Information Systems) and expert system technologies for the management of the drainage basin?

Furthermore, three secondary objectives of the pilot study are implied by these questions, which foresee:

- development of a basin-wide database (or information system) including not only hydrological, but also geographical and demographic characteristics;

- establishment of coordination between various institutions that are involved in basin management;
- application of new technologies (e.g. GIS, expert systems) in management.

These objectives are to be achieved by separate sub-projects incorporated into the macro-level management plan. In essence, the pilot study planned for the Yesilirmak basin is an integrated river basin scheme which will eventually assume the form of a regional development plan. Similar schemes will be applied to other basins in Turkey once the above objectives are efficiently realised in the case of Yesilirmak. The results of the pilot study are expected to delineate basic guidelines for management of the nation's natural resources using new technologies (e.g. GIS and expert system methodologies) and to initiate cooperation between various institutions related to management. Realisation of the latter issue is particularly significant as the water resources and related agencies in Turkey have long lacked the interactive communication and collaboration which are essential for development of sound management policies and actions.

PROPOSED PILOT STUDY

Various levels of activities in the form of sub-projects are planned for the Yesilirmak basin, as outlined in Figure 1. Among these, two stages, the statistical analysis of river runoff (step 1) and the development of a mathematical model for plankton (step 6) were completed in 1991.

Levels 4 and 7 are the current stages being undertaken to satisfy one of the specific objectives of the project, the establishment of a regulatory system to control and protect the quality of the environment for purposes of maintaining favourable conditions required by fishery production. Such a system can be developed by resolving the following issues:

- determination of seasonal variations in the physical, chemical and biological properties of the river using water quality simulation program packages and delineation of required databases;
- establishment of coordination among various institutions to decide who should monitor which variables;
- determination of compliance with receiving media standards in the case of each polluting source;
- assessment of the share of each polluting source in the variation of water quality;
- determination of the assimilation capacity of the river and those water quality variables whose discharge must be prevented.

The two water quality simulation models, SALMON-Q and QUAL2E-UNCAS, were selected, on the basis of their suitability for project purposes, to realise the first two issues above. The results of model applications are expected to delineate space-time distributions of both hydrodynamic and water quality variables. Currently, the existing monitoring system in the basin is being revised and modified to develop a database required by the models. Such modifications include selection of new

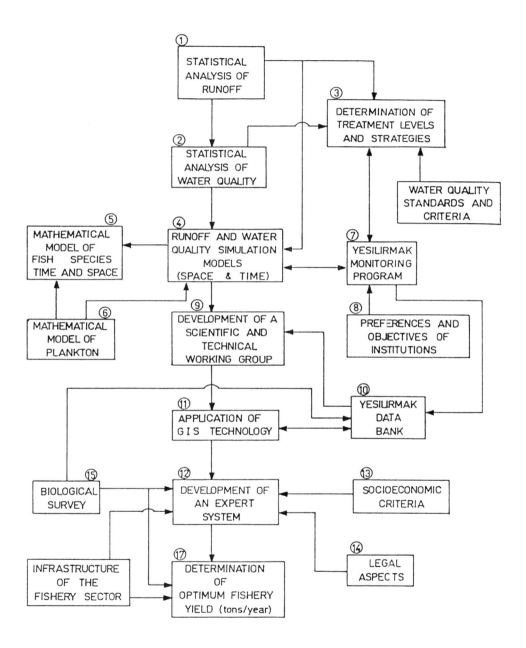

Figure 1 Stages of the pilot study proposed for the Yesilirmak basin

sampling sites, sampling frequencies, and new water quality variables to permit effective model applications.

An investigation of the current monitoring practices in the basin has revealed that redundant data are collected in some regions by the three monitoring agencies, DSI, TKB and EIE, whereas other regions reveal a complete lack of information. Thus, a coordination between agencies was deemed necessary to optimise monitoring efforts. Finally, the required coordination was initiated in 1993 when the three agencies agreed to support a regional cooperative approach toward Yesilirmak basin management. Accordingly, the guidelines for joint monitoring practices have been developed to define each agency's share in the total sampling program. EIE has undertaken the work of measuring streamflow and river cross sections at required points. DSI and TKB have started to assess their sampling and laboratory facilities on a joint basis, so as eventually to decide upon who should sample which variables. They have also initiated field trips to investigate the variability of hydraulic conditions along the river. As a result of these studies, the initially defined river reaches will be specified more precisely to permit effective model applications together with model calibration and verification processes.

As a prerequisite for the new data collection program, preferences and monitoring objectives of various agencies in Turkey have been investigated by the university using a questionnaire. In view of the current lack of coordination among these agencies, this study has constituted a significant initial step towards establishment of cooperation. The responses to the questionnaire have revealed that most agencies have a positive attitude towards cooperative approaches to monitoring practices. This result indicates possibilities for the development of a national data bank for management of river basins, and the Yesilirmak pilot study is expected to initiate the efforts towards progress, as indicated in level 10 of Figure 1.

In addition, the laboratory and computer facilities of each monitoring agency in the Yesilirmak basin are to be updated and modified. The university research group started training courses for DSI and TKB regional staff so that they can apply SALMON-Q and QUAL2E programs and evaluate model simulation results without the need for expert help. These courses are supported by reports and user manuals of program packages prepared by the university (Harmancioglu et al., 1993). Such training programs constitute the basis for level 9 of Figure 1, i.e. the development of a scientific and technical working group that will actively contribute to basin management.

TKB and the university have also initiated investigations to identify the pollutant loads in the basin. Numerous types of industries, settlements, and their discharges are analysed for the amounts and types of water quality constituents they involve. Such information is then transferred onto discharge maps which are used in revising and modifying the water quality monitoring scheme planned by concurrent studies.

The new sampling program foresees the monitoring of those water quality variables that can be simulated by both SALMON-Q and QUAL2E. The formerly irregular patterns of sampling frequencies have been revised into systematic programs. Thus, water quality will be observed on a regular monthly basis to permit an evaluation of seasonal changes. The new sampling locations are identified on the basis of model requirements to provide data on initial conditions for particular simulation runs. Furthermore, additional sampling sites are selected to observe the effects of significant pollutant discharges.

The investigations described will eventually constitute inputs to the basin management model which will be developed at further stages of the pilot study. This model will help to identify the required levels and locations of treatment as a part of the regulatory system to control and protect the quality of the Yesilirmak river.

APPLICATION OF WATER QUALITY SIMULATION MODELS

Of the two program packages selected for modelling water quantity and quality, QUAL2E-UNCAS is supplied to every laboratory in the basin, whereas SALMON-Q is used only by the university for basic research to develop the framework for the Yesilirmak management plan. In the future, SALMON-Q will be installed at one of the regional offices of TKB, which will eventually serve as a reference laboratory.

Application of these models constitutes an integral part of the pilot study as shown in level 4 of Figure 1. Both models simulate water quality processes in space-time dimensions, thereby making it possible to estimate the contribution of each polluting source to the variability of water quality along the river. Information provided by simulation studies will then constitute inputs to the management model at the final stage.

QUAL2E is a program package that permits the simulation of 15 water quality constituents in a branching river system. It applies a finite difference solution to the one-dimensional advective-dispersive mass transport and reaction equation (Brown & Barnwell, 1987). The program can be operated as a steady-state model to investigate the effects of pollutant loads upon river quality and to identify the magnitude and structure of non-point discharges. As a dynamic model, QUAL2E analyses the effects of diurnal variations in meteorological data on water quality. Diurnal dissolved oxygen (DO) variations caused by algal growth and respiration can also be studied by the dynamic model. Recently, the program has been enhanced by: (a) the addition of the QUAL2E-UNCAS model for uncertainty and risk analyses; (b) the inclusion of an option for reach-variable climatology input for steady-state temperature simulation; and (c) the capability to plot observed DO data on the line printer plots of estimated DO concentrations. The QUAL2E-UNCAS model provides the capability to assess the effect of model sensitivities and of uncertain input data on model forecasts so that the risk of a water quality variable being above or below an acceptable level can be evaluated.

QUAL2E is highly comprehensive, flexible and able to model a wide spectrum of physical conditions. However, this flexibility may be a disadvantage to users unfamiliar with the structure of the package since several decisions have to be made to adjust this structure to the specific problem investigated. Nevertheless, QUAL2E has been applied for many years both in the United States and worldwide and is considered to be a proven model in the analysis of DO balance in streams. Barnwell *et al.* (1989) claim that the program is also suitable for incorporation into expert system methodologies.

SALMON-Q is a one-dimensional mathematical model of flow, mud transport, water quality, and saline intrusion in tidal and riverine channel networks. It can simulate the oxygen and nutrient balance of looped and dendritic networks using an implicit finite difference approach to the solution of advection-dispersion equations. The basic assets of the model comprise its capability to assess: (a) impacts of effluent discharges; (b) effects of water abstraction and flow regulation schemes on water quality, sediment transport and saline intrusion; (c) effects of engineering schemes on hydraulic and water quality processes; and (d) alternative pollution control schemes (Wallingford Software, 1993). Although SALMON-Q is a new release by HR Wallingford, it involves process modules which are developed as a result of extensive studies over many years. Thus, it is supported by a long history of experience and applications. Like QUAL2E, it is highly flexible and capable of modelling a wide spectrum of physical conditions. It permits the evaluation of particular river management plans to control whether or not specified water quality objectives can be met by proposed actions.

At present, calibration and verification of the two models for the Yesilirmak basin cannot be realised due to lack of necessary field data. Although water quality observations are sufficient for preliminary simulations, data on hydrodynamic variables do not meet the requirements of either of the models. Thus, a monitoring program has been initiated to define flow conditions at necessary points along the river for a realistic description of river reaches. At present both models can be run with the available data; however, the physical system of the basin cannot be realistically introduced to the computer models.

Although the two models have not yet been calibrated, the results of preliminary applications with available data reveal specific characteristics of the programs. The hardware requirements of SALMON-Q are more extensive than those of QUAL2E. This is to some extent related to the dimensional limitations imposed on each model. In SALMON-Q, the number of reaches that can be defined is approximately three times as much as that allowed by QUAL2E. Similarly, the maximum number of computational elements is twice that specified for QUAL2E. Thus, the user can describe the actual physical system more precisely with SALMON-Q provided that sufficient data are available. This advantage is enhanced by the capability of SALMON-Q to accept variable lengths of computational elements in different reaches, whereas QUAL2E requires uniform lengths throughout all reaches. Furthermore, SALMON-Q introduces a calibration file which guides the user with respect to the information required for calibration. QUAL2E involves only an observed DO data file to assist in calibration and verification procedures.

Evaluation of the SALMON-Q program as applied to the Yesilirmak Basin
One of the most powerful aspects of the SALMON-Q program is its user-friendly graphical interface which comprises data capture and edit, data visualisation, model

run management, and model results review. The menu system with its main commands and windows of subcommands is easy to use as it permits access to all sections of the program and controls their operation. The interface also has a comprehensive help system which guides the user at various levels of modelling. The context sensitive spreadsheet type editor is again fairly easy to use since it operates like an expert system checking input errors and parameter suitability.

The setting up module is another advantageous feature of SALMON-Q, where input cross-section and topography data are used to generate the geometric data required by the model. When this module runs, the river network comprising reaches, nodes and junctions is schematised on the screen, allowing the user to check whether the physical system is correctly defined in the program. For complex systems such as the Yesilirmak river, particular reaches and nodes can be easily identified and accessed through the labelling and zoom options provided by the module. These features are not available in QUAL2E where it is difficult to visualise the network system defined. Furthermore, the graphics module of SALMON-Q is much more advanced than that of QUAL2E as it includes coloured time history and longitudinal profile plots of any parameter simulated. In essence, this module involves several options to produce output plots of any kind and quality. Evaluation of outputs is further enhanced by the statistical post-processor which again is not available in QUAL2E.

Several difficulties are encountered in applying both models to the Yesilìrmak basin. The major difficulty is related to availability of required data as discussed earlier. Other problems pertain to various modelling details such as the handling of steep slopes and incorporation of specific characteristics of the basin into the program. One major difficulty in model applications is the inclusion of a large number of existing reservoirs to model structures. At present, it has not been clarified how such large structural elements should be treated except to apply the models at discrete groups of river reaches. Such an approach increases the number of reaches to be investigated, leading to further data requirements and surveying efforts.

At further stages of the Yesilirmak pilot study, an expert system will be developed for management of water quality. Accordingly, the possibility and conditions necessary for incorporating SALMON-Q into this system will have to be investigated. Such an analysis has already been realised for QUAL2E and the results have shown the applicability of this model within an expert system methodology (Barnwell et al., 1989). Similarly, elements of an expert advisor will need to be developed for SALMON-Q as part of future studies.

On the other hand, it is known that the SALMON-Q program package is regularly updated and modified by ongoing research at HR Wallingford. As part of future developments, inclusion of such topics as migration of toxic substances and risk and uncertainty analyses may be suggested. In particular, the latter topic will significantly enhance model features since it will then be possible to assess the effects of model sensitivities and of uncertain input data on model forecasts. Consequently, the model will be much more suitable as a management tool.

FURTHER STAGES OF THE YESILIRMAK MANAGEMENT PLAN

Additional stages of the Yesilirmak pilot study will be initiated in 1994, to be used eventually in the formulation of a management model. These stages are:
- Biological surveys;
- Development of a data bank for the basin;
- Development of a geographical information system (GIS);
- Development of an expert system.

Biological surveys (level 15 in Figure 1) will be carried out to collect data on existing fish species, to analyse the current fishery practices, to define the fishery potential of the basin, and to investigate the possibility of increasing the current freshwater production to an optimum yield level.

The data bank to be developed for the basin (level 10 in Figure 1) will comprise all types of data including the hydrology, topography, geology, geography, hydraulic structures, lakes, reservoirs, land use, vegetation, coastal regions, settlements, industry, economy, and various other characteristics of the basin. It will provide the sources of these data in the form of maps, reports, images, and field surveys. If there are legal regulations relevant to any data type, these will also be stated. Such a data bank will eventually constitute the basis of basin management activities and will be accessible to all relevant institutions.

Another stage planned for 1994 is the initiation of a Geographic Information System for the basin (level 11 in Figure 1). The preliminary studies will cover the collection of all types of information (e.g. geographical, topographic, meteorological, hydrologic, demographic, land use, etc.) to be incorporated into a GIS system. For this purpose, a GIS database will be developed and by the end of 1994, the system will be ready for experimenting. As the final stage of the pilot study (level 12 in Figure 1), an expert system will be developed to incorporate the GIS system and the simulation models together with other relevant information regarding the basin. This system will then be used for management purposes; in 1994, preliminary studies for development of such a system will be initiated.

ACKNOWLEDGMENT

The research leading to this paper was sponsored by the Turkish Scientific and Technical Research Council (TUBITAK) and the Ministry of Agriculture (TKB) in the form of Grants DEBAG-71G and DEBAG-113G. This support is gratefully acknowledged.

REFERENCES

Barnwell, T.O., Brown, L.C. & Marek, W. 1989. Application of expert systems technology in water quality modelling. *Wat. Sci. & Technol.* **21**, 1045-1056.

Brown, L.C. & Barnwell, T.O. 1987. *The Enhanced Stream Water Quality Models QUAL2E and QUAL2E-UNCAS: Documentation and User Manual.* EPA-600/3-87/007, Environ. Research Lab., Environmental Protection Agency, Athens, GA 30613, USA.

DSI, 1991. *Statistical Bulletin for the Year 1991* (In Turkish). General Directorate of the State Hydraulic Works Authority of Turkey, Ankara.

Harmancioglu, N., Alpaslan, N., Ozkul, S., Saner, E. & Fistikoglu, O. 1993. *Project on Determination of Water Pollutant in Yesilirmak Basin: Water Quality Simulation Program Packages* (in Turkish). Report prepared by Dokuz Eylul University (DEVAK) for TKB as part of TUBITAK Project No. 71G. Vol. I: General Guidelines, Vol. II: QUAL2E User Manual, Vol. III: SALMON-Q User Manual.

Ministry of Environment, 1991. *Turkey: National Report to UNCED 1992.* Ministry of Environment, United Nations Conference on Environment and Development, Ankara, Turkey.

TKB, 1992. *Project on Determination of Water Pollution in Yesilirmak Basin: Final Report for the Year 1991* (in Turkish), Ministry of Agriculture (TKB) and The Scientific and Technical Research Center of Turkey (TUBITAK), Project No. DEBAG-11G, Samsun, Turkey.

Wallingford Software, 1993. *SALMON-Q User Documentation Version 1.0.* HR Wallingford Ltd, Wallingford, UK.

PART III

GROUNDWATER/ SURFACE WATER INTERACTION

11 Aquatic weed control options evaluated with the integrated surface and groundwater flow model MOGROW

E P QUERNER
DLO Winand Staring Centre for Integrated Land, Soil and Water Research, Wageningen, The Netherlands.

INTRODUCTION

Aquatic weed control is carried out to maintain an acceptable transport capacity in water courses during the growing season. Traditionally, weed control in water courses in the Netherlands has been stipulated by the requirements for farming. This implies increasing the agricultural benefits and minimizing the cost of weed control. The timing of such weed control often conflicts with other interests such as nature conservation. The frequency of weed control has been determined by rule of thumb. Therefore an integrated approach must be used to establish the schedule and method of weed control that best satisfies all interest groups. Having so many variables complicates the process of determining operational rules. Only limited experiments in the field are feasible, because of the possible risk of flooding.

The timing of weed control influences the levels of surface water and of groundwater. Therefore a hydrological model would be a tool to analyse weed control. Such a model must therefore include surface and groundwater flow. To achieve this, a groundwater flow model was combined with a surface water model. To calculate the surface water levels properly, information is needed on the weed growth and the corresponding flow resistance.

The model MOGROW (**MO**delling **GRO**undwater flow and the flow in surface **W**ater systems) was developed to give solutions for such regional water management problems (Querner, 1993). In the Netherlands particularly the relation between the required hydrological situation for agriculture and nature conservation is difficult to quantify. Other situations are the extractions for water supply or the management of surface water, such as aquatic weed control. For these type of situations it is necessary to simulate the flow of water in the saturated zone, the unsaturated zone and the surface water system in an integrated manner. The aim was to simulate the rather complex processes involved in such a way that it is accurate enough without requiring too much input data and computer time. The model MOGROW is physically-based and can therefore be used in situations with changing hydrological conditions.

The combined surface and groundwater flow model MOGROW was tested and verified in a drainage basin of 6.5 km², named Hupselse Beek (Querner, 1993). This paper describes the application of the MOGROW model to a larger area, Poelsbeek and Bolscherbeek, in which the main problem was focused on the necessity of weed clearing by the local water board. Different weed control options were simulated numerically. The results were evaluated by means of a cost-benefit analysis and a multi-objective decision analysis.

CONCEPT USED TO EVALUATE WEED CONTROL

Weed control primarily influences the water levels in the water courses. The secondary effects are differences in groundwater levels and in evapotranspiration. These hydrological changes affect crop production, soil workability and trafficability. The ecological impact of aquatic weed control is great. The integrated approach used for evaluating the impact of weed control is shown in Figure 1. Based on 'professional judgement' or the present weed control programme different weed control options are proposed as numerical experiments (Figure 1). A combined surface and groundwater

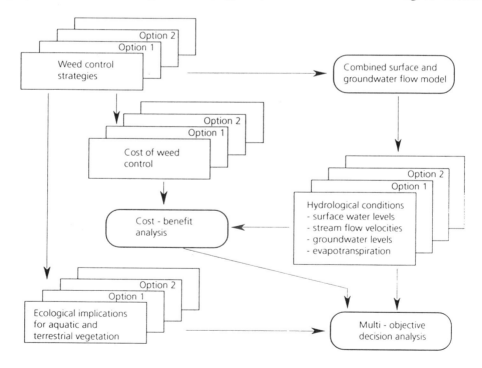

Figure 1   Approach to evaluate weed control options, using a cost-benefit analysis and a
           mutlti-objective decision analysis

flow model was used to quantify the hydrological conditions. This enables the weed control programme to be included and its effects to be ascertained. Simulations with these conditions in the model yield the repercussions on surface water levels, stream flow velocities, groundwater levels and evapotranspiration (Figure 1).

It is rather difficult to analyse all the results of the numerical experiments and choose the best option. An evaluation technique was necessary to find acceptable options. The method also had to consider the environment. This complicates the decision process even more, because these environmental criteria cannot easily be translated into costs or benefits at the same scale as, for example, in the agricultural sector. This type of problem therefore requires a technique involving multi-objectives. Each objective has its own scale of expressing its appropriateness for a given management option. The different options were therefore evaluated using a cost-benefit analysis and a multi-objective decision analysis.

THE COMBINED SURFACE AND GROUNDWATER FLOW MODEL MOGROW

The hydrological model MOGROW consists of the surface water flow model SIMWAT and the regional groundwater flow model SIMGRO. The model SIMWAT (**SIM**ulation of flow in surface **WAT**er networks) describes the water movement in a network of water courses, using the Saint Venant equation (Querner, 1986). The model SIMGRO (**SIM**ulation of **GRO**undwater flow and surface water levels) simulates regional groundwater flow in relation to drainage, water supply, sprinkling, subsurface irrigation and water level control (Querner, 1988).

Surface water flow
In the Netherlands the surface water system is often a dense network of water courses. It is not feasible to explicitly account for all these water courses in a regional computer simulation model. The surface water levels in the major water courses are important for the flow routing and to estimate the drainage or subsurface irrigation. Therefore in SIMWAT the major water courses are modelled explicitly as a network of sections; the other water courses are treated as reservoirs and connected to this network. Also the model includes special structures such as weirs, pumps, culverts, gates and inlets, necessary for the modelling of all water movements within a certain region.

Groundwater flow
To model regional groundwater flow, as in SIMGRO, the system has to be schematized geographically, both horizontally and vertically. Land use is schematized in the first aggregation level. The second aggregation level deals with subregions describing moisture vertically in the unsaturated zone. The third level covers various subsurface layers for saturated groundwater flow by means of a finite element network (Figure 2).

The unsaturated zone is considered by means of two reservoirs, one for the root zone

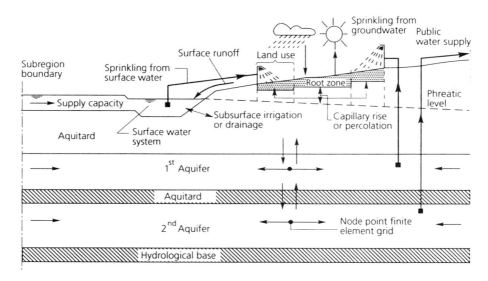

Figure 2 Schematisation of water management in the model MOGROW. The main feature is the integration of the saturated zone, unsaturated zone and surface water systems within a subregion.

and one for the subsoil (Figure 2). If the equilibrium moisture storage for the root zone is exceeded, the excess water will percolate to the saturated zone. If the moisture storage is less than the equilibrium moisture storage, then the result will be an upward flow from the saturated zone. The height of the phreatic surface is calculated from the water balance of the subsoil, using a storage coefficient which is dependent on the depth of the groundwater level below soil surface. The unsaturated zone is modelled one-dimensionally per subregion and land use (Querner & Van Bakel, 1989). Special processes are included in the unsaturated zone model, i.e.: surface runoff, perched water tables, hysteresis and preferential flow. This part of the model is more simplified than for instance the SHE model (Abbott *et al.*, 1986).

Evapotranspiration is a function of the crop and moisture content in the root zone. The measured values for net precipitation and potential evapotranspiration for grassland and woodland must be available. The potential evapotranspiration for other crops or vegetation types are derived in the model from the values for grassland by converting with known coefficients. The potential evapotranspiration for pine-forest is calculated as the sum of transpiration and interception. The model allows for a simulation of sprinkler irrigation according to a rotation scheme, depending on the water requirements of a crop.

The intensity of the network of water courses, is an important influence on the interaction between surface water and groundwater. The smaller water courses are assumed to be spread evenly over a finite element or subregion. These water courses are primarily involved in the interaction between surface water and groundwater: commonly called the secondary water courses, the tertiary water courses and the trenches. A so-called channel system can be present as well, but in specific nodes. It should represent the larger channels, also considered in the surface water model. In this way four drainage subsystems are used in the model for the interaction between surface and groundwater. This interaction is calculated for each subsystem using a drainage resistance and the difference in water level between groundwater and surface water (Ernst, 1978).

Link SIMWAT-SIMGRO

The geographical link between both modules is that a nodal point of the finite element grid in the groundwater module is assigned to a nodal point of the surface water module. The model MOGROW has a groundwater part that reacts slowly to changes, plus a surface water part with a quick response. Therefore both parts have their own time step. The result is that the surface water module performs several time steps during one time step of the groundwater module. The groundwater level is assumed to remain constant during that time and the interaction between groundwater and surface water is accumulated using the updated surface water level. The next time the groundwater module is called up the accumulated drainage or subsurface irrigation is used to calculate a new groundwater level.

APPLICATION OF MOGROW

Study area and schematisation

The Poelsbeek and Bolscherbeek catchments, 64 km² in size, are located in the east of the Netherlands (Figure 3). The ground surface slopes from about 30 m above NAP (reference level in the Netherlands) in the south-east to about 12 m above NAP in the north-west. The area consists of sandy soils. Land use is predominantly agricultural; about 48% is pasture, 20% is arable land (mainly silage maize), 24% is woodland and 8% residential. Both catchments are part of a bigger area controlled by the water board 'Regge en Dinkel'. This board is responsible for managing water quantity and quality.

For the model MOGROW the groundwater system needs to be schematised by means of a finite element network. The network, comprising 437 nodes spaced about 500 m apart, is shown in Figure 3. For the surface water module only the major water courses present were considered, requiring 151 nodes, 25 typical cross-sections and 61 weirs.

The physical properties for modelling the flow of water in the unsaturated zone module were obtained from Wösten *et al.* (1987). For each of the four units distinguished the actual characteristics, i.e. upward flux, storage coefficient and equilibrium moisture storage of the root zone, were calculated using the model CAPSEV (Wesseling, 1991). The basin was subdivided into 78 subregions using the physical soil properties and the layout of subcatchments (level for modelling the unsaturated zone). For the saturated zone the transmissivity varied between 100 and 700 m² day⁻¹.

The interaction between groundwater and surface water is characterised by a

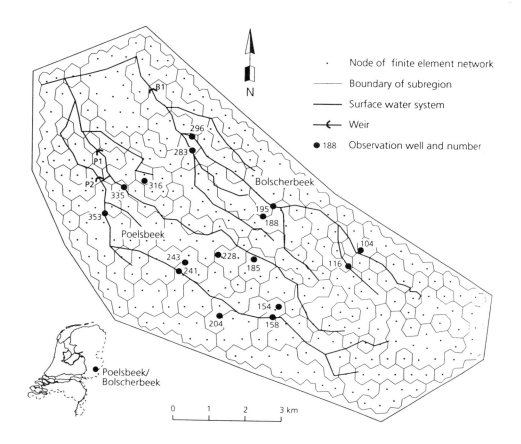

Figure 3 The Poelsbeek and Bolscherbeek drainage basins showing the schematization for groundwater and surface water. The subregions are needed for the schematization of the unsaturated zone.

drainage resistance (Ernst, 1978). This resistance is derived from hydrological parameters and the spacing of the water courses. The density of the water courses controlled by the water board was obtained from maps and the smaller ditches not controlled by the water board were measured in the field.

The network of water courses controlled by the water board is shown in Figure 3. The difference in height of about 1 m from south-east to north-west means that a number of weirs are needed to control water level and flow. Most of the weirs are adjustable, so that the target water level in summer can be raised. The Poelsbeek mainly drains agricultural land and the discharge is very irregular. The Bolscherbeek receives effluent water from two sewage treatment plants. The water board does some weed control and contracts out the remainder. In the research area the major water courses have access roads, i.e. a maintenance path 1.5 m wide on either side. In general, the weed is cleared mechanically three times a year.

Flow resistance and weed growth
Field measurements and laboratory experiments suggested that it was best to use the unobstructed part of the cross-section only and exclude the part covered by weeds (Querner, 1993). The obstructed part has a very low discharge capacity and can be ignored. The roughness coefficient k_M for the unobstructed part was derived from these field measurements and laboratory experiments and was found to vary between 30-34 $m^{1/3}$ s^{-1}.

Weed obstruction was measured throughout the growing season. The data were obtained from water courses without any weed control and also in sections where weeds were cleared twice during the growing season. The data were grouped in three classes based on the average water depth in summer (Querner, 1993).

Comparison of calculated and observed groundwater levels and discharges
The phreatic groundwater levels from June 1989 onwards were available for 15 locations (Figure 3). The discharge from the two catchments were measured at the locations P1, P2 and B1 (Figure 3). The input parameters were calibrated by comparing measured and calculated groundwater levels and discharges. The computations were carried out over the period June 1989 until September 1990 for the calibration and from October 1990 until September 1991 for the verification. The time step was one day for the groundwater module and 15 minutes for the surface water module. It took approximately 2 hours CPU time to simulate one year on a VAX 4200.

Figure 4 gives the groundwater hydrographs for two arbitrary locations, showing the typical differences between calculated and measured groundwater levels. The groundwater levels calculated for node 204 were slightly too high for the summer period and for node 228 were lower than the measured levels (Figure 4). The mean standard deviation (root mean square) between calculated and measured groundwater levels for the 15 locations varied between 0.08-0.31 m (on average 0.17 m). For nine locations the mean standard deviation was less than 0.20 m (Querner, 1993).

Figure 5 shows the measured and calculated discharges for the main stream of the Poelsbeek (weir P1). Both discharges compare well, but small differences were noted, such as for autumn 1989, where the calculated discharge was slightly higher than measured. Some peak discharges were underestimated, which may be because special processes, such as preferential flow and surface runoff were not included in the simulations. The mean standard deviations between calculated and measured discharges for this weir P1 was 0.08 m^3 s^{-1}.

EVALUATION OF WEED CONTROL

Simulations with model MOGROW of weed control
Because weed control is costly and is ecologically disruptive, the options formulated all involved less weed removal than the present situation. The alternative schedules included different periods and frequency of weed control, ranging from twice during

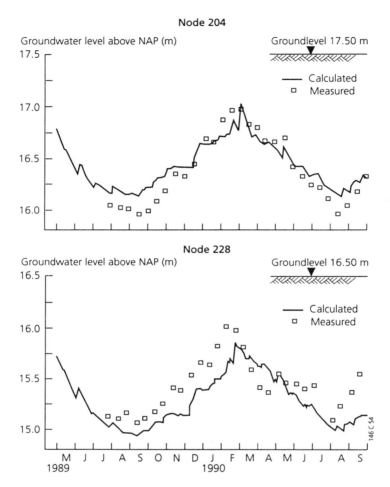

Figure 4 Calculated and measured water table for locations 204 and 228 (for locations see
Figure 3)

the growing season to no weed control at all. To clear only part of the cross-section,
such as the bottom only, was also included.

For the simulations with the model MOGROW, it was necessary to select a number
of meteorological years. These were selected on the basis of extreme wet conditions
during summer, having recurrence intervals of five and ten years. The period during
which these wet conditions occurred was either early, middle or the end of summer.

The impact of two weed control options, calculated with the model MOGROW for
the surface water levels and groundwater levels of node 185, is shown in Figure 6 (for
location of node 185, see Figure 3). The levels shown are for two extreme options,
being weed control twice during the growing season (option 1—present situation) and
no weed control at all (option 6). The meteorological year (1954) was characterised
by extreme wet conditions from mid-July onwards, resulting in higher surface water

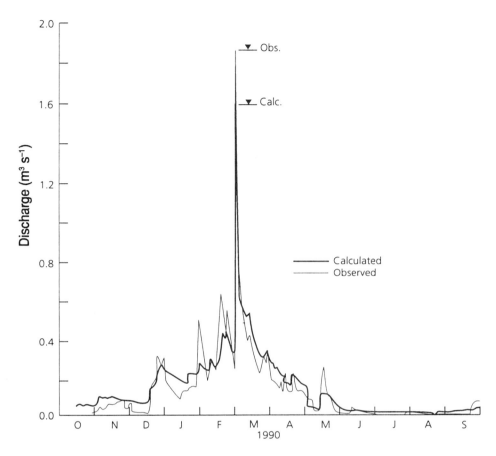

Figure 5   Hydrograph showing the measured and calculated discharges for the Poelsbeek at
weir P1 (for location see Figure 3)

levels and also higher groundwater levels (Figure 6). For option 6, the weed clearing
at the end of summer only, resulted in a sudden fall of the surface water level, but the
groundwater level changed much more slowly (Figure 6). The results shown for node
185 show one of the extreme differences in calculated levels: most nodes showed
considerably smaller differences.

Cost-benefit analysis
The weed control options were evaluated by comparing the costs of weed control
against the financial damage to agriculture. The financial loss caused by high surface
water levels, high stream flow velocities and high groundwater levels was calculated
on the basis of data from various sources (Querner, 1993). The meteorological
conditions causing a variation in discharge characteristics have a distinct effect on the
results. In the catchment the damage is primarily caused by high groundwater levels

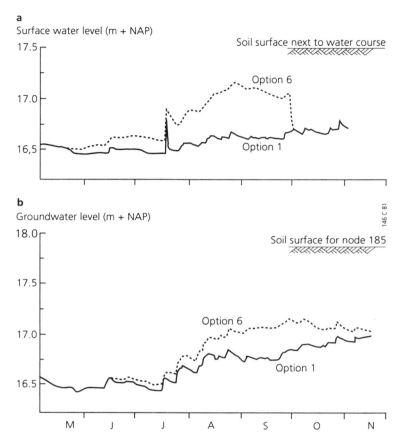

Figure 6   Difference in surface and groundwater levels between two week control
alternatives using meteorological data for a wet summer: (a) Suface water levels
(b) Groundwater levels

in combination with a few high surface water levels. In most years a weed control
frequency of twice a year or even once a year causes a small increase in 'damage',
compared to the present situation of weed clearing twice during summer.

Multi-objective decision analysis
The integrated water management approach for weed control incorporates the
ecological impact, using the ELECTRE II method (Goicoechea *et al.*, 1982). The
ELECTRE II method was selected because it enables the evaluation of discrete
options and uses quantitative values to represent the criteria. Each criterion is also
assigned a weight to express its relative importance. The procedure is focused on the
ranking of alternatives.

For the analysis of weed control options the important goals are: adequate drainage,
flood protection, scour protection, minimize cost of weed control and obtain desirable
ecological impact. Such goals were translated into specific criteria for which a

numerical value is used for each option and each criterion to express how 'good' or 'bad' this option is. The hydrological criteria were expressed quantitatively. The simulations carried out with the model MOGROW gave the number of high surface water levels, high stream flow velocities and high groundwater levels. The financial losses calculated using these extreme conditions in the cost-benefit analysis were input in ELECTRE II. For the ecological values a number of water boards and scientists from research institutes were consulted. These values are to some extent arbitrary, but the best available.

The analysis with the ELECTRE II method was carried out for three options. The first option was the present clearing, the second focused on agricultural benefits and in the third option the ecological aspects are paramount and agriculture is less important. The strategy focusing on agriculture gave various options as the best depending on weather conditions. For example, in summers with moderate wet conditions no weed control during summer was the best, because of the saving on maintenance cost. For nature conservation, weed control options avoiding weed clearing of the side slopes seem to be the best choice. A compromise is required between maintaining sufficient conveyance capacity for drainage whilst simultaneously preserving weeds that provide shelter for birds, amphibians and other fauna.

CONCLUSIONS

The surface water model SIMWAT and the groundwater model SIMGRO were combined into one hydrological model MOGROW. The direct link between the unsaturated zone, the saturated zone and the surface water, enables the model to predict all the physical processes within these systems, such as drainage, capillary rise and seepage in an integrated manner. The model MOGROW was kept fairly simple in order to handle practical applications in which simulations over several years are required.

In the Poelsbeek and Bolscherbeek catchments the groundwater levels and discharges could be successfully modelled over long periods with the integrated model. In this catchment the discharges do not react very quickly to rainfall events, but nevertheless cannot be modelled exactly in the same way as the measured discharges. Some small disparities between measured and calculated values were noted.

To date the water boards determine the frequency of weed control often by rule of thumb. This may result in too much weed control or the failure to carry out the weed control during periods in which ecological disturbance is less damaging. The analysis of weed control options by means of simulation models was successful. The purpose of such evaluation is to identify the options which meet the objectives stipulated by the water board. The accuracy of the input data is more questionable for the multi-objective decision analysis than for the cost-benefit analysis. The ecological impact was largely based on expert judgement. The multi-objective decision analysis must be regarded as a first step towards an integrated approach for the scheduling of weed control.

REFERENCES

Abbott, M.B., Bathurst, J.C., Cunge, J.A., O'Connell, P.E. & Rasmussen, J. 1986. An introduction to the European Hydrological System, "SHE". *J. Hydrol.* **87**, 45-77.

Ernst, L.F. 1978. Drainage of undulating sandy soils with high groundwater tables. *J. Hydrol.* **39**, 1-50.

Goicoechea, A., Hansen, D.R. & Duckstein L. 1982. *Multi-objective decision analysis with engineering and business applications.* John Wiley & Sons, New York.

Querner, E.P. 1986. An integrated surface and groundwater flow model for the design and operation of drainage systems. *Int. Conf. on Hydraul. Design in Water Res. Engg: Land Drainage,* Southampton, 101-108.

Querner, E.P. 1988. Description of a regional groundwater flow model SIMGRO and some applications. *Agric. Wat. Manag.* **14**, 209-218.

Querner, E.P. 1993. Aquatic weed control within an integrated water management framework. Wageningen Agricultural University. Doctoral thesis. Also as: Report 67. *DLO Winand Staring Centre,* Wageningen, The Netherlands.

Querner, E.P. & Van Bakel, P.J.T. 1989. Description of the regional groundwater flow model SIMGRO. Wageningen, *DLO Winand Staring Centre.* Report 7.

Wesseling, J.G. 1991. CAPSEV; Steady state moisture flow theory; Programme description; User manual. Wageningen, *DLO Winand Staring Centre.* Report 37.

Wösten, J.H.M., Bannink, M.H. & Beuving, J. 1987. Water retention and hydraulic conductivity characteristics of top- and subsoils in the Netherlands: The Staring Series (in Dutch). Wageningen, *ICW.* Report 18.

12 Integrated development of Shiyang river basin, Ganshu Province, China

LIN XUEYU and YANG YUESUO
Institute of Applied Hydrogeology, Changchun University of Earth Sciences, China

INTRODUCTION

Shiyang river basin is a structural deposit basin located in the arid desert area of north-western China. Its total area is 30 000 km². Eight rivers with a total runoff of 15.43×10^8 m³a⁻¹ flow through this area (see Figure 1).

Shiyang river basin is divided into Wuwei basin in the south and Minqin-Chaoshui basin in the north. It was divided into 16 subdivisions for water management (see Figure 2). The water from the Qilian mountains replenishes the aquifers of Wuwei basin and overflows as spring groups on the forward position of the diluvial fan of the Wuwei basin. The Shiyang river is formed by these spring groups. To the north is the lowest area of Wuwei basin where the groundwater level is quite close to the ground surface and forms the evaporation zone in the study area.

Further northwards, the aquifer in Minqin-Chaoshui basin is recharged with the waters from the lower reaches of Jinchun river and Hongyashan reservoir. In short, the river water leaks down to groundwater from banks upstream and then the groundwater flows out to supplement the river somewhere lower down: turn and turn about. The river-aquifer system forms an integrated water resources system in the river basin.

The main aquifer of the basin is gravel and sand gravel deposits of the late Pleistocene and Holocene periods. The large unconsolidated deposit which forms the basin is the source of natural storage of water resources and the inter-exchange system of surface water and groundwater.

Since the 1960s, in the mountainous areas of Huangyang, Nanying and Jinta, many reservoirs were built to intercept river water at the gaps bordering the Qilian mountains. These all suffer from evaporation from the reservoir surfaces. According to the data, annual evaporation is more than 2868 mm, i.e. about 270×10^4 m³ a⁻¹ km⁻² of water will be evaporated from the surface of reservoirs. The surface of reservoirs that are inlaid over the branches of Shiyang river is many hundreds of square kilometres and the evaporation from reservoirs is about 3×10^8 m³. That is to say, the reservoirs play an important role in agriculture in the upper course of rivers, but they have also created a lot of problems in the middle and lower reaches.

For example, the water decrement in rivers (about 5×10^8 m³ in the 1950s and

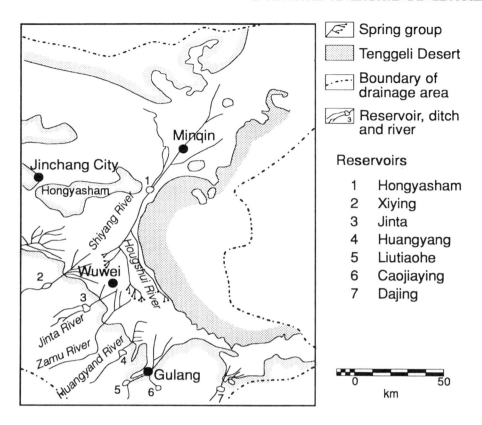

Figure 1 The Shiyang Basin

2×10^8 m³ in 1980s) decreased the water replenishment to groundwater and springs and also caused the farmers in the middle and lower reaches to overpump groundwater for irrigation. The rate of groundwater drawdown is 0.25-0.57 m a⁻¹. The average groundwater level depth in Shiyang river basin has dropped to 4-7 m since the 1960s.

The water pumped from the confined aquifer in the Minqin area has induced the upper salty groundwater to flow vertically to the lower aquifer and has thereby degraded the water quality. In addition, the average total dissolved solids (TDS) in the water of most wells has reached 0.12 g l⁻¹. 1.049×10^8 m³ of fresh water (accounting for 44% of the total groundwater in this area) has turned into salty water (the TDS is more than 2.3 g l⁻¹). All of this accelerates and aggravates soil salinisation.

The development of vegetation has a close relationship with groundwater level in the arid area. The optimal groundwater level for vegetation development is 3-5 m depth: when the groundwater level drops to 7 m the vegetation will mostly die off. Therefore, the fact that the regional groundwater level is falling rapidly and continuously, and already exceeds 7 m, is threatening the vegetation in most of the

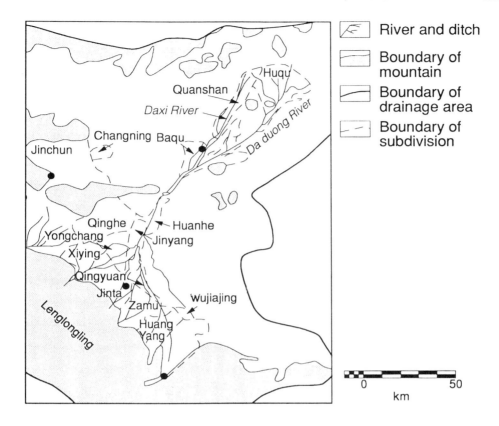

	River and ditch
	Boundary of mountain
	Boundary of drainage area
	Boundary of subdivision

Figure 2   Subdivisions of the Shiyang Basin

area. In some areas of the Minqin basin, the groundwater depths are 2-3 m, but the TDS exceeds the optimal value of <1 g l⁻¹. Based on the data in this area, when the TDS is 1-1.5 g l⁻¹, the cover rate of vegetation is 60-70%; when it is 2-2.5 g l⁻¹, 30-60%, and when more than 10 g l⁻¹, it has mostly died off. All of these factors have made, and will make, the oasis desertification occur rapidly.

MULTI-OBJECTIVE PROGRAMMING MODEL OF WATER SYSTEM

To improve the eco-environmental problems outlined above, a multi-objective programming model of water resources system was built using an objective programming method (Lin *et al.*, 1988). The method of linear programming was used in the solution.

The basic principle of objective programming is to construct a set of multi-objective functions. Under the conditions of satisfying the objective and the water resources constraints, the solutions for decisive variables can be obtained by minimizing the deviation of the decisive value and the given objective value. In order to satisfy the objectives, or to compromise between them, the efficiency of some will be reduced when others are improved: this is an inherent conflict in the use of multi-objective planning.

As described above, the water resources system of Shiyang river basin is a complex system and the water resources management of this basin is multi-objective and multi-decision.

There are nine objectives, p1 - p9:
P1: that surface water demand should be less than the surface water supply; and groundwater pumpage is smaller than the allowable groundwater pumpage, i.e:

$$\text{Min } (DSJ^+ + SSM^+ + DSW^+ + DGJ^+ + DGM^+ + DGW^+)$$

where DSJ^+, DSM^+, DSW^+ are the positive deviations of allowable surface water developing quantity (the unit is "10^4 m³") in Jinchang basin, Minqin basin and Wuwei basin; DGJ^+, DGM^+, DGW^+ are the positive deviations of allowable groundwater pumpage in Jinchang basin, Minqin basin and Wuwei basin.

P2: to satisfy the water needs of human beings and livestock, i.e.:

$$\text{Min } \sum_{i=1}^{16} (DRG_i^- + DRS_i^-)$$

where DRG_i^-, DRS_i^- are the negative deviations of groundwater and surface water supply for human and livestock use in i subdivision.

P3: to satisfy the industrial water use:

$$\text{Min } \sum_{i=1}^{16} (DGG_i^- + DGS_i^-)$$

where DGG_i^-, DGS_i^- are the negative deviations of groundwater and surface water supply for industrial use in i subdivision.

P4: to satisfy agricultural water use:

$$\text{Min } \sum_{i=1}^{16} (DNG_i^- + DNS_i^-)$$

where DNG_i^-, DNS_i^- are the negative deviations of groundwater and surface water supply for agriculture in i subdivision.

P5: to satisfy forestry water use:

16

$$\text{Min} \sum_{i=1} (DLG_i^- + DLS_i^-)$$

where DLG_i^-, DLS_i^- are the negative deviations of groundwater and surface water supply for forestry business use in i subdivision.

P6: to maximise the output of industries and agriculture:

$$\text{Min } DE^-$$

where DE^- is the negative deviation of desired total output index (The unit is "10^4 yuan", the same below) of industries in Shiyang river basin.

P7: to minimise the investment in water development:

$$\text{Min } DC^+$$

where DC^+ is the positive deviation of desired investment index of water development in the basin.

P8: to make the water level drawdown in the optimal range of water level fluctuation:

$$\text{Min} \sum_{i=1}^{16} (DS_i^- + DS_i^+)$$

where DS_i^-, DS_i^+ are the negative and positive deviations of water level drawdown in i division; a_i is the weight of water level objective.

P9: to make the concentration increment of water quality (here, it denotes the TDS) the minimum:

$$\text{Min} \sum_{i=1}^{16} b_i (DM_i^+)$$

where DM_i^+ is the positive deviation of the TDS at controlled point; b_i is the weight of water quality objective.

The objective function is
$$\text{Min } Z = P_1 + P_2 + P_3 + P_4 + P_5 + P_6 + P_7 + P_8 + P_9$$

where the "+" symbol means satisfying the objectives in their given order.

The decision variables are:
XGG_i: the groundwater quantity for industrial use in i subdivision.

XNG_i: the groundwater quantity for agricultural use in i subdivision.
XRG_i: the groundwater quantity for human and livestock use in i subdivision.
XLG_i: the groundwater quantity for forestry business use in i subdivision.
XGS_i: the surface water quantity for industrial use in i subdivision.
XNS_i: the surface water quantity for agriculture use in i subdivision.
XRS_i: the surface water quantity for human and livestock use in i subdivision.
XLS_i: the surface water quantity for forestry business use in i subdivision.

The constraints of surface water balance are:

$$\sum_{i=1}^{2}(XGS_i + XNS_i + XRS_i + XLS_i) + DSJ^- - DSJ^+ = SJ$$

$$\sum_{i=3}^{5}(XGS_i + XNS_i + XRS_i + XLS_i) + DSM^- - DSM^+ = SM$$

$$\sum_{i=6}^{16}(XGS_i + XNS_i + XRS_i + XLS_i) + DSW^- - DSW^+ = SW$$

where SJ and DSJ$^-$ are the allowable surface water developing quantity and its negative deviation in Jinchang basin; SM and DSM$^-$ are the allowable surface water developing quantity and its negative deviation in Minqin basin; SW and DSW$^-$ are the allowable surface water developing quantity and its negative deviation in Wuwei basin.

The constraints of the groundwater balance are:

$$\sum_{i=1}^{2}(XGG_i + XNG_i + XRG_i + XLG_i) + DGJ^- - DGJ^+ = GJ$$

$$\sum_{i=3}^{5}(XGG_i + XNG_i + XRG_i + XLG_i) + DGM^- - DGM^+ = GM$$

$$\sum_{i=6}^{16}(XGG_i + XNG_i + XRG_i + XLG_i) + DGW^- - DGW^+ = GW$$

where GJ and DGJ$^-$ are the allowable groundwater pumpage and its negative deviation in Jinchang basin; GM and DGM$^-$ are the allowable groundwater pumpage and its negative deviation in Minqin basin; GW and DGW$^-$ are the allowable groundwater pumpage and its negative deviation in Wuwei basin.

The constraints of water supply objectives are:

$$\begin{aligned}
XGG_i + DGG_i^- - DGG_i^+ &= GG_i \\
XGS_i + DGS_i^- - DGS_i^+ &= GS_i \\
XNG_i + DNG_i^- - DNG_i^+ &= NG_i \\
XNS_i + DNS_i^- - DNS_i^+ &= NS_i \\
XRG_i + DRG_i^- - DRG_i^+ &= RG_i \\
XRS_i + DRS_i^- - DRS_i^+ &= RS_i \\
XLG_i + DLG_i^- - DLG_i^+ &= LG_i \\
XLS_i + DLS_i^- - DLS_i^+ &= LS_i
\end{aligned}$$

where GG_i and GS_i, DGG_i^+ and DGG_i^-, DGS_i^+ and DGS_i^- are the groundwater and surface water supply for industrial use and their positive and negative deviations in i division respectively; likewise NG_i and NS_i, DNG_i^+ and DNG_i^-, DNS_i^+ and DNS_i^- are the equivalent variables for agriculture use; RG_i and RS_i, DRG_i^+ and DRG_i^-, DRS_i^+ and DRS_i^- are for human and livestock use; LG_i and LS_i, DLG_i^+ and DLG_i^-, DLS_i^+ and DLS_i^- for forestry business use.

The constraints of the economic objectives are:

$$\sum_{i=1}^{16} (\varepsilon G_i \,(XGG_i + XGS_i) + \varepsilon N_i \,(XNG_i + XNS_i) + \varepsilon R_i \,(XRG_i + XRS_i)$$
$$+ \varepsilon L_i \,(XLG_i + XLS_i)) + DE^- - DE^+ = E$$

$$\sum_{i=1}^{16} (CG_i(XGG_i + XNG_i + XRG_i + XLG_i) + CN_i(XGS_i + XNS_i + XRS_i + XLS_i)$$
$$+ DC^- - DC^+ = C$$

where εG_i, εN_i, εR_i, εL_i are the efficiency coefficients caused by unit water use of industries, agriculture, human and livestock, and forestry business respectively in subdivision i (the unit is "10^4 yuan 10^{-4} m^{-3}"); CG, CS_s are the operation fees for developing unit groundwater and surface water quantity respectively in i subdivision; E, DE^+ are the desired total output index of industries and agriculture and its positive deviation (the unit is "10^4 yuan") in Shiyang river basin; C, DC^- are the desired investment index of water development and its negative deviation in the basin.

The constraints of water level are:
(i) For maintaining the existence of springs:
$$[A].(XGG + XNG + XRG + XLG) + DS^- - DS^+ = -[S_y]$$
(ii) For water development constraint:
$$[A].(XGG + XNG + XRG + XLG) + DS^- - DS^+ = [S_{max}] - [S_0]$$
(iii) For eco-environment constraints:

Table 1 (a) The results of the calculations: Near future plan (before 1995)

		WUI	WUHL	WUF		WUA	
		NY	NY	NY	DY	NY	RY
Jingchang	GW		423	1031	4333.3	5510	5510
	SW	9281	1163	843	4604	4604	4604
Changning	GW		14	48	862.9	999	999
	SW						
S.	GW		437	1079	5196.2	6509	6509
	SW	9281	1163	843	4604	4604	4604
Huqu	GW		51	102	1278	1882	1882
	SW		98	590	11790	13054	13054
Baqu	GW	42	190	368	6032	6032	6032
	SW			404	3876.3	6610	6610
Quanshan	GW		115	271	4853	5169	5169
	SW			404	4260	7660	7660
S.	GW	42	356	741	12164	13083	13083
	SW		98	1398	19926	27324	27324
Huanhe	GW		41	161	1080	1980	1980
	SW						
Yongchang	GW	67	124	219	6236	9018	9018
	SW	5	9	16	641	641	641
Jinyang	GW	44	143	369	4603	5820	5820
	SW	16	39	140	2205	2205	2205
Qingyuan	GW	61	135	400	5219	5828.6	8984
	SW	11	24	71	1594	2594	1594
Qinghe	GW		256	1547	4745	5124	6922
	SW		62	377	1688	2604	1688
Gulang	GW		6	12	180	180	180
	SW	130	265	497	5481	7429	7429
Zhamu	GW	31	32	43	1169	1169	1169
	SW	479	488	652	15622	17881	17881
Jinta	GW	51	43	36	612	930	930
	SW	500	419	351	9129	9129	9129
Xiyeng	GW	68	56	79	1101.8	1963	2471
	SW	192	584	834	20796	25990	25990
Wujiajing	GW		16	96	639	953	953
	SW						
Huangyang	GW						
	SW	566	434	880	6594	10929	10929
S.	GW	272	812	2962	25584	32966	38427
	SW	1079	2324	3818	66009	79403	77486
T.S.	GW	314	1608	4782	83369	52558	58019
	SW	11180	3589	6059	9053	111330	109414

Table 1 (b) The results of the calculations: The specified future plan (2000)

| | | WUI | WUHL | WUF | | WUA | |
		NY	NY	NY	DY	NY	RY
Jingchang	GW		1676	1413	1008	4742	5940
	SW	20007	1706	1437		5470	6042
Changning	GW		73	91	451.15	530	860
	SW						
S.	GW		1749	1504	3459.2	4772	6800
	SW	20007	1706	1437		5470	6042
Huqu	GW		85	428	3429	2562	3629
	SW		176	1653	9356	10202	12356
Baqu	GW	42	322	836	5132	3884	5132
	SW			1732	1129.3	8524	10532
Quanshan	GW		200	801	2028	5062	5767
	SW			1381	5995	5846.5	4537
S.	GW	42	607	2065	10589.9	11508	14328
	SW		176	4766	16480	24572.5	27625
Huanhe	GW		68	315	1703	1703	1703
	SW						
Yongchang	GW	80	175	322	6728	7728	7728
	SW	6	14	27	623	623	623
Jinyang	GW	50	142	481	3766	4746	4746
	SW	22	63	213	1579	2102	2102
Qingyuan	GW	71	189	531	690	7443	7443
	SW	15	39	110	1290	1537	537
Qinghe	GW		329	179	6664	6394	6394
	SW		109	505	1412	2123	2123
Gulang	GW		14	21	233	233	233
	SW	132	441	669	529	5627	7269
Zhamu	GW	6	7	11	180	180	180
	SW	606	668	1036	13928	15004	15826
Jinta	GW	38	37	32	532	532	532
	SW	622	607	515	6625	7411	8616
Xiyeng	GW	8	27	41	862	862	862
	SW	243	797	1196	15771.8	22721	24148
Wujiajing	GW		120	120	721	721	721
	SW						
Huangyang	GW						
	SW	679	574	1115	6990	9272.7	9587
S.	GW	253	1008	2053	26316	30542	28839
	SW	2325	3312	5386	53509	66421	71792
T.S.	GW	295	3364	5622	40366	46822	49967
	SW	22332	5194	11589	68669	96463	105258

Notes: WUI: Water use for industries. WUHL: Water use for human & livestock.
WUF: Water use for forestry. WUA: Water use for agriculture. DY: Dry year. NY: Normal
year. RY: Rainy year. S.: Sum. T.S.: Total sum. GW: Groundwater. SW: Surface water.

$$[A].(XGG + XNG + XRG + XLG) + (DS^-) - (DS^+) = [S_{max}] - [S_0]$$
$$[A].(XGG + XNG + XRG + XLG) + (DS^-) - (DS^+) = [S_{min}] - [S_0]$$

(Lemoine *et al.*, 1986)

where [A] is the matrix of water level responding; $[S_0]$ is the matrix of additional water level drawdown; $[S_{max}]$ is the matrix of maximum water level drawdown at the controlled points; $[S_{min}]$ is the matrix of minimum water level drawdown at the controlled points.

The constraint of the water quality objective is:

$$B_i (XGG_i + XNG_i + XLG_i + XRG_i) + DM_i^- - DM_i^+ = C_{maxi} - C_{0i}$$

where DM_i^- is the negative deviation of TDS at controlled point i; C_{maxi}, C_{0i} are maximum TDS controlled index and original TDS at controlled point i.

Non-negative constraints

All the above objective function and constraints equations are formed as part of the multi-objective programming model of the Shiyang river basin.

CONCLUSION AND ANALYSIS

Based on the designed plans of the near future and the specified future plan, the result of the calculation is shown in Table 1 (previous page).

The following conclusions can be drawn from the results:

- The groundwater pumpage and surface water developing quantity do not exceed the allowed value when the optimal plans are operated.
- In the near future (before 1995), water supply cannot satisfy water demands in a normal and dry year: the water deficit is about 36077.98×10^4 m³ in a dry year and 3544.72×10^4 m³ in a normal year. The water deficit areas are mainly in Qinyuan, Qinhe and Xiying but the water supply is bigger than the water demand in a rainy year.
  At a specified future plan (2000), the water supply cannot satisfy water demands in normal, rainy and dry years: the water deficit is 20399.8×10^4 m³ in a dry year and 6746.5×10^4 m³ in a rainy year.
- From the economic point of view, the total output value of industry and agriculture is lower than the desired value in the normal and dry year according to the near future plan. The deficit values are 208.58×10^4 yuan and 1068.89×10^4 yuan respectively. In a rainy year, the investment in water development exceeds the desired cost of the near future plan.
  For the specified future plan, the total output value of industries and agriculture is lower than the desired value of the plan in the year of 2000 and the water developing investments are within the desired investment budget.
- The over-development of existing groundwater has caused continuous

groundwater level drawdown, has caused vegetation to wither, and has led to soil desertification and salinisation.

Because of the regulation of well distribution and water pumping quantity after optimization, the groundwater drawdown is limited at the optimal condition and water quality degradation is controlled, which benefits the eco-environment.

Based on the above analysis, the environmental efficiency is best with the most severe deficit between water supply and water demand in a dry year. In a rainy year, the water demand can be fully satisfied with water supply, but the environmental efficiency is the worst. Only in a normal year can the water demand and environmental efficiency be satisfied. Therefore it is suggested that the plan in a normal year should be the adopted plan from a long-term point of view. Under this condition, the surplus water from a rainy year can be stored to satisfy water demand in a dry year.

Furthermore, importing water from outside the basin and accepting water saving measures are essential ways to solve both the environmental problem and the problem of conflicting aspects of water supply and demand.

REFERENCES

Lin Xueyu and Hou Yinwei 1988. *Computer programs of groundwater quantity and quality.* Scientific book.

Lemoine, P.H., Reichard, E.G. and Remson, I. 1986. Matrix method for coupling a groundwater simulator and a regional agriculture management model. *Water Resources Bulletin,* **22.**

13 Review of the Southern Okavango integrated water development project

R E MANLEY
Consultant in Engineering Hydrology
E P WRIGHT
Hydrogeological Consultant

BACKGROUND

The integrated management of a river basin can mean different things in different basins. This paper considers the use of the Okavango river and delta for water resources while at the same time maintaining its attraction to tourists and its ecological integrity as one of the world's great wetlands.

Botswana is one of Sub-Saharan Africa's richest countries—its wealth comes mainly from diamonds, but also from the sale of beef and from tourism (see Figure 1). For the tourists, one of the main attractions is the Okavango delta which seasonally floods up to 20 000 km² of the Kalahari desert, which is why the Okavango is sometimes called the Jewel of the Kalahari. The inflow to the delta comes from the River Okavango which rises in Angola where the average rainfall is over 1200 mm a⁻¹. At its northern end it consists of perennial papyrus swamps, with channels interconnecting *lediba*, oxbow lakes from an earlier stage of the delta's development. Many of the channels are poorly defined, being little more than tracks made by animals through the floating papyrus beds. Further south more of the delta is seasonal swamp, but still with permanent pools; permanent islands are characteristic features of both perennial and seasonal swamps and have developed in a variety of ways, through geomorphological changes, tectonic activity and geochemical precipitation. There are several distributaries from the delta, the main one being the Boteti which flows south-eastwards.

The Okavango delta is not a true delta but an alluvial fan whose primary origin and, to some extent, evolution has been controlled by regional earth movements and land subsidence. The delta overlies a sedimentary basin which has been in existence with progressive downwarping and faulting since the late Mesozoic period. There is a notable occurrence of NE-SW faults on the southern peripheral margin (Figure 2). The basin infill includes mainly unconsolidated sediments (Kalahari Beds—Tertiary to Recent) with thicknesses up to some 300 m and probably overlying older, more consolidated, sedimentary rocks (Karoo Age).

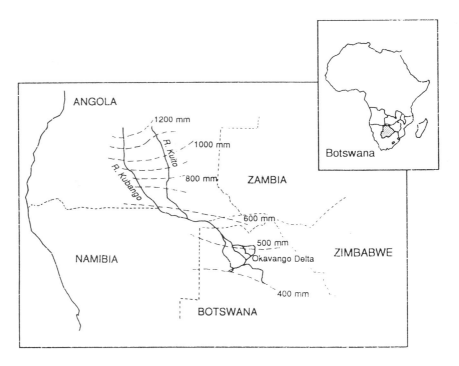

Figure 1   The Okavango delta — location map

The Okavango delta is probably a relatively recent feature but the proto-Okavango river may have had a significantly different course(s) with theories linking it to the Zambezi, Limpopo or Orange rivers (Thomas and Shaw, 1991). Groundwater occurrence in the underlying sedimentary infill is not well known other than to shallow depths in the southern and western marginal delta areas. Fresh groundwater in these locations corresponds with current distributary occurrence. Increasing salinity at depth is thought to be a reflection of an absence of flushing and is not necessarily representative of conditions below the main delta. Subsurface groundwater outflows appear constrained — low hydraulic gradients to the NE and SW; geological boundaries to the SE beyond the main marginal fault. Recharge to the Kalahari Beds along the Boteti river valley is however an important factor in available water resources.

The river system in the delta is not stable (Wilson, 1973; Shaw, 1984; Scudder *et al.*,1993). When David Livingstone first visited the delta in the mid-nineteenth century the Thaoge river to the west was the main distributary. Flows then became concentrated in the Gomoti to the east (Figure 3). From around 1950 the main outflow from the delta has been the Boro River in the centre of the delta. The river Boro discharges into the Thamalakane and then into the Boteti. The reasons for the instability are not completely understood. At least part of the process is common to all deltas, namely the deposition of silt in river channels which then become higher than surrounding land, with at some point instability resulting in the river taking a new

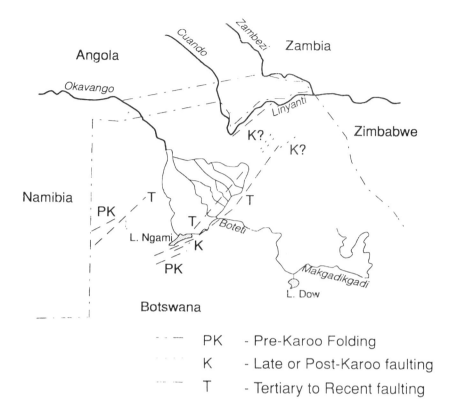

PK   - Pre-Karoo Folding
K   - Late or Post-Karoo faulting
T   - Tertiary to Recent faulting

Figure 2 Tectonic map of Ngamiland

course. Another reason is that vegetation, principally floating beds of papyrus, breaks away and blocks river channels; it has also been suggested that earthquakes and channel clearance by man have been factors (McCarthy (1992)).

Some 200 km down the Boteti is the Orapa diamond mine, which earns a large percentage of the country's foreign exchange and which currently needs around 5 MCM (million cubic metres) of water each year for processing the ore. Near to the Thamalakane/Boro confluence is the rapidly growing town of Maun which is the District's administrative centre and also services the tourist industry in the delta.

To meet these two water needs, and other proposed needs such as irrigation and riparian agriculture, a project called the Southern Integrated Water Development Project (SOIWDP) had been under consideration from 1985 onwards (Figure 4). The final proposal was to deepen and widen the Boro river for 42 kilometres upstream of its confluence with the Thamalakane and to construct three reservoirs, one to control water levels for flood recession agriculture and the other two for water storage. Despite efforts by the government to inform local residents of their proposals, when engineering plant arrived on site to start work, strong opposition to the scheme rapidly developed. Part of the opposition was from a deep-seated belief of the local people

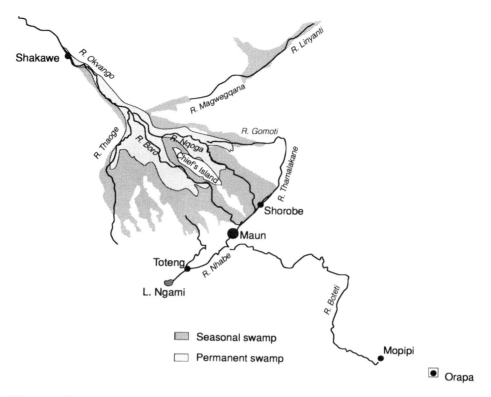

Figure 3   The Okavango delta

that changing rivers was against nature and could only be for the worse (earlier river works had served only to establish this fear) and also from the international community which was concerned with the possible environmental impacts. Following a *khotla*, an open air public gathering when anybody can participate and express their views, which the citizens of Botswana do with vigour, the Government cancelled the engineering contract and suspended the project.

The World Conservation Union (IUCN) was invited by the Government of Botswana to conduct an independent review and to advise on alternative proposals, if deemed more appropriate. The review was financed by SIDA (the Swedish International Development Agency), NORAD (Norwegian Development Co-operation) and the Government of Botswana.

THE REVIEW

In October 1991 the multi-disciplinary team started work. There were thirteen members from five countries covering a range of disciplines (biology, botany, economics, engineering, geography, hydrogeology, sociology and water resources and land use planning). The team leader was a Professor of Developmental

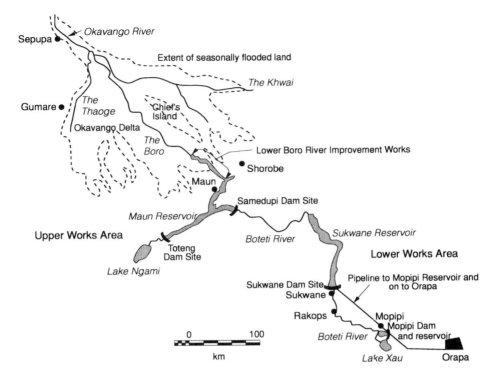

Figure 4 The proposed engineering works

Anthropology and his deputy was a hydrologist/engineer. These two were the only full time members of the team during the eight-month study. To ensure the team's independence none of the members had had any connection with the development of the project nor any recent connection with Botswana. Three of the team had worked in Botswana previously, one was fluent in Setswana, one of the principal local languages, and another had done geological mapping in the region of the delta while the country was still a protectorate. To compensate for the lack of recent experience in Botswana the team was allowed to recruit 'local experts' with specialised knowledge.

The Government of Botswana, principally through its Department of Water Affairs but also other Departments and Ministries, fully supported the review. Not only did they provide the data and reports necessary for an understanding of the decision making process leading up to the study, they also accepted that the study was an independent one and never at any stage tried to influence the team's thinking or appraisal. Good co-operation was also obtained from the NGOs, some of whom had been critical of the SOIWDP.

The review started with a visit to the delta by many of the team and the collection of data and reports. After this the team members started pursuing studies in their own disciplines.

A hydrological model of the delta had been developed during a project sponsored for the Food and Agriculture Organisation of the United Nations in the 1970s (Dincer, 1985). The original model had simulated the delta as a single cell. This model was further developed and used by the team designing the SOIWDP (Dincer, Child and Khupe (1987). Whilst this model gave a good representation of the annual flows its representation of monthly flows was less good. It was therefore decided to develop a model of the delta for comparison purposes. The main differences introduced were better representation of the depth/area relationships and a parameter to take account of the water loss as the delta filled up and previously dry areas became inundated. This model gave the hoped for improvement in monthly flow simulation. The main data were inflows at the northern end of the delta (10,000 MCM/annum or 316 cumecs), rainfall at Maun and Shakawe and outflows at the lower end of the delta (less than 5% of the inflows).

During this period other components of the study were also under way. These included a socio-economic study of the potential benefits for the riparian population and a review of the hydrogeology.

THE OUTCOME

At the end of the study the team unanimously proposed that the SOIWDP be terminated. The conclusions covered four main areas:

- The socio-economic study had confirmed the deep-seated belief of the local people that altering rivers was flying in the face of nature. Also, by studying how people lived and what activities they carried out to get food, the team was able to demonstrate that, even if the project had worked as intended, the local people who were supposed to benefit would in fact be worse off. One of the main reasons for this was that the reservoirs would have eliminated a large proportion of the area of river bed currently used for flood-recession agriculture.

- The main aspects of concern to the hydrogeologist were the water supplies to Maun, the riparian populations and the Orapa diamond mine. In the case of Maun, it was considered that the potential for additional groundwater development had been discounted in the original study with insufficient justification. Groundwater for rural populations and livestock in the riparian areas is an important resource and becomes increasingly so in drought periods and in the lower Boteti which floods at less frequent intervals. It was considered that the proposed modifications to the pattern of flows and related recharge would put these supplies at risk with little or no compensating advantage. In the case of the diamond mine the operating company had developed a mathematical model of the aquifer which had indicated adequate supplies for the foreseeable future. The team concluded that the results of this study were valid. Although this water was being mined it was considered that alternative uses for the highly saline water meant that the mine was not depleting a source of great future value. It also became apparent that as the surface water to be released under the proposed scheme would not constitute a guaranteed continuous supply of water to the mine, additional investment in a groundwater system would still be necessary.

- An important environmental concern was that the regraded channel of the Boro River would leave a very unnatural looking river in an area much used by tourists. The Government had been aware of this concern and the project had included plans for revegetation. It was felt that it had not been adequately demonstrated that this method would work. Another environmental concern was that of the health of the population living so close to an open body of water in a region where malaria and bilharzia are endemic.

- In the area of water resources the team demonstrated that, at the very least, there were serious doubts that the project would have worked as intended. There were four reasons for this:

(1) A flow record for the delta inflows was available for 1933 onwards. From 1933 to 1946 water levels had been recorded at a site whose whereabouts are now known only approximately; from 1947 to the present flow measurement is of good quality. For the original study it was assumed that the site of the early levels was close enough to the current ones for the present day flow rating to be used. When these flows were used with the simulation model of the delta they indicated that in 1941 and 1942 there had been no outflow from the delta. This was investigated further using rainfall data from Angola, where the Okavango receives most of its runoff, details of two gaugings carried out in 1937 and a comparison of water levels and average flows for two periods of more than a decade which had similar rainfall. None of the studies in itself could be considered conclusive but all led to a similar conclusion, namely that the earlier study had in fact underestimated flows at the entry to the delta by about 20% during the early years and that water would have flowed out of the delta in both drought years. The earlier project had calculated that, with the current river system in the delta, there would have been 30 months with no flow at Maun during the 1940/42 drought. Using the modified inflow record the team calculated that there would only have been nine months with no flow.

(2) It had proved impossible to get a set of parameters for the earlier model which was able to simulate the whole of the recorded outflow of the delta from 1969 onwards. It had therefore been concluded that the delta responded differently during different periods, or to use the terminology adopted for that project, there were three different 'regimes'. At first the review team was not able to explain these differences other than by a similar hypothesis to that of the earlier study until it looked in more detail at the data. A double-mass plot of the two rainfall stations used for the study was produced which exhibited two kinks in the line. By plotting each of the stations against groups of other stations it was possible to identify which of the two was in error and at which date. As the rainfall is affected by movement of the inter-tropical convergence zone, whose variable movement might have caused some anomalies, two separate tests were carried out using nearby and more distant groups of gauges which showed the same changes. The effect of errors in rainfall was magnified in both the models as there is no long term evaporation data so this was estimated by correlation with rainfall over a twenty

year period and the rainfall was then used to calculate evaporation. When the rainfall had been corrected it was found that it was no longer necessary to introduce the concept of 'regimes' and that one set of parameters with corrected data gave almost as good a simulation as three sets of parameters and uncorrected data. The fact that it was possible to obtain a good simulation of flows with a single stable parameter set was a further indication that the model was a valid tool for studying the delta.

Whilst the above were important findings they did not answer the question "Would the SOIWDP have increased flows down the river?". After all, the people living along the Boteti are not very concerned about whether the average flow for the last 60 years has been 200 or 300 MCM; what interests them is whether work on the river will improve water availability. This leads to two other hydrological findings.

(3) One of the most contentious proposals was for channel excavation of 42 km of the lower reaches of the Boro river. The aim of this was to reduce water levels in the excavated reach for a given flow, thereby reducing flooded areas and in turn evaporation. With less evaporation there would be increased outflow from the delta. The original study calculated that there would be an almost constant annual increase in flow 40 to 55 MCM. This would have been ideal as there would be good benefit in dry years - when total natural outflow could be as low as 30 MCM - but at the same time minimal effect in wet years - when flow would be up to 800 MCM. Unfortunately it is difficult to measure flows within the delta itself; in dry years particularly access is not easy except during the height of the flood and the channels are poorly defined by banks of reeds. Nevertheless three gaugings were made at the upper and lower end of the proposed works. For the original study these values were used with the delta model, which was linear in its response, to estimate the gain in yield at other times. The team's approach was more direct. For the three gaugings the downstream flow was always less than the upstream flow; the difference represented the evaporation loss between the two points. It was reasoned that the maximum theoretical gain in flow would occur if all these losses were eliminated. From a curve of these losses against flow the envelope of this maximum theoretical gain was defined. When this envelope was applied to the gains calculated at the design stage it was found that the average gain was reduced by a half and that in dry years the gain was very small; a very different conclusion to the original study.

(4) The reservoirs proposed for the project were long, thin and shallow and in the bed of the existing rivers. The furthest upstream of the two had an average depth of 2.08 m (compared with the average open water evaporation of 2.17 m), a width of 504 m and a periphery of 258 km. At present the rivers are fringed with trees for a depth of 100 to 200 m. Beyond the trees there is only scrub and bushes. Therefore the trees are getting their water from the river. The amount of this loss which will continue after reservoir construction is difficult to calculate, our best estimate was around 20 MCM/annum, but in any case it will be higher than the zero which was assumed by the original study. It is likely that the trees would be

affected by the drowning of their roots. However, even if some of them died off seepage would still continue.

When all these factors were taken into account it was estimated that construction of the reservoir would lead to a reduction on flows downstream of the Samedupi Dam of 26 to 46 MCM. The original study had calculated an increase of 4 MCM. Since one of the criteria for the scheme was that there should be no reduction in downstream flow then the review's studies indicated that this criterion would not have been met.

THE ALTERNATIVE

In view of the team's conclusions that the SOIWDP would not meet the objectives claimed, consideration of an alternative was needed. A major component of the SOIWDP was the provision of an assured water supply for Maun. At present Maun gets all its water supplies from shallow, small diameter boreholes which currently yield in excess of 1 MCM/year. The existing supplies are drawn from two well fields, one in a buried sand channel which borders the Shashe river and the second in the sand-clay sequence below the Thamalakane river (Figure 5). The Shashe aquifer is recharged during the intermittent flows in the Shashe river (approximately once every three years); the relation of the aquifer in the sand-clay sequence below the Thamalakane to surface water is not clearly understood and there is also some deterioration of quality at around 30-40 m depth. The resources of the Shashe aquifer depend on recharge intervals, natural outflow rates and the manipulation of storage. Calculations indicate a possible aquifer full condition of 6 to 12 MCM and a modelling study of the initial well field (BRGM, 1986) indicated the feasibility of yields reducing from 0.66 MCM to 0.33 MCM during a three year drought. The resources of the Thamalakane aquifer have not been evaluated in the same degree of detail.

Demand projections for 2010 are between 2.5 and 4.0 MCM/year, the range reflecting some difference in the controlling assumptions. Groundwater development in a conjunctive use scenario would allow maximum manipulation of available storage with potential amounts in excess of 4.5 MCM from the Shashe well field alone in an aquifer full condition. Artificial recharge would assist control of the storage and pilot studies in 1986/87 were encouraging. The yields of the Thamalakane boreholes appear more consistent although constrained by high drawdowns and a deterioration of water quality with depth. Collector wells could improve both abstraction rates and water quality and may demonstrate better interaction with ground and surface water storage.

Further groundwater development would have significant financial and other advantages over the proposed channel excavation and reservoir construction. In addition to substantially lower costs, effective management is more feasible. The evaporation loss from the proposed Maun reservoir would average 68 MCM/year with additional seepage losses of up to 20 MCM/year which together greatly exceed projected demands (2.5 to 4.0 MCM/year). In a protracted drought it would be

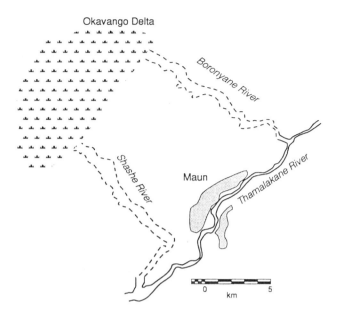

Figure 5 Existing well fields

impossible to effect any significant conservation of resources and, once the reservoir had dried up, supply would cease. On the other hand, demand management by controlling borehole abstractions and bringing additional supplies on-line from more distant well fields is a more practical option. Potential also exists for regional aquifers below the main delta but as yet there has been no exploration. Preliminary conclusions are therefore that a conjunctive use scenario is feasible for the 2010 demand projections, even with the existing well fields and assumed drought durations. Fuller information on both ground and surface water resources would improve planning and management.

THE PRESENT SITUATION

This type of study — looking at the work of other professionals - is always difficult. The first requirement is to remain objective but this is not always easy. This is particularly true in a case like this where so much hinges on the interpretation of data of doubtful accuracy. Throughout the study the team was able to maintain good relations with both the government and NGOs.

The situation was complicated by the fact that in 1993 the inflow to the delta, regardless of whichever data series is correct for the 1933 to 1946 period, was the lowest on record. So consequently the outflow was also very low. The situation in Maun would have been difficult whatever had been done but did lead to questions concerning the best course of action. On the other hand there were some encouraging signs regarding potential locations for new well fields in the vicinity of Maun. One of the aquifers we considered for development was successfully tested by exploration drilling. Another aquifer, at present unused, which had been evaluated in some detail, and which we had thought was recharged from surface flows only once or twice in a decade, was at least partly recharged even during this extreme drought. Detailed studies, which we recommended, will of course be needed to confirm the potential of these resources.

Over the last year or so there have a number of meetings, usually in different contexts, which have enabled the team's conclusions to be discussed. The Department of Water Affairs has started to implement works which are in line with the recommendations, by installing a water treatment plant to use the water in the river for example. Recent press reports from Botswana have indicated that future development will concentrate on conjunctive use of surface and groundwater.

Although - as the terms of reference required - the team indicated the outline of an alternative which seemed to offer advantages to the original proposal, they would not claim that they had provided a definitive solution. First, whilst there are positive signs, more detailed studies of both the groundwater and the hydrology are necessary to prove that the resources are there and can be used as efficiently as expected. Second, the interaction between the delta and the groundwater in the surrounding areas remains poorly understood. At the moment the major loss mechanism in hydrological models of the delta is evapotranspiration and the additional loss to groundwater appears to be small enough to be within the range of error in evapotranspiration estimates. Both these aspects require further clarification.

CONCLUSIONS

- The Government of Botswana took a very brave decision in inviting an international team to review their proposals for the Okavango. One wonders how many major water developments in the UK or elsewhere would have survived such detailed scrutiny. The government not only invited the review but they also decided to cancel the project when the team's findings were published. The reason for the cancellation was at least as much the response of a democratic government to the local opposition as to the team's recommendations.

- The need to maintain independence during the review meant that the team was always working at arm's length from those who had been responsible for the promotion of the SOIWDP. Although, as far as possible, the calculations were cross checked within the team it was not possible during the study to submit them to detailed appraisal by the Department of Water Affairs or others. Whilst the team tried to be as constructive as possible in its review of the project and in making recommendations, a further phase in which its premises, arguments and conclusions could have been discussed in an open way would have been useful.

ACKNOWLEDGEMENTS

The authors wish to thank their colleagues for the help and support during the study and all those who contributed data and information. The views expressed in this paper are those of the authors and do not necessarily represent those of the IUCN or the Government of Botswana.

REFERENCES

BRGM, 1986, *Maun Water Supply - hydrogeological survey*. 2 volumes. Report to the DWA.

Dincer, T. 1985. *Okavango Swamp Model - Phase II*. Department of Water Affairs, Gaborone.

Dincer, T., Child, S. and Khupe, B.B.J. 1987. A simple mathematical model of a complex hydrologic system. *J. Hydrol*, **93**, 41-65.

McCarthy, T.S. 1992. Physical and Biological Processes Controlling the Okavano - A Review of Recent Research. Dept of Geology, Univ. of the Witersrand, Johannesburg, S.A.

Scudder, T. *et al*. 1993. The IUCN review of the Southern Okavango Integrated Water Development Project. IUCN Wetlands Programme, Gland, Switzerland.

Shaw, P. 1984. A historical note on the outflows of the Okavango Delta system. *Botswana Notes and Records, 16*.

Thomas, D. S. G. and Shaw, P.A. 1991. *The Kalahari Enviroment*. Cambridge University Press.

Wilson, B.H. 1973. Some natural and man-made changes in channels of the Okavango delta. *Botswana Notes and Records, 5*.

PART IV

MANAGEMENT OF LOW RIVER FLOWS

14 Low flow studies of the Khimti River in Nepal

PRAVIN KARKI

BPC Hydroconsult, Kathmandu, Nepal

INTRODUCTION

Water is Nepal's greatest natural resource. In a drive to utilise this resource, the 60 MW Khimti River Hydroelectric Project is being proposed (see Figure 1). This will take advantage of a gross head of 684 m and will have a tunnel length of 10 km; the intake will be a low diversion weir with limited pondage.

The Khimti basin has an area of 443 km^2, set in the High and Low Himalayas. Extreme seasonal variations in river discharges will significantly affect the amount of water diverted at the dam site for ultimate power development. The tributaries of the Khimti River do not contribute enough water to the Khimti during low flow periods to sustain the various water users. Therefore, there will be considerable effects on water users in the areas below the proposed dam site. Optimal use of the dam will require increased power generation in the monsoon period to make use of the large monsoon discharges and to compensate for the flow diverted during the period of low flow to sustain the various water users. The Khimti Valley has a biologically diverse and rich ecosystem, including some rare and endangered species, which construction of the dam could significantly affect.

REFERENCE HYDROLOGY

Existing Department of Hydrology and Meteorology (DHM) gauges, are generally located at some distance from the tributaries of the Khimti, as is the case elsewhere in Nepal. The records from these existing gauges are not therefore directly applicable to the ungauged sites under study. The low flow studies necessary for the project require the development of reference hydrology appropriate for the ungauged sites. It is therefore imperative to consider how best to transpose the hydrological records from the nearest gauging station to a particular tributary and to generate the required reference hydrology for ungauged sites in the Khimti basin. The estimation of the flows contributed by these ungauged tributaries to the Khimti flows below the proposed intake site will help to assess the impacts on the environment of the area.

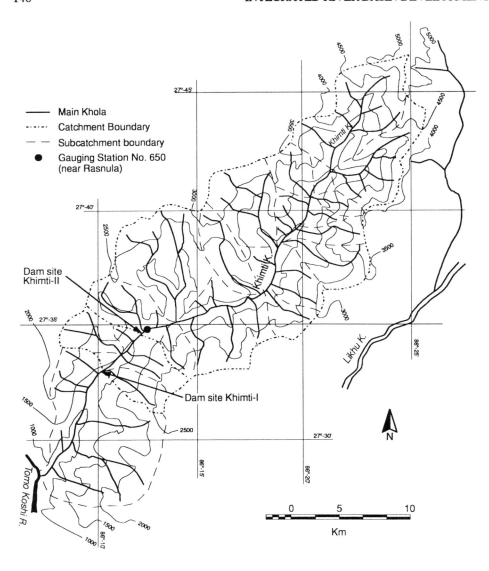

Figure 1 The Khimti catchment, showing the Khimti Khola Hydroelectric Project

METHODOLOGY

Field work

In April and May 1992, discharge measurements were conducted with a current meter. The April visit of the field workers coincided with the occurrence of the lowest discharges at the end of a year drier than average. The May visit, on the other hand, was already affected by the pre-monsoon showers. The results of the measurements are given in Table 1.

Table 1 Detailed list of field measurements

Distance from intake (km)	River Name	Location	Area below 5000 km²	April		May	
				Flow ($l\,s^{-1}$)	Spec. flow ($l\,s^{-1}\,km^{-2}$)	Flow ($l\,s^{-1}$)	Spec. flow ($l\,s^{-1}\,km^{-2}$)
-5.2	Khimti	Phulping	302	2981	9.9		
-5.0	Hosinga	Phulping	16.2	139	8.6	218.5	13.5
-0.4	Palati	intake site	7.9			81.5	10.3
+0.5	Hanba	Hanba	5.7			5.6	0.98
+1.0	Khimti	Bhangja	365	3218	8.8		
+1.5	Chhahare	Hanba	1.16	6	5.2	8.2	7.1
+2.0	Gairi	Chyama	3.1			11	3.5
+2.8	Kagune	Chyama	1.8			6.3	3.5
+5.6	Haluwa	Banchare	22	140	6.4	190	8.6
+7.1	Pharpu	Gobantar Besi	18			93.2	5.2
+7.8	Khahare	Gobantar Besi	3.3			11.7	3.5
+9.6	Khimti	Pharpu	430	3356	7.8	4271	9.9
+9.7	Chatiwane	Pharpu-Khimti	7.7	7	0.91	21.2	2.8

Theoretical approach

Typically, the transposition of hydrological records involves the pro rata adjustment of the hydrological data on the basis of the total drainage area at the proposed intake site compared with the total drainage area at the nearest gauge on the river. This method is only appropriate where both basins are hydrologically the same, but does give satisfactory results if both of the basins exhibit a reasonable similarity in characteristics.

To estimate yield from ungauged catchments, several regional techniques are available to predict flows. Three such methodologies were employed to predict flows in the ungauged catchments.

Medium irrigation project method (MIP)

The MIP method presents a technique for estimating the distribution of monthly flows throughout a year for ungauged locations. The MIP methodology uses a database consisting of DHM spot measurements. The occasional wading gaugings conducted by DHM include only low flows and these flows do not represent the natural conditions since they are residual flows remaining after abstraction for different purposes like irrigation. MIP presents non-dimensional hydrographs of mean monthly flows for seven different physiographical regions. These hydrographs present monthly flows as a ratio of the flow in April (lowest annual flow). For application to ungauged sites, it is necessary to obtain several low flow discharge estimates by conducting wading discharge measurements at particular sites. These measurement values are then used with the regional non- dimensional hydrograph to synthesize an annual hydrograph for a particular site. The discharge measurements conducted in April and May 1992 have been applied to the MIP methodology to generate a hydrograph.

WECS/Department of Hydrology and Meteorology (DHM) method
The Water and Energy Commission Secretariat (WECS) and DHM (1991) method is based on a series of regression equations that are derived from analyses of all the hydrological records from Nepal. These take into account different catchment parameters such as basin area, basin shape factor, length of main channel, main channel slope, land-use data, hypsometric curve data and wetness indices. The findings of this regression analysis have been used to produce equations for predicting flows and in the case of monthly mean flows, these depend on basin area, the area below 5000 m and the monsoon wetness index. The WECS method was used to determine the low flows, flood flows, long term average discharges and the flow duration curves in the Khimti tributaries.

Area ratio method
The product of the catchment area ratio (km^2) of the Khimti Khola and the tributaries and the specific discharge ($l\ s^{-1}\ km^2$) ratios give different sets of coefficients. These coefficients, when multiplied by the flows for the Khimti Khola at Station No. 650 yield flows for the tributaries. The results from all these techniques are presented in Table 2.

Comparisons of WECS, MIP and the area ratio methodologies
In the low flow period from November to May all three methods yield different flows but do not vary significantly in their respective magnitudes. However, the variations in the three methodologies for the wet season months (from June to October) are quite striking. The basic principle in producing a hydrograph of long-term average discharges for the low flow period lies in selecting the lowest figures of the three methodologies but trying to keep them close to the measured values. As far as the monsoon figures are concerned, either the WECS or the area ratio methods can be employed. Note that the MIP Design Manual suggests that "the WECS figures should more accurately represent total flow" in the wet season.

The major drawbacks in applying these methods for predicting flows are:
WECS: Delineation of drainage basins and elevation contours are often distorted on the available maps; also, regressions were derived on the basis of observed flows for catchments ranging in size from 4 up to 54 100 km^2. Therefore, for flows on the smaller tributaries of the Khimti, the results would prove to be unreliable.
MIP: The occasional measurements carried out by DHM are all residual flows, i.e. after abstraction for different purposes such as irrigation, and they do not represent the natural conditions: they cannot be expected to give a good estimation of total flow

Table 2 Generated long-term average discharges $(l\ s^{-1})$ for the tributaries of Khimti Khola

River	Method	Jan	Feb	Mar	Apr	May	June	July	Aug	Sep	Oct	Nov	Dec
Hanba	WECS	90	80	70	60	70	280	960	1210	960	420	180	120
	MIP	13	9	7	5	9	16	67	124	103	52	25	19
	area-ratio	96	82	75	86	132	575	1406	1431	917	395	192	124
Haluwa	WECS	310	260	230	210	260	1070	3330	4130	3230	1410	640	420
	MIP	314	218	160	116	218	362	1567	2893	2410	1206	579	434
	area-ratio	238	204	186	214	326	1426	3486	3546	2274	979	475	308
Khahare	WECS	60	50	40	40	40	160	620	780	620	270	110	70
at	MIP	21	15	11	8	15	25	106	196	163	82	39	29
Gobantar	area-ratio	56	48	44	50	77	336	821	836	536	231	112	73
Pharpu	WECS	250	220	190	170	210	880	2750	3410	2670	1170	530	350
	MIP	169	117	86	62	117	195	842	1555	1295	648	311	233
	area-ratio	304	261	237	274	417	1824	4458	4536	2909	4252	608	393
Chatiwan	WECS	120	100	90	80	90	370	1240	1560	1230	540	230	150
	MIP	37	25	19	14	25	42	183	338	282	141	68	51
	area-ratio	10	9	8	9	14	61	150	152	98	42	20	13

in the monsoon months.

AREA RATIO: About 1% of Khimti's catchment area is covered by permanent snow, and up to 25% by seasonal snow. The snow melt contribution to the Khimti flows have not been taken into consideration when applying this method. The Khimti Khola flows themselves are not totally reliable to start with, particularly in the monsoon period. It is also inherently assumed that rainfalls are similar over different parts of the catchment, which is clearly not the true situation. However, there are no better data available.

The final hydrograph of the flows contributed by the individual tributaries to the Khimti Khola is given in Table 3. When plotted, the results give some discontinuity between dry season and monsoon because of the different methods used but in the present situation this is of no concern. The monsoon flows are high, and no significant project impact is expected. The dry season flows given here are conservatively low, and the impact should be less severe than the results suggest.

PROJECT IMPACTS

The diversion of the Khimti Khola at the proposed intake site will cause a reduction of flows below the proposed intake site.

Table 3 Generated flows (l s⁻¹) of the tributaries

River	Jan	Feb	Mar	Apr	May	June	July	Aug	Sep	Oct	Nov	Dec
Hanba	13	9	7	5	9	280	960	1210	960	420	25	19
Haluwa	238	204	186	116	218	1070	3330	4130	3230	1410	475	308
Khahare	21	15	11	8	15	160	620	780	620	270	39	29
Pharpu	169	117	86	62	117	880	2750	3410	2670	1170	311	233
Chatiwane	10	9	8	9	14	370	1240	1560	1230	540	20	13
other river	22	20	16	17	20	150	600	720	610	250	35	27
Total	473	374	314	217	393	2910	9500	11810	9320	4060	905	629

REDUCTION OF DISCHARGES IMMEDIATELY BELOW THE INTAKE SITE

Water diversion at the proposed intake site will significantly reduce the amount of water available to the Khimti Khola downstream. Estimated diversion for ultimate power development ranges from a low of $3 \text{ m}^3\text{s}^{-1}$ in the low flow period to a high of $10.6 \text{ m}^3\text{s}^{-1}$ during the monsoon. Optimal use of the project will require increased power generation in the monsoon period to make use of the large monsoon discharges. The basic shape of the hydrograph will thus be significantly altered. In the low flow period, maximum available water will be diverted for power generation. A residual flow immediately below the intake of $0.35 \text{ m}^3\text{s}^{-1}$ has been estimated as being necessary to meet the needs of water users downstream during this period, while between 0.5 and $1.0 \text{ m}^3\text{s}^{-1}$ is estimated as necessary for the fishery. The release flow will govern the shape of the hydrograph from November to May, but for now a release of $0.35 \text{ m}^3\text{s}^{-1}$ has been assumed, which with the expected seepage through the intake weir of about $0.15 \text{ m}^3\text{s}^{-1}$, gives a residual flow of $0.5 \text{ m}^3\text{s}^{-1}$. During the monsoon period, the maximum diversion will be $10.6 \text{ m}^3\text{s}^{-1}$. Table 4 gives the expected discharge immediately below the project diversion site.

Table 4 Hydrograph of Khimti Khola flows ($\text{m}^3\text{ s}^{-1}$) immediately below the intake site

Jan	Feb	Mar	Apr	May	June	July	Aug	Sep	Oct	Nov	Dec
0.5	0.5	0.5	0.5	0.5	25.1	77.3	78.8	46.6	13.8	11.7	0.5

REDUCTION OF DISCHARGES AT KHIMTI BESI

The combined contribution of the tributaries to the Khimti flows up to Khimti Besi, near to the confluence with Tama Koshi has been added (as shown in Table 3) and the resulting flow after diversion is presented in Table 5.

Table 5 Hydrograph of Khimti flows (m^3s^{-1}) at Khimti Besi

Jan	Feb	Mar	Apr	May	June	July	Aug	Sep	Oct	Nov	Dec
0.97	0.87	0.81	0.72	0.89	28.0	88.8	90.6	55.9	17.9	2.01	1.13

EFFECTS ON WATER USERS

The effects of diversion of water at the proposed intake site relates mainly to impacts on aquatic life, irrigation, drinking water and wildlife habitats.

IMPACT MITIGATION

The most undesirable features of the diversion of water at the proposed intake site include the deficiency of water to the water users. The need to mitigate morphological and flow alterations in the stretch from the intake to the confluence relates mainly to impacts on irrigation, drinking water, fish and wildlife habitats.

CONCLUSIONS

The more significant conclusions derived from the low flow hydrological studies related to the diversion of the Khimti Khola at the proposed intake site and its consequential impacts on the environment below the proposed intake site. These are summarized as follows:

- Although the hydrological data such as stream flow and precipitation for parts of the basin are inadequate, the overall results of the data analyses conform reasonably with stream flow measurements at the hydrometric station No 650 at Rasnalu.
- The spot measurements conducted on the tributaries gave the best estimate of the contribution of these tributary flows in the low flow period to the Khimti Khola below the proposed intake site.
- The generated long-term average discharges for the tributaries is a result of application of several regional techniques based on theoretical assumptions. Their accuracy therefore, is subject to certain reservations.
- During the low flow period, a compensation release of approximately 0.35 m^3s^{-1} has been assumed to sustain aquatic life, irrigation, drinking water and other water users below the proposed intake site.
- After the contribution of the tributaries to the Khimti Khola below the proposed intake site and a compensation release, the minimum flow of about 720 ls-1 is estimated to be available to meet the needs of water users at Khimti Besi in an average year.

REFERENCES

DHM. Climatological records of Nepal. 1971 to 1986. Department of Hydrology and Meteorology, HMGN.

DHM. Surface water records of Nepal. 1971 to 1976. Department of Hydrology and Meteorology, HMGN.

DHM, 1991. Detailed gauging station data records. Department of Hydrology and Meteorology, HMGN.

MIP, 1990. Design manuals for irrigation projects in Nepal. M.3 Hydrology and Agro-meteorology manual. Sir M. MacDonald and Partners for HMGN Ministry of Water Resources, Department of Irrigation. UNDP (NEP/85/013)/World Bank. February 1990.

NEA, 1988. Khimti Khola hydroelectric project. Prefeasibility study (2 volumes). Nepal Electricity Authority, July 1988.

WECS/DHM,1990. Methodologies for estimating hydrologic characteristics of ungauged locations in Nepal. (2 volumes). HMGN Ministry of Water Resources, Water and Energy Commission Secretariat and Department of Hydrology. July 1990.

15 A model for the assessment of minimum instream flow in Italian scenarios

T BAGNATI and S PARINI
CISE Tecnologie Innovative SpA, Milan, Italy
I SACCARDO
ENEL/CRIS, Mestre, Italy

INTRODUCTION

It is only recently that Italian Rules have stated the need for integration and coordination of land planning and water resource management in hydrographic basins, and in this context the minimum instream flow problem must be considered. However, it should be noted that in the last century Italian laws relating to water management were conceived or even modified mainly to take into account the varying priorities given to the competing uses of water (irrigation, land reclamation, industry, hydroelectric power production, etc.) or even as a mere consequence of these different priorities.

In 1989 Law No.183 stated the need for the integrated development of hydrographic basin planning and the "Regulations on water resources management", approved by the Italian Parliament on 21st December 1993, gave the necessary approval.

The Law No. 183/1989 coordinates the different items involved (the hydrological cycle, land protection, water quality improvement and conservation, drinking water resources management, environmental safeguards, etc.) within a single regulation linking both "water" and "land" intervention and management strategies from a formal and a legal viewpoint. Furthermore, the Regulations define basic items relevant to Administrations and Authorities for hydrological systems management (waterworks, wastewater treatment plant, etc.).

Both Law No.183 and the "Regulations" consider—explicitly and for the first time in Italy — the minimum instream flow problem as a basic environmental concern which must be preserved to assure the optimal development of human activities with respect to the availability of the resource and to water quality characteristics.

In Italy, as in the other industrialised countries, social and economic development has resulted in a great increase in water demand for different competing uses in recent decades, and therefore more and more hydrological resources have been required. Even in areas previously characterised by great water availability, increasing demand together with pollution and irrational uses have determined an evident shortage and inevitable rising conflicts between different competing users.

The growing problems connected with shortages or poor quality water available for the final users underlined the need for a global revision both of control policies and management procedures, so that a growing political and social interest was addressed to the reorganisation and rationalisation of water services.

Within this framework, the Fifth Environmental Action Program of the European Economic Community states as a main EEC strategy the granting of water availability and demand matching through properly planned hydrological resources management, while the LIFE fund programme considers integrated basin management as one of its main priorities.

In this framework and with respect to the new laws which have determined the Italian approach to hydrographic basin planning, ENEL, the main energy producer in Italy, has undertaken a research programme, together with CISE SpA, aimed at the definition of a proper methodology to determine the minimum instream flow which must be assured in rivers.

Being at first mainly a research programme to validate hydraulic and biological methodologies aimed at the minimum instream flow evaluation for aquatic life preservation, the analysis performed has shown the global complexity of the problem.

In the present Italian social-economic and territorial scenario, the need to assure a minimum streamflow in the rivers may be seen as a typical management problem, relevant to streamflow definition and sharing of the basin, rather than a real issue concerning the impact of new power plants.

The minimum instream flow problem is therefore a system of complex linkages, related both to many natural trends and to socio-economic and territorial parameters and constraints. It therefore calls for considerable skill to implement (and manage) complex models that will tackle the physical, as well as the political, economic, social and biological aspects.

THE ITALIAN SCENARIO

Thirty-one hydrographic basins, both at a national and at a regional level, have been defined by the coming into force of Law 183/1989, governing a complex territorial framework consisting of mountainous land (about 106.000 km²), hills (about 125.000 km²) and plains (about 69.000 km²).

In this territory 526 artificial basin may now be found, if we consider only those with an active capacity greater then 0.1 million m³ (34 have an active capacity greater than 100 million m³). About 180 artificial basins are more than 50 years old and almost 50 basins are now being built or tested (Figure 1).

The Italian situation is a particularly interesting one owing to its general complexity and to the high population density (187 inhabitants per km²), which, together with the historical evolution of land use, highly diversifies the hydrological resources as regards the use and quantities involved. Furthermore, large urban and industrial sites

Figure 1 Regional distribution of artificial basins over 0.1 million m³

have been developed near, or even inside, the main hydrological basins.

The main characteristics of the hydrological system depend on different litho-morphological features relevant to plutonic and metamorphic rocks in the mid-western Alps, carbonate and dolomite rocks (highly karstified with considerable infiltration and active underground flows) in the Pre-Alps and in the eastern Alps. In the Appennini the hydrological system is mainly determined by the linkage between underground flows and tectonic fractures, largely diffused arenaceous sandstones and impermeable clays.

The river flow system is highly diversified owing to the different geographic and orographic characteristics and the various hydrological features of the country; the

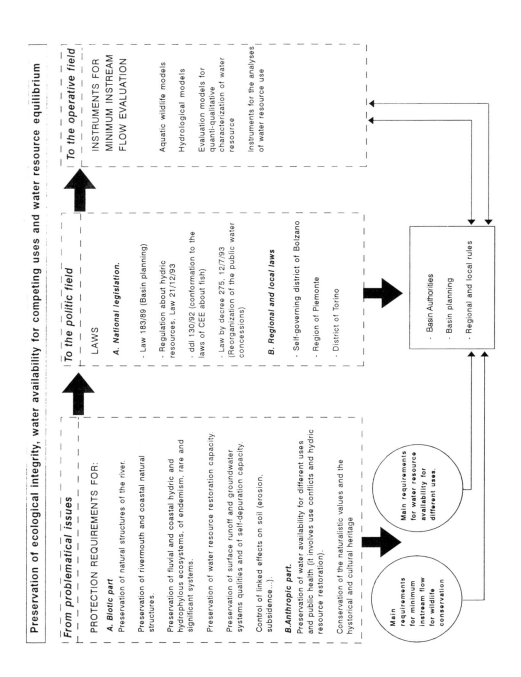

Figure 2 Theoretical approach to the determination of minimum streamflow in the Italian context

general streamflow characterisation, particularly in low flow conditions, is therefore strongly dependent on such a situation.

Natural low flows may be found at different periods of the year and with different magnitudes. In the Alps, where a mainly snow-glacier system is present, winter low flows are recorded. Impermeable basins in the Appennini and the basins of the islands have long lasting summer low flows characterised by almost zero flows, whereas mainly permeable basins, with high sources contributions, have rather short summer natural low flows.

The multiple water uses do not generally fit the yearly trends of natural water flow and therefore different interactions and problems arise. The complexity of the global river system, which stems both from the physical-geographical configurations and from the varying flow rates and trends, is therefore increased by these different uses and by the "quality" requirements of the water itself.

Overall water availability and acceptable carrying capacity seem to be at their limits today in Italy. Even if the human settlements and activities have not basically been increased, water quality appears to be worsening, owing to the increased number of diversions and uptakes.

The specific problem of the definition and evaluation of the minimum instream flow which must be assured is therefore only a part of the more general "water problem", linked both to the management of hydrological resources for different uses and to the development of rules and laws in the Italian situation.

On the basis of the aforementioned assertion it may be easily understood that the use of only traditional hydraulic/biological simulation models (PHABSIM, HQI, etc.) may not be a completely adequate approach to examining the global problem. An integrated theoretical approach has been developed, therefore, to single out the most significant indicators, relevant both to the strictly environmental field (hydraulic/ ecological factors) and to socio-economic aspects; this approach allows a minimum streamflow determination based on optimal sharing of water resource, quality requirements and definition of required amounts for different uses.

The theoretical scheme of the approach is aimed at ecological integrity preservation, optimal economic use and water resource balance; from the different problematical issues and the national and local political trends the scheme derives the methodological procedures and instruments to evaluate the minimum streamflow (Figure 2).

METHODOLOGY

ENEL produces 21% of its energy output via hydroelectric power plants: the need for a methodological approach to minimum streamflow analysis in the different Italian scenarios is therefore evident. In the context of cooperation between ENEL and CISE, a research activity is in progress to develop the methodology described here.

This methodological approach has been defined in three steps:
- An adequate general approach scheme for the "Environment" has been defined. In the present work the "Environment" is thought of as the whole system linking

(for structure and functions) biotic, abiotic and human factors to the river in question or, more generally, to flowing water.
- A linkages network has been implemented taking into account environmental parameters and main effects of flow reduction.
- Evaluation matrices have been created.

The general approach, developed by E.P.Odum and properly modified for the minimum streamflow problem, is shown in the following scheme.

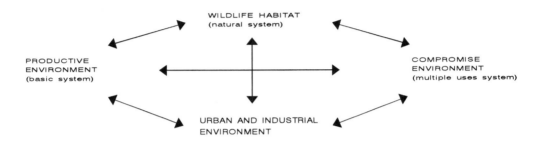

The minimum instream flow must be defined as the result of a complex system of linkages among different environmental concerns. This evaluation phase — still a theoretical one — takes into account not only hydrological, hydrogeological and biological parameters (such aspects may be properly defined on the basis of a wide and rich international scientific literature), but also the "optimal share" among different uses, water quality requirements and the different quantities being considered, together with the conservation of environment and the natural system as a whole.

The general assessment must be performed taking account of both the effects of a minimum "vital" flow not actually authorised and, on the contrary, the effects of a situation in which this minimum vital flow may not be assured.

The implementation of the following step requires:

(a) a preliminary characterisation of the possible conflicts which may be connected to the realisation of water reservoir or derivation plants (particularly for different competing uses); environmental unacceptability and economical losses may be caused by neglecting these items.

(b) the proper definition of "the minimum vital streamflow" is in some way the threshold which must be reached to begin the general analysis and evaluation.

A space-time linkages system has been built corresponding to the previous theoretical model; possible impacts on each environmental item are considered (the "linkages network" in Figure 3). The linkages system, though not directly an evaluation tool, shows the evaluation parameters to be used in impact assessment for water flow reductions.

The third step in the procedure needs a proper analytical procedure to be used for the assessment and for the definition of the evaluation matrices.

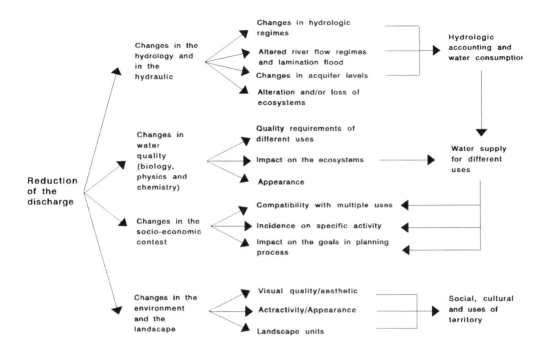

Figure 3   General approach to determine instream flow assessment

OPERATION DIAGRAM AND EVALUATION MATRICES

The proper analytical procedure to be used for the assessment of instream flow is shown in Figure 4.

This diagram consists of three steps, each of which contains a number of specific procedures:

- definition of the aim of the specific instream flow evaluation; analysis design and planning.
- general framework and description of evaluation fields.
- instream flow evaluation.

As for steps two and three, the approach model has been implemented through the definition of two complex matrices to be used in the assessment of the minimum instream flow in Italian basins.

Previous to evaluation matrices definition, two further phases had been developed. After the general situation analysis has been performed, the development of an assessment approach model has been tackled, aimed at the minimum instream flow definition. The following steps have been performed:

I.   Definition of more widely used water stores and derivation technologies in the Italian scenario, interacting with the minimum instream flow problem, as different

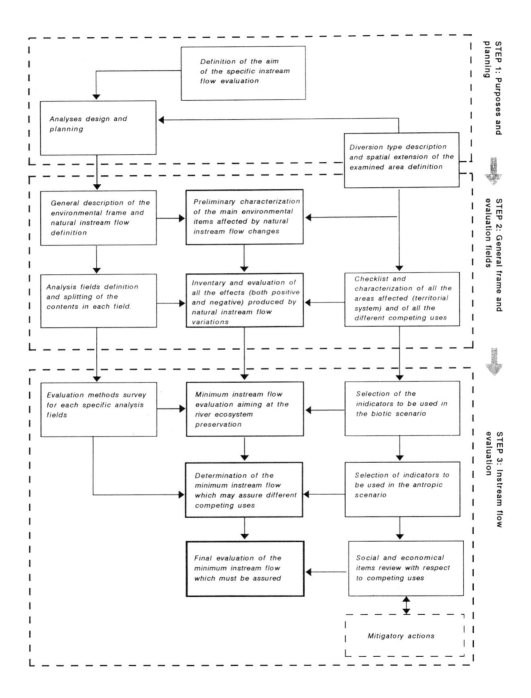

Figure 4 Flowchart describing actions and studies to be performed for minimum instream flow evaluation

plants produce different effects.

II. Definition of the hydrographic basin, not only as a physical concept, but mainly as an institutional reference for the proper balancing of the different water requirements for competing uses. The network matrices comprise a complete set of actions and environmentally important effects (both positive and negative) linked (in the upper part of the matrix) to water retention and derivation plants, to components relevant to the plant itself, and linked (in the lower part of the matrix) to the different receptor types (physical, biological, environmental factors, and more generally flowing water factors) and uses of the hydrological resources.

Owing to the large number and complexity of the variables involved, and considering the relevant Italian law, the matrices have been implemented differently for human factors (relevant to the uses of the hydrological resource) and for the biotic-abiotic scenario (relevant to the effects on ecosystems, the physical environment, etc.). Elementary impacts have been characterised in each macro-area (or matrix) so that every impacting action and/or evaluation factor together with specific receptors may be clearly singled out. As is obvious, a number of elementary impacts are considered in both matrices. Figures 5 and 6 show the two complete evaluation matrices relevant to the human and biotic/abiotic macro-areas respectively.

In Figure 5 the human scenario matrix has been shown; in this matrix 36 possible effects may be seen which appear also in the biotic-abiotic scenario. The latter is represented through a different matrix comprising further possible effects, adding up to a total of 54 (Figure 6).

How to use the evaluation matrices may be summarised as described in the following five steps:

- the types of utilisations, i.e. derivations, regulation types and final destination of the derived flows, must be identified in the matrix upper column;
- the environmental parameters which may be impacted must be identified in the upper part of the matrix; in this phase the matrix plays mainly a "checklist" role; a first inventory of positive and/or negative effects is obtained. Figure 6 gives an example of a diversion with active capacity and the possibility of flow modulation and restitution in the same river.
- the lower matrix part is considered and the impacts as defined at the previous step are linked to relevant receptors (lower column);
- an evaluation step in which evaluation criteria may not be (at the present stage in the research) at their final definition: every impact may be evaluated differently according to the receptors considered and/or the different second order effects (secondary impacts) relevant to the geographic scenarios. At each impact/cross-point the matrix will be provided with a proper description of the corresponding parameters (scientific or legislative indexes or indicators), checklists, models and quantitative/qualitative parameters. An example is given in Figure 6;
- finally, many mitigations are taken into account for each considered impact.

As a formal and conceptual connection with step 4, the evaluation matrices have been provided with a further analysis tool aimed at the definition of specific indicators relevant to each individual effect. Such a tool, consisting of numbered operational

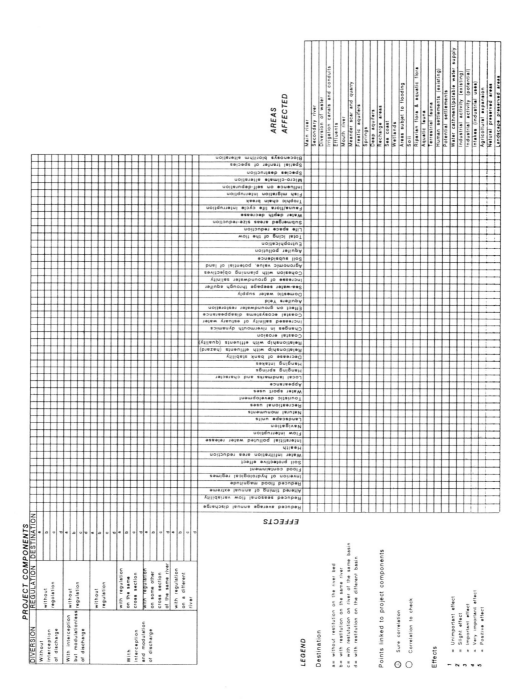

Figure 5 Matrix of potential effects to check human impacts

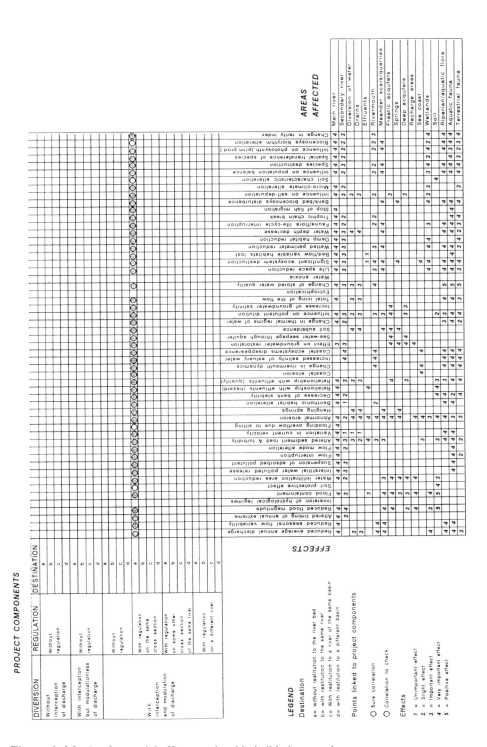

Figure 6 Matrix of potential effects on the abiotic/biotic scenario

tables which are linked to every considered effect, allows the implementation of a proper procedure defining (according to the reason of the choice of the particular effect) the contents of the analysis, the studies which must be performed, the basic information needed, the data sources, and the connections with the other effects considered in the matrices.

CONCLUSIONS

A matrix model has been illustrated which may be considered a working tool aiming at the proper management of the various items and disciplines which must be considered and compared in the definition of streamflow requirements. However, this is not a quantitative algorithm to evaluate the minimum streamflow to be assured but it may prove to be useful for compatibility statements for competing needs.

BIBLIOGRAPHY

Bagnati, T. & Parini, S. 1992. *Analisi delle problematiche inerenti i deflussi minimi vitali nella gestione di bacino della risorsa idrica, per la predisposizione di un capitolato di studi.* Rapporto Topico CISE 7017, Milano
Dracup, J. 1980. On the definition of droughts. *Wat. Res. Research*, 2.
ENEL-DSR, 1993. *Criteri di approccio alla trattazione del minimo deflusso costante vitale negli alvei sottesi.* Venezia.
EPRI, Electric Power Research Institute 1990. *Evaluating Hydro Relicensing Alternatives: Impacts on Power and Nonpower Values of Water Resources.* Los Altos, California.
Ghetti, P.F. 1993 *Manuale per la difesa dei fiumi.* Fondazione Giovanni Agnelli, Torino.
Linsley, R.K. 1975. *Hydrology for Engineers.* McGraw-Hill. New York.
Marchetti, R. Cotta Ramusino, M. & Crosa, G. 1988. Determinazione delle portate minime necessarie per la tutela della vita acquatica in corsi d'acqua soggetti a derivazione o ritenute. *Acqua Aria* **7,** 839-850.
Milhous, R.T. Wagner, D.L. & Waddle, T. 1984. *Users Guide to the Physical Habitat Simulation System (PHABSIM).* Instream Flow Information Paper 11 Rep. FWS/OBS-81/43. U.S.Fish Wildlife Service, Washington D.C.
Ministero dell'Ambiente, 1992. *Seconda relazione sullo stato dell'ambiente.* Roma.
Odum, E.P. 1971. *Fundamentals of Ecology*, Saunders, Philadelphia & London.
Pedroli, G.M. Willem Vos, Dijkstra, H. Rossi, R. (ed) 1988. *The Farma River Barrage Effect Study.* Progetto Toscana, Marsilio Editori, Venezia.
Rossi, G. 1991. *Caratteristiche delle magre nei corsi d'acqua italiani ed effetti delle utilizzazioni idroelettriche.* Rapporto Topico CISE n.6987. Milano
Saccardo, I. Bagnati, T. Parini, S. & Menajovsky, S. 1993. Minimum Streamflow Assessment in Italian Scenarios. In: ASCE (ed) *Engineering Hydrology*, San Francisco, 300-305.

16 The impact of inter-basin water transfers on invertebrate communities of the River Wear

C N GIBBINS
Department of Environment, University of Northumbria at Newcastle, UK
C SOULSBY
Department of Geography, University of Aberdeen, UK
R F MERRIX
National Rivers Authority, Northumbria and Yorkshire Region, Newcastle Upon Tyne, UK

INTRODUCTION

The National Rivers Authority (NRA) is currently developing a national Water Resources Development Strategy to redress regional water imbalances in the UK (NRA, 1992b) and it is likely that inter-basin water transfers will be used increasingly in the future. While the impact of impoundments on downstream river ecology has been widely researched, little is known about the implications of inter-basin transfers (Meador, 1992).

The Kielder water scheme in north-east England transfers water from the River Tyne catchment southwards to the rivers Wear and Tees (Figure 1). Transfers to the River Wear are made to support a statutory Minimum Maintained Flow (MMF) in its lower reaches. This ensures that potable water abstractions for the Sunderland conurbation can be made during most low flow periods and prevents adverse water quality conditions developing during summer droughts. Aquatic communities are considered to benefit from transfers through the prevention of extreme summer low flows.

Concerns over the physico-chemical effects on the River Wear and the possible transfer of biota through the Kielder tunnel have been voiced. Moreover, the ecological benefits of avoiding low flows on the river have never been demonstrated through empirical work. To date, transfers to the Wear have only been used intermittently as daily pumped mine-water discharges to the river and its tributaries have acted to augment summer low flows (Younger, 1993). However, the imminent closure of the Durham coalfield is likely to lead to the cessation of this pumping, and the increased use of the Kielder transfer scheme.

This paper is concerned with the hydro-biological impacts of the Kielder scheme on the River Wear. The regulation system and its impacts on the river's low flow

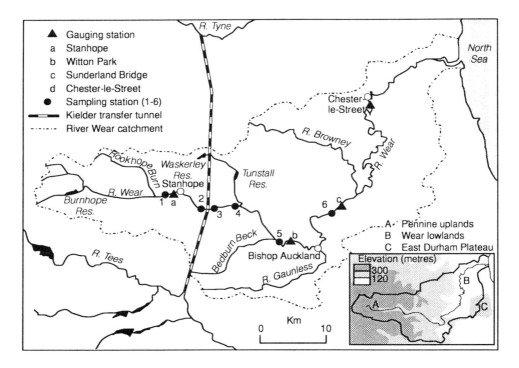

Figure 1   The Kielder water transfer scheme and the location of gauging stations and study
sites within the River Wear catchment, north-east England

hydrology are described, and the extent to which macro-invertebrate assemblages
have been impacted by regulation is examined. The results of one season's
fieldwork are presented and their implications for future work and the management
of the Wear catchment are discussed.

STUDY AREA

The River Wear drains an area of 1163 km². Carboniferous limestones and mill-
stone grit dominate the upper catchment with coal measures in the east. Land-use
is dominated by livestock rearing in the upper Wear valley, through arable farming
in the Wear lowlands to urban/industrial areas in the lower catchment. Soil types
within the catchment parallel the geological regions (Figure 1), with peat and
stagnogleys in the upper catchment and brown earths in the Wear lowlands. The
Durham coalfield underlies the mid and lower reaches of the river.

   Mean annual precipitation ranges from over 2000 mm in the headwaters to
650 mm at the coast. The flow regime of the River Wear is flashy because peaty soils
and steep channel slopes predominate in the headwaters (Table 1).

The Kielder transfer system, completed in 1982, provides great flexibility in the

Table 1 Selected flow statistics for gauging stations on the River Wear.

	Catchment area (km^2)	Mean daily flow ($Ml\ d^{-1}$)	Q_{95} ($Ml\ d^{-1}$)	Mean annual flood (cumecs)
Stanhope	172	316.3	43.4	121.9
Witton Park	455	672.4	107.4	196.1
Sunderland Bridge	657	961.1	172.9	221.3
Chester-le-Street	1008	1257.6	260.5	256.8

management of water resources in north-east England. Previous regulation of the River Wear was restricted to compensation releases of 9.1 Ml d^{-1} and 5.6 Ml d^{-1} respectively from the small Burnhope and Tunstall reservoirs in the catchment headwaters (Figure 1). The mine-water discharges of up to 105 Ml d^{-1} constitute a large source of regulation downstream of Bishop Auckland. Pumping mainly occurs during the night, resulting in a diurnal flow pattern which may vary by up to 1 cumec at Chester-le-Street. Upstream from Chester-le-Street, discharges from 18 sewage treatment plants also augment flows. The statutory 2 cumec MMF measured at Chester-le-Street is the main mechanism for managing low flows in the river. An instantaneous MMF is set rather than a mean daily flow due to the diurnal flow variations.

WATER QUALITY

The water quality of the Wear from its headwaters to Chester-le-Street was Good (Class 1a or 1b in the 1990 NRA survey) although the reach from Chester-le-Street to the tidal limit fell to Fair (Class 2) where it receives substantial inputs of treated sewage effluent. Serious problems during the droughts of the mid-1970s occurred in the lower reaches of the river where high temperatures and excessive weed growth due to nutrient enrichment resulted in deoxygenation and a number of fish kills. However, improvements in waste water treatment over the last two decades have greatly improved the quality of the river.

METHODS

Six research sites were established in April 1993. At the catchment scale, these were selected to fall both upstream and downstream of the Kielder water release point and to extend into the region influenced by mine-water discharges. Sites with well-defined riffle-pool sequences and an absence of excessive engineering or other channel disturbance were chosen. Where possible, sites close to existing NRA gauging stations were used (Figure 1).

Physical habitat features of the six sites (each approximately 1 km in length) were mapped using River Corridor mapping techniques (NRA, 1992). On the basis of these maps, five individual riffles were randomly selected at each site. A single, one minute kick sample was taken from each of the five riffles for each site. In this way, individual riffles were used as the replicate sample unit.

In a pilot study, community data from five replicate one minute samples was found to be similar (Goodness of fit test: $\chi^2 = 6.60$; df 9; p<0.05) to that generated by the same number of three-minute samples. (Species rank abundance distributions from five one-minute samples were compared to a predicted distribution calculated from that of the three minute sample data.) Further, applying the Percentage Community Similarity Index (PCS) (Brock, 1977) to the pilot study data, 90% of invertebrate sample comparisons from different riffles were found to fall within the range of PCS coefficient values exhibited by samples taken from within a single riffle. This precision in intra-site sample replicates forms the basis of comparisons between sites. Samples were collected on 28th June, nine months after the last 1992 release. In the laboratory, samples were sieved (500 µm mesh size) and preserved in 4% formalin. Identification of invertebrates was to species level, except for oligochaetes (true worms) (identified to Subclass) and dipterans (true flies) which were taken to either Family (Chironomidae, Ceratopogonidae and Simulidae) or Genus (Tipulidae, Empididae) level.

RESULTS AND DISCUSSION

THE HYDROLOGICAL IMPACTS OF WATER TRANSFERS

The Kielder transfer scheme was used extensively during the 1989-92 drought (Table 2). These transfers have a significant impact on the low flow hydrology of the Wear. Figure 2 shows the daily volume of water released at Frosterley in relation to the mean daily flow at Stanhope, Witton Park and Chester-le-Street gauging stations over summer 1989. Transferred water accounted for up to 67% of flow in the river at the tunnel discharge point and although the relative proportion of transferred water in the river decreased downstream, it could represent as much as 30% of the mean daily flow at Chester-le-Street. Without regulation during the 1989 drought, the instantaneous MMF of 2 cumecs at Chester-le-Street would have been violated for prolonged periods between July-October.

Table 2 Annual volume of water transferred through the Kielder tunnel into the River Wear (1982-1992)

Year	1984*	1989	1990	1991	1992
Volume (Ml d⁻¹)	505	7910	5270	1052	1195

* No transfers were made or required during the years 1985-1988 inclusive.

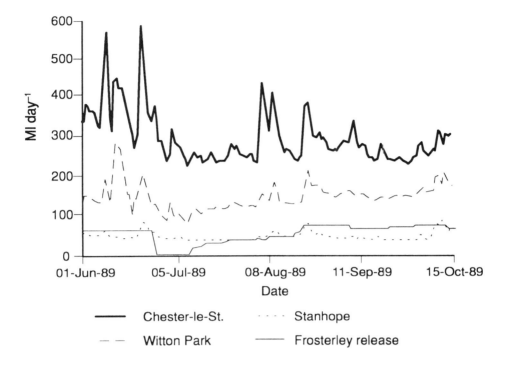

Figure 2 River Wear flows June-Oct 1989 showing the volume of transferred water released at Frosterley

Increased flows resulting from regulation may be compared for the years 1976 and 1989 when droughts of similar severity were experienced. At Sunderland Bridge (see Figure 1) the minimum mean daily flow in 1976 was 83.8 Ml d^{-1} compared to 165.3 Ml d^{-1} in 1989. The increased flows also increase water velocities in the river. For example, mean velocities at Witton Park gauging station increase by around 0.1 m s^{-1}. The magnitude of such velocity changes will vary in different parts of the channel and have important, but unclear, implications for the hydraulics of aquatic habitats and may affect macroinvertebrate assemblages.

The chemistry of the source water in the River Tyne is similar to that in the Wear due to its similar catchment geology. The increased dilution and flow velocities of the regulated low flow regime have alleviated deoxygenation problems in the lower part of the Wear (Johnson, 1988). Less clear is the impact of water temperature changes that occur in the transfer tunnel due to the influence of geothermal heat. Limited data are available but measurements taken in 1982/83 indicate that tunnel water may be up to 8°C higher than river water, with the effect being greatest in the spring and autumn. Deoxygenation of release water may also occur in the transfer tunnel.

IMPACTS OF REGULATION ON MACRO-INVERTEBRATE ASSEMBLAGES

A combination of classification, ordination and predictive techniques were used to
examine the structure of invertebrate communities in the River Wear. The purpose
of this was to look for changes in stream fauna which could be attributed to
regulation in previous years, and also to form a baseline for comparison with future
data. Sample numbering follows a downstream progression, with sample 1 being
the uppermost in the sequence (Stanhope) and sample 30 the furthest downstream
(Page Bank).

Classification and Ordination
Figure 3 presents a Two-Way Indicator Species Analysis (TWINSPAN) classification
of invertebrate replicate sample data for each of the study sites (Hill, 1979a).

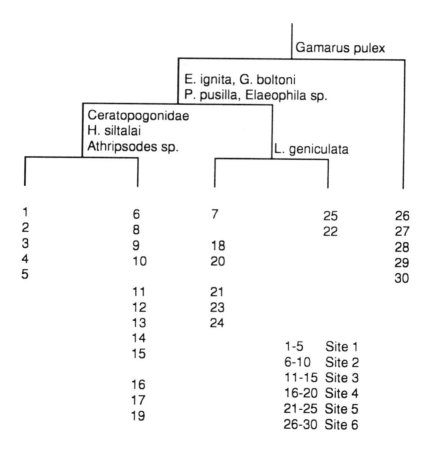

Figure 3   TWINSPAN Classification of 30 invertebrate samples collected from 6 sites on the
River Wear. Numbers in the TWINSPAN end-groups represent individual samples
and indicator species are given for each division. Abundance levels of 1, 10, 100 and
1000 were employed for the TWINSPAN pseudospecies facility.

Indicator species for each of the three divisions used in the analysis are given. Apparently distinctive communities exist at Stanhope (the uppermost site, samples 1-5) and Page Bank (the most downstream site, samples 25-30). The Amphipod *Gammarus pulex* (L.) characterises the Page Bank samples, with the beetle *Brachyus elevatus* (Panzer) and the mayfly *Caenis lactuosa* Burmeister also important in distinguishing this site. Invertebrate assemblages of sites 2, 3, 4 and 5 are separable from Stanhope and Page Bank, yet the differences between these middle reach sites are less apparent. In particular, samples from certain sites fall into TWINSPAN end groups dominated by samples from other sites. Using three divisions (resulting in five end-groups), TWINSPAN failed to differentiate the sites immediately up and downstream of the Kielder discharge point.

Figure 4 shows a DECORANA (Detrended Correspondence Analysis) ordination of the invertebrate data, with samples grouped according to the associations produced by TWINSPAN. The Stanhope and Page Bank samples remain as distinctive communities, but while the Stanhope samples are tightly grouped, there is greater intra-site variability of samples collected from Page Bank. There is considerable overlap between the other four groups.

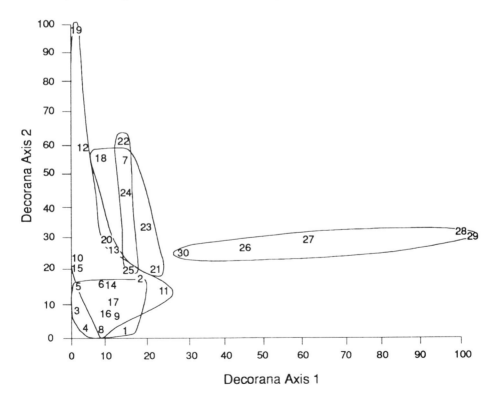

Figure 4 Detrended Correspondence Ordination of invertebrate sample data from six sites on the River Wear. Associations are based on a TWINSPAN classification using three divisions.

The TWINSPAN and DECORANA analyses of sample data indicate the overall similarity of sites 1-5. Downstream changes occur in a progressive fashion, with no discontinuities apparent from the data. Site 6 however remains distinctive: its communities are dominated by *G. pulex,* while differences between individual riffle habitats within the site are more marked than elsewhere.

Prediction

The River Invertebrate Prediction And Classification System (RIVPACS) uses a range of environmental data to predict the probability with which given species (or families) could be expected at a given site (Wright *et al.*, 1989). In the present study, nine environmental variables (mean annual air temperature, altitude, slope, distance from source, channel width, mean depth, discharge, substrate classification, and alkalinity) were used to predict the invertebrate taxa for each of the sites. Predictions for spring, summer and autumn were generated. For each site, seasonal predictions proved very uniform and this consistency was used as a basis from which to investigate specific prediction or sample anomalies, and assess the current state of the site faunas. In interpreting the results, it should be borne in mind that RIVPACS predictions are based on three-minute kick samples.

To compare 'predicted' to observed fauna, attention was focused on taxa whose probability of occurrence was greater than 50%, following Moss *et al.* (1987). Species distributions for June 1993 sample data for each site were compared with RIVPACS 'summer' predictions by means of observed to expected ratios. The expected values were calculated from the sum of the individual predicted probabilities of occurrence of the species listed in the RIVPACS output file.

Given the likelihood of reduced sample catches due to the shortened sampling duration, the ratio of observed to expected for all samples and sites (Figure 5) compares favourably with those published previously (Moss *et al.*, 1987; Wright *et al.*, 1989). Overall the macro-invertebrate communities conform to predictions, with no significant differences in observed to expected ratios between sites (tested by Analysis of Variance, $P<0.05$) This supports the interpretation of a downstream continuum in invertebrate assemblages taken from the DECORANA ordinations, a pattern known to occur in lotic systems (Voelz & Ward, 1990) and one implicit in the RIVPACS model.

IMPLICATIONS AND FUTURE WORK

Initial results imply that Kielder water releases have no long term impact on the invertebrate assemblages of the River Wear. Observed communities at the sampling stations are similar to those predicted by RIVPACS while the change in community structure downstream occurs in a continuous manner, with no discontinuity in the faunas up and downstream of the transfer outlet. If this is the case, then with the resultant water quality improvements, regulation of the river through Kielder releases appears beneficial. Any such interpretation however would be premature.

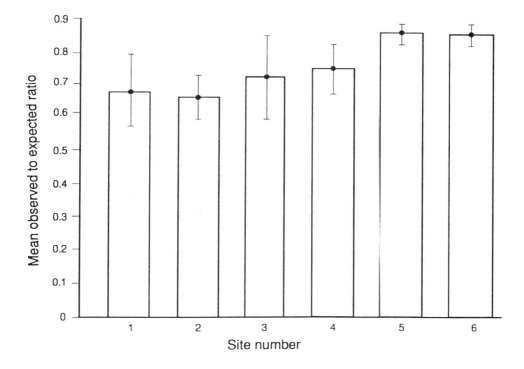

Figure 5 A comparison between the observed and expected number of taxa predicted to occur
with probability P 0.5 for a summer sampling occasion on the River Wear. Mean and
standard deviation for the five individual sample ratios calculated for each site are
given.

The data described here represent only the first in a series relating to 1993 Kielder
releases, with the river having had nine months to recover from previous release
episodes. Moreover, with possible increased use of the Kielder system due to mine
closures, the impacts of regulation may change in the future as the frequency and
duration of transfer releases increase. The current data therefore establish a baseline
from which to assess the implications of these operational changes.

Of the six study sites, Page Bank is the most distinctive. As the site is 66 km from
the source, the invertebrate communities may be entirely consistent with those
expected in the river potamon, where diversity is known to be greatest (Statzner &
Higler, 1986). The invertebrate fauna of riffles habitats within the site varies
greatly, as indicated by the DECORANA ordination. Mine-water releases of up to
11.8 Ml d^{-1} are discharged into the mid-part of the Page Bank sampling site, and
being the most downstream its fauna may also be influenced by the cumulative
impact of sewage inputs to the river. However, further work is needed to test
whether these patterns persist through time, and to look for causes. Such intra-site
variability in invertebrate communities has been explained in terms of habitat
heterogeneity resulting from regulation (Petts, Armitage & Castella, 1993).

It is also important to understand both the short and long term impacts of transfer releases. The processes through which changes in invertebrate community structure occur and the mechanisms and timing of recovery are therefore central to the research; invertebrate drift, an important element of stream invertebrate dispersal and recovery, is known to be sensitive to physical cues such as discharge. For this reason, the relative importance of release intensity and duration for invertebrate drift and recovery will be examined through a series of experiments in 1994, along with simultaneous physico-chemical sampling of release water.

The difficulty of equating instream flows with the ecological requirements of invertebrate taxa pervades many river management issues. Instream hydraulic information for use in the Physical Habitat Simulation (PHABSIM) component of Instream Flow Incremental Methodology (IFIM) has already been collected from the Wear, and this may allow the availability of suitable habitat under different discharge regimes to be quantified. Taxonomic groups, such as mayflies and stoneflies, with known sensitivities to flow changes can be used in the modelling process. Hopefully, this will contribute to an understanding of how invertebrate groups may be affected by different catchment management scenarios.

ACKNOWLEDGEMENTS

Thanks are due to Dr Michael Jeffries and David Turnbull for comments throughout the preparation of the manuscript. The research was funded by the National Rivers Authority. Views expressed are those of the authors and not necessarily the NRA.

REFERENCES

Brock, D.A. 1977. Comparison of community similarity indices. *J. Wat. Pollut. Control Fed.* **49**, 2488-2494.

Hill, M.O. 1979a. TWINSPAN — A FORTRAN program for arranging multivariate data in an ordered two-way table by the classification of the individuals and their attributes. *Ecology and Systematics*. Cornell University, Ithaca, New York.

Hill, M.O. 1979b. DECORANA — A FORTRAN program for detrended correspondence analysis and reciprocal averaging. *Ecology and Systematics*. Cornell University, Ithaca, New York.

Johnson, P. 1988 River Regulation: a regional perspective — Northumbrian Water Authority. *Regulated Rivers: Research and Management* **2**, 233-255.

Meador, M.R. 1992. Inter-basin water transfers: ecological concerns. *Fisheries (Bethesda)* **17**, 17-22.

Moss, D., Furse, M., Wright, J.F. & Armitage, P.D. 1987. Prediction of the macro-invertebrate fauna of unpolluted running water sites in Great Britain using environmental data. *Freshwat. Biol.* **17**, 41-52.

NRA. 1992a. *River Corridor Surveys: Methods and Procedures*. Conservation Technical Handbook No. 1. National Rivers Authority.

NRA. 1992b. Water Resources Development Strategy. National Rivers Authority, Bristol.

Petts, G.E., Armitage, P. & Castella, E. 1993. Physical habitat changes and macroinvertebrate responses to river regulation in the River Rede, UK. *Regulated Rivers: Research and Management* **8**, 167-178.

Statzner, B. & Higler, J. 1986. Stream hydraulics as a major determinant of benthic invertebrate zonation patterns. *Freshwat. Biol.* **16**, 127-139.

Voelz, N.J. & Ward, J.V. 1990. Macroinvertebrate responses along a complex environmental gradient. *Regulated Rivers: Research and Management* **5**, 365-374.

Ward, J.V. & Stanford, J.A. 1983. Insect species diversity as a function of environmental variability and disturbance in stream systems. In: Barnes, J.R. & Minshall, G.W. (eds) *Stream Ecology* Plenum, New York. 265-278.

Wright, J.F., Armitage, P.D., Furse, M.T. & Moss, D. 1989. Prediction of invertebrate communities using stream measurements. *Regulated Rivers: Research and Management* **4**, 147-155.

Younger, P.L. 1993. Possible environmental impact of the closure of two collieries in County Durham. *J. Instn Wat. &. Environ. Manag.* **7**, 521-540.

PART V

IMPACT OF RURAL LAND USE CHANGE

17 Influence of vegetation, land use and climate on streamflow in The Philippines

SANTIAGO R BACONGUIS
Department of Environment & Natural Resources, Laguna, The Philippines

INTRODUCTION

The increasing population of The Philippines and rising per capita consumption of water, together with the destruction of forests, leads to nationwide concern about the adequacy of water supplies. This concern is further heightened by the increasing demand for water to meet the needs of agro-industrial expansion, domestic consumption and the generation of hydroelectric power.

It has been observed that the water in our river system does not arrive in an ideal quantity, quality or at the best time. While some regions suffer acute shortages, many areas sustain major damage from severe floods. Thus it is important to determine the role of the different types of forest vegetation in regulating streamflows from catchments, as well as the effects of varying land uses. These factors among others greatly influence streamflow characteristics and sedimentation. With proper management, the forest cover and land use could be adjusted to minimise streamflow extremes, provide more water during the dry season or reduce flood and erosion hazards during the rainy season.

The study described here was conducted to evaluate streamflow characteristics as influenced by the major forest vegetation and climatic types in The Philippines, and suggest ways by which the forest cover may be manipulated to improve water yield, quality and timing without prejudice to the management objectives of the other forest resources.

LITERATURE REVIEW

The need to control or improve water yield arises from the increasing demands for quality water in sufficient quantity and delivered at the right time. This is vital for the developing industries and increasing population centres which draw their water supplies from streams draining from the highlands (Bibal 1974). Moreover, Pablico (1971) observed that important catchments supplying water to industries and domestic establishments are now in a critical condition resulting in shortage of water supply during dry seasons and flash floods during rainy days. In places where seasons are not distinct, water yield can be improved by proper manipulation of vegetation (Coleman, 1953). However, the question of what vegetation type or species to use, and how it should be manipulated under varying local conditions, is the subject of research.

Several studies in developed countries have shown that water yield can be altered by manipulation of transpiring vegetation (Lull, 1962; Hibbert, 1967). Dortignac (1967) reported several aspects of the effects of catchment management practices on water yield in the U.S. such as (1) converting Coweeta catchment from hardwood to white pine greatly reduced streamflow; (2) converting hardwood covered catchment to grass significantly increased streamflow; (3) timing of streamflow could be improved. Practices which increased annual flow produced extra water during low-flow periods, and those that decreased annual flow also reduced streamflow during low months.

Hydrological studies in The Philippines have produced some useful results. Baconguis (1980) found that of the total annual rainfall from 1975 to 1979 in a secondary dipterocarp forest, 62% was streamflow. In a study on grassland uses within the upper Magat River Basin, Jasmin (1976) found that undisturbed burned plots had the highest increase of surface runoff followed by corn and rice plots. Runoff was greatly reduced in the moderately grazed and protected plots. With cultivation on slopes, the frequency of runoff increased even with decreasing rainfall.

Bocato (1981) studied the effect of differing vegetal covers on runoff and found highly significant differences between treatments. Runoff percentages of total rainfall are as follows: sweet potato plot, 2.69%, natural vegetation, 3.38%; para grass + centrosema mixture, 3.77%; para grass, 4.56%; guinea grass, 6.61%; centrosema, 7.04%; stylo, 11.13%, corn + stylo, 21.58%; corn, 36.30%; and bare plot, 44.86%.

MATERIALS AND METHODS

The data used in this study were taken from the rainfall and streamflow hydrographs of the gauging stations maintained and operated by the Coastal Zone and Freshwater Ecosystems Research Division (CZFERD) of the Ecosystems Research and Development Bureau (ERDB), College, Laguna, The Philippines. Different authors on climatology have classified climate based on rainfall occurrences, rainfall amounts, temperature and other parameters. There are four climatic systems in use in The Philippines: (1) the Mohr system; (2) the Koppen system; (3) the Corona system and (4) the Hernandez (Q ratio) system. At present, the Corona system is the most widely used classification. Fr. Jose Corona devised a climatic classification based on rainfall characteristics. The prevalence of rainfall from month to month is considered in consonance with the northeast and southwest monsoons, as follows:

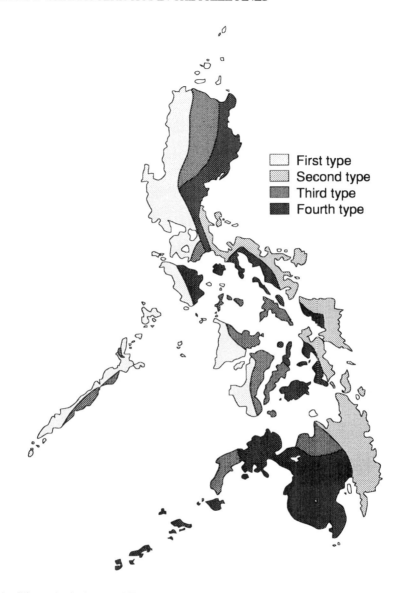

Figure 1 Climatological map of The Philippines

Type I   Two pronounced seasons, dry from November to April;
            wet during the rest of the year.
Type II   No dry season with a very pronounced rainfall from
            November to January.
Type III   No very pronounced maximum rainy period with a short dry season
            lasting only from one to three months.
Type IV   Rainfall more or less evenly distributed throughout the year.

Brief description of catchments under study
Climatic Type I

A *small grassland catchment* was chosen for this study. The area is located in the undulating grassland at the western portion of the Angat Pilot Forest, So. Paco, San Mateo, Norzagaray, Bulacan.

The study site has an average elevation of 225 metres above sea level at the flume site. The topography varies from slightly to moderately undulating, with a mean slope of 25%. Soil type varies from clay to clay loam with pH values ranging from 4.9 to 5.6. The vegetative cover is composed of a mixture of cogon (*Imperata cylindrica* (L.) Beauv.), talahib (*Saccharum spontaneum* Linn.) and a few small alibangbang trees (*Bauhinia malabarica* Roxb.). This was burned annually in March during the period of calibration although the whole catchment is quickly covered with grass at the start of the rainy season in April or May.

The *secondary dipterocarp forest catchment* which is the site of the dipterocarp stream gauging station is located downstream of the Angat Hydroelectric Dam and about 2 km from the former Angat Pilot Forest Meteorological Station. The highest elevation above the weir site is 160 metres. The topography is from moderately undulating to very steep, with an average slope of 50%. The irregular configuration of the catchment is orientated to the south-east with the majority of the slopes facing south to south-east.

The soil is generally silt loam, yellowish red to fairly dark grey in places where vegetative matter is abundant. Because of the rugged relief, the soil is thin on the sides of the slopes and the better soils are accumulated in pockets and at the base of slopes. Soil in such locations is loamy, with a high humus content. The vegetative cover is composed of a mixture of dipterocarp and non-dipterocarp broad-leaf species.

The *pine forest catchment* under study is located within the Ambuklao catchments at Yangyang, Bokod, Benguet. Elevation ranges from 800 metres at the lowest to 1440 metres above sea level at the highest divide of the catchment. The texture of the soils ranges from clay to clayloam to sandy loam. Bulk density ranges from 1.09 - 2.50 g cm^{-3} which is somewhat high for surface soils of forested areas. *Pinus kesiya* dominates the vegetation.

The *mossy forest catchment* under study lies in the Central Cordillera Mountain at Monamon Sur, Bauko, Mt. Province. The catchment drains towards the Chico river. Elevation ranges from 2260 m a.s.l. at the weir site to 2435 m a.s.l. at the highest divide of the catchment. The soil is characterized by thick humus reaching a depth of about 30 cm. The soil type is sandy loam. The parent material is composed of andesite, deorite and basalt, all originating from igneous rock. This type of parent material exhibits fair to excessive surface and internal drainage as well as the presence of boulder outcrops. Vegetation is a mixture of *Quercus, Dacridium Eugenia* and miscellaneous species. Ferns cover all open areas.

Climatic Type III

The *secondary dipterocarp forest catchment* under study is located in Central Mindanao at Musuan, Bukidnon and drains towards Pulangi river. The soil is clayey.

Elevation ranges from 320 to 1200 metres above sea level. Residual or secondary growth forest is very evident. Slash and burn agriculture (kaingin) is prevalent in the upper part of the catchment due to encroachment of settlers.

The *pine forest catchment* under study is located in Central Mindanao at Malaybalay, Bukidnon and drains towards the Pulangi river. The soil is a moderately acidic clayey type. Elevation ranges from 650 to 950 metres above sea level. The pine forest is a man-made pine plantation. Vegetation under the pines consists of grasses, mostly *Themeda* and *Imperata* species.

RESULTS AND DISCUSSION

The streamflow characteristics of the different types of catchment were assessed based on their trend, monthly average and monthly variations. Streamflow values were converted to millimetres to conform with the unit of rainfall.

Climatic Type I
For the grassland catchment over the eight-year period, mean monthly rainfall varied from 7.68 mm in February to 560.22 mm in August with a monthly mean of 260.70 mm and a standard deviation of 207.29 mm. Rainfall peaked at the mid-rainy season (August). Streamflow was similar to the rainfall pattern with mean monthly values ranging from 0 in February to April to a maximum of 134.84 mm in August with a monthly mean and standard deviation of 47.25 mm and 49.57 mm, respectively. Annual values of these particular variables fluctuated periodically from year to year with an annual average of 1807.27 mm and a standard deviation of 252.62 mm. The lowest mean streamflow (runoff) coefficient is 1.20% (Table 1) which occurs in January to a maximum of 24.4% in the mid-rainy season (August). Mean annual streamflow coefficient for the grassland catchment is 18.12%.

From 1975 to 1992, the average monthly rainfall for the *secondary dipterocarp forest catchment* in the Angat District ranges from 10.06 mm in February to 635.89 mm in August with a monthly mean of 267.63 mm and standard deviation of 228.82 mm. Streamflow pattern was similar to the rainfall pattern with mean monthly values varying from 0.42 mm in March to 403.72 in August with a monthly average and standard deviation of 129.11 mm and 143.63 mm, respectively.

Generally, the streamflow (runoff) coefficient is lowest during the dry season due to the fact that more water is absorbed in dry conditions. The lowest mean monthly streamflow (runoff) coefficient is 0.0125 which occurs in April to a maximum of 0.6589 in September. Mean annual streamflow coefficient for the secondary diptero-carp forest catchment is 0.4824 (see Table 1).

For the *pine forest catchment*, the mean monthly rainfall during the three-year period varied from 3.97 mm in December to 718.73 mm in August with a monthly mean of 275.14 mm and a standard deviation of 285.98 mm. Rainfall peaked in the mid-rainy season (August). Streamflow pattern was similar to the rainfall pattern with mean monthly values ranging from 14.09 in April to a maximum of 382.32 mm

Table 1 Mean monthly runoff coefficients of the catchments with varying vegetation/land-use and climatic types in The Philippines

Months	Grassland	Secondary Dipterocarp	Pine	Mossy Forest
Climatic Type I				
January	0.0120	0.1880	0.9269	1.2571
February	0.0	0.1204	2.9576	0.8875
March	0.0	0.0136	4.1501	0.3052
April	0.0	0.0125	0.2032	0.2018
May	0.1323	0.1694	0.0527	0.2769
June	0.1639	0.2048	0.5890	0.5876
July	0.2427	0.4697	0.5793	0.5563
August	0.2442	0.6327	0.2801	0.7203
September	0.2065	0.6589	0.4128	0.7845
October	0.1616	0.6297	0.4901	0.7803
November	0.1411	0.6273	2.4318	0.7359
December	0.0539	0.2482	10.2519	0.8180
Mean	0.1812	0.4824	0.4584	0.6250
Climatic Type III				
January		0.0610	0.0056	
February		0.0440	0.0124	
March		0.0312	0.0068	
April		0.0577	0.0062	
May		0.0335	0.0557	
June		0.0618	0.0350	
July		0.1036	0.0492	
August		0.1677	0.0312	
September		0.1533	0.0576	
October		0.2206	0.1281	
November		0.1740	0.0447	
December		0.1936	0.0159	
Mean		0.1158	0.0507	

in June with a monthly mean and standard deviation of 126.14 mm and 135.56 mm, respectively. The lowest mean streamflow (runoff) coefficient is 0.0527, which occurs in May to a maximum of 10.2519 in December. The mean annual streamflow coefficient for the pine forest catchment is 45.84%. Table 1 presents the average streamflow coefficient for a pine forest catchment in the Ambuklao District.

During a five-year period, the average monthly rainfall for a *mossy forest catchment*, at the source of the Chico river, varies from 42.50 mm in February to 562.52 mm in July with a monthly mean of 297.94 mm and a standard deviation of 187.94 mm. The rainfall pattern had two maxima (peaks), the highest peak in July

Table 2   Mean annual runoff coefficients for the major vegetation types under two climatic types in The Philippines

	Rainfall	*Runoff*	*Coefficient*
Climatic Type I			
Grassland	3128.51	567.04	18.12
Dipterocarp	3211.42	1549.62	48.25
Pine	3301.70	1513.64	45.84
Mossy	3575.22	2234.57	65.30
Climatic Type III			
Dipterocarp	1908.67	221.08	11.58
Pine	2096.32	106.32	5.07

and the second peak in September. The streamflow pattern was more or less similar to the rainfall pattern although it had only one peak (in September). The absence of the first peak in the streamflow was due to the availability of soil water storage space. The mean monthly values ranged from 27.90 mm in March to 425.45 mm September with a monthly average of 186.21 mm and standard deviation of 135.57 mm. Annual values of rainfall and streamflow fluctuated periodically with an annual standard deviation of 408.62 mm and 370.45 mm, respectively.

The streamflow coefficient for the mossy forest catchment showed an increasing trend from the lowest point of 20.18% in April to 125.71% in January and then decreased to the lowest point (Table 1). Streamflow could also be affected by fog drips, particularly during the cold season, apart from rainfall in the high elevations. The highest streamflow coefficient which increased abruptly in January could be influenced by the coldest time of the year when the Siberian winter affects the region. The mean annual values for the streamflow coefficient was 62.50%.

The mean rainfall and streamflow had a linear relationship with a coefficient of determination equal to 78.57%. This means that 78.57% of the variation in streamflow was caused by rainfall and only about 31.43% was associated with factors (most probably fog drips or high relative humidity) other than rainfall.

Climatic Type III

For the four-year period, mean monthly rainfall in the *secondary dipterocarp forest catchment* varied from 25.33 mm in March to 345.33 mm in June with a monthly mean of 159.06 mm. Rainfall peaked at the early rainy season (June). The streamflow pattern was similar to the rainfall pattern, with mean monthly values ranging from 0.62 in April to a maximum of 52.51 mm in August with a monthly mean of 18.42 mm. The lowest mean streamflow (runoff) coefficient was 0.0312 which occurred in March; the maximum was 0.2206 in the late rainy season (October). The mean annual streamflow coefficient for the secondary dipterocarp forest catchment with climatic type III was 11.58%. Table 1 presents the average streamflow coefficient for a secondary dipterocarp forest in Central Mindanao.

For the four-year period, mean monthly rainfall in the pine forest catchment varied from 46.77 mm in February to 369.85 mm in August with a monthly mean of 174.69 mm. Rainfall peaked at the mid-rainy season (August). Streamflow pattern was similar to the rainfall pattern with mean monthly values ranging from 0.43 in April to a maximum of 28.77 mm in October with a monthly mean of 8.86 mm. The lowest mean streamflow (runoff) coefficient is 0.0527 which occurs in April to a maximum of 0.1281 in the late rainy season (October). Mean annual streamflow coefficient for secondary dipterocarp forest catchment with climatic type III was 3.74%. Table 1 presents the average streamflow coefficient for a pine forest catchment in Central Mindanao.

Water balance computation

The water balance computation is one technique used in analysing the hydrologic response of a catchment to rainfall in terms of streamflow, evapotranspiration, soil moisture storage and groundwater recharge and storage. In this study, only one vegetation type with two climatic variations was discussed and compared.

In the 11 years of data from the undisturbed secondary dipterocarp forest with Type I climate, annual rainfall was broken down as follows: 49% was streamflow, 34% was evapotranspiration and 17% was recharged to the groundwater. In comparison with the secondary dipterocarp forest with Type III climate, annual rainfall was broken down as follows: 14.23% was streamflow, 59.09% was evapotranspiration and 26.69% was recharged to the groundwater.

SUMMARY AND CONCLUSION

Streamflow characteristics of the major forest vegetation types (grasses, dipterocarps, pine and mossy forest species) with two climatic types (Type I and III) in the Philippines was assessed using rainfall and streamflow data from the six stream gauging stations maintained by the Coastal Zone and Freshwater Ecosystems Research Division, Ecosystems Research and Development Bureau (formerly Forest Research Institute). Results from this study indicate that:

- From the available data of the six gauging stations, rainfall and streamflow were observed to be lowest during the first four months of the year.
- Streamflow pattern follows that of the rainfall.
- The average monthly maximum rainfall for the four vegetation types per climatic type does not vary significantly but their corresponding average monthly maximum streamflow does show a significant difference.
- In terms of the annual and monthly mean rainfall, the amount received by the vegetation type catchments also reveals no significant difference, but the corresponding mean monthly streamflow differs significantly. The mean monthly streamflow from the grassland, secondary dipterocarp forest and mossy forest is

47.25, 129.11, 126.14 and 186.21 mm, respectively. This means that the secondary dipterocarp, pine and mossy forest has 173.28, 166.94 and 294.08% more streamflow than the grassland catchment. This difference was caused primarily by the inherent characteristics of each vegetation type to store and utilize rainfall and regulate streamflow. From the above data, it appears that grassland has the lowest and mossy forest has the highest (Table 2).

- The streamflow (runoff) coefficient reveals an interesting aspect of the streamflow characteristics of the major forest vegetation types. For the grassland catchment and the secondary dipterocarp forest in climatic type I, the coefficient has an increasing trend up to the mid-rainy season and decreases thereafter. For pine and mossy forest, there was a general increasing trend for the coefficient from April to January and for the pine forest up to March. This could be due to the effect of fog drips especially during the months when cold weather as influenced by Siberian Winter affect the region.
- Water balance computations revealed that for same vegetation type, its response to rainfall differs in different types of climate.

Based on the above results, it is recommended that studies should be pursued on the appropriate management practices which would enhance sustained water yield and minimize peak flood flows during the rainy season. The development of prediction models for water yield with given management schemes is very desirable. Such information would provide catchment planners, managers, engineers and researchers with the vital tool to improve water yield and minimise high flood flows.

REFERENCES

Baconguis, S.S. 1980. Water balance, water use and maximum water storage opportunity of a dipterocarp forest catchment in San Lorenzo, Norzagaray, Bulacan. *Sylvatrop* 5, 73-98.
Bibal, J.N. 1974. Permafor-Forest Farms and Gardens, Quezon City.
Bocato, F.C. 1981. Effects of different vegetal cover on runoff and soil losses. M.S. Thesis, University of The Philippines at Los Baños, College, Laguna (Unpublished).
Bruce, J.P. & Clark, R.C. 1965. *Introduction to hydro-meteorology*. Pergamon Press, London.
Coleman, E.A. 1953. *Vegetation and Watershed Management*. Ronald Press Co., New York.
Dortinac, E.J. 1967. Forest water yield management opportunity. In: *Int. Symposium on Forest Hydrology*. Pergamon Press, New York.
Hibbert, A.R. 1967. Forest treatment effects on water yield. In: *Int. Symposium on Forest Hydrology*. Pergamon Press, New York.
Jasmin, B.B. 1976. Grassland uses: effects on surface runoff and sediment yield. *Sylvatrop* 1,156-172.
Lull, H. & P.W. Flietcher. 1962. Comparative influence of hardwood trees, litter and bare area on soil moisture regimen. *Res. Bulletin 800*.

18 The hydrological effects of clear-felling established coniferous forestry in an upland area of mid-Wales

GARETH ROBERTS, JIM HUDSON, GRAHAM LEEKS and
COLIN NEAL
Institute of Hydrology, Wallingford, UK

INTRODUCTION

Uplands are defined as those areas over 300 m in altitude. In the UK, they cover 7.3×10^6 ha, or approximately one third of the total land area, concentrated in Scotland, northern England and Wales. They are characterised by high rainfall, low temperatures and short growing seasons. In the past, this has restricted agriculture to marginal sheep farming, with stocking densities ranging from 0.25 to 1.5 ewes per hectare, and the management of heather moorland. As a result of this abundant rainfall and low agricultural activity, upland runoff is plentiful, and generally of a high quality, needing little treatment before being suitable for potable water supply. Many reservoirs have been built in these upland areas, and provide the greater part of the nation's water supply.

During the course of this century, many of the upland areas of the UK have been subject to land use change. In particular, large areas of land were afforested in the 1920s to overcome a timber shortage following the First World War. Since then, the rate of afforestation, mainly with softwoods, has increased, with Forestry Commission plantations being supplemented by those of other concerns, following the granting of tax relief on private forestry investment in 1988. Afforestation, mainly with Sitka spruce, is presently the most rapidly expanding land use change in the British uplands, particularly in Scotland. At the same time, there has been a smaller increase in the rate of upland grassland improvement, particularly when subsidies became available following the Second World War and under the EEC Common Agricultural Policy in the early 1970s.

Many upland forested areas have now reached felling age, typically 40-60 years. There is a growing concern reported in the literature, mainly from overseas studies (see, for example, Bosch & Hewlett, 1982; Likens *et al.*, 1970; Packer, 1965), that clear-felling may significantly reduce the quality of upland water supply.

A number of studies have been initiated since the 1970s in the UK into various aspects of upland land use change. The particular study reported on here concerns clear-felling of established forestry, and its effects on the quantity and quality of streamflow. The initial results, covering the first five years after felling, are reported, and the processes contributing to these results described. The extent of felling is determined by remote sensing, so that direct comparisons with other studies may be made, and to permit the results to be put into context with regard to changes within the UK uplands.

STUDY AREA

The study is being conducted within the forested catchment of the Institute of Hydrology's area of research in the Plynlimon range of hills in mid-Wales (Kirby *et al.*, 1990). The catchment of the upper Severn under study is 870 ha (see Figure 1) in area and forms a small part of the Hafren forest. Prior to the clear-felling, the catchment was approximately 70% afforested with softwoods, mainly Sitka spruce, planted between 1937 and 1964. The altitude range in the catchment is 320 to 740 m, with slopes varying between 0° and 15°. Most of the forested areas are on the steeper valley sides and the flat valley bottoms. The high altitude areas which are relatively flat support Nardus and Agrostis species on acid moorland. Many of the forested

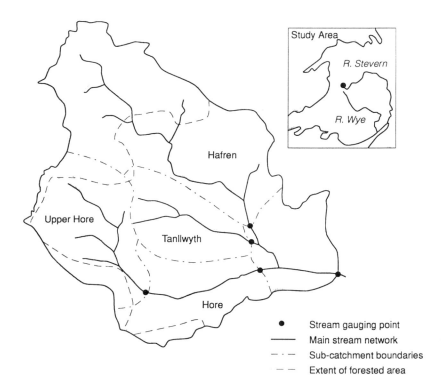

Figure 1   The Upper Severn experimental catchment

areas, particularly those in the flat valley bottoms, were drained prior to planting. The soils are thin (<1 m depth), and consist of a mixture of stagno-podzols, stagno-gleys, brown earth and peat types, overlaying shales of Silurian and Ordovician age, mudstones, sandstones and grit bedrock. Average 1972-88 annual rainfall was 2469 mm, of which 564 mm was lost as evapotranspiration. A more detailed description of the upper Severn catchment is given in Newson (1976).

CLEAR-FELLING

The clear-felling was started in late spring 1985 and took five years to complete. It was mainly confined to the lower reaches of the Hore sub-catchment (Figure 1 and Table 1). A number of removal techniques were employed, including skidding, forwarding and skylining. Whole tree harvesting was not used and extensive brash and tree stumps were left to decompose on site. Soon after harvesting, the slope areas were re-planted with juvenile Sitka spruce. A total area of 95 ha of the Hore was scheduled for felling. The areas affected, and the timing and method of extraction are shown in Figure 2.

Table 1 Details of the gauged areas in the upper Severn catchment

		Total area (ha)	Forest (ha)	(%)	Grassland (ha)	(%)	Clearings (ha)	(%)
Severn	1985	870	554	64	205	24	111	13
	1989	870	439	50	308	35	123	14
Hafren	1985	355	166	47	133	38	56	16
	1989	355	149	42	179	50	27	8
Tanllwyth	1985	92	84	92	2	2	5	6
	1989	92	79	86	8	9	5	5
Hore	1985	315	204	65	68	22	43	4
	1989	315	117	37	116	37	82	26
Upper Hore	1985	184	113	61	53	29	18	10
	1989	184	98	53	81	44	5	3
Lower Hore	1985	131	91	70	15	12	25	19
	1989	131	19	14	35	26	77	59

Gauged data (1976-84)	Rainfall P (mm)	Runoff Q (mm)	P–Q (mm)	Altitude range (m)
Severn	2470	1924	546	320 – 740
Hafren	2460	1950	510	330 – 640
Tanllwyth	2550	1980	570	350 – 560
Hore	2258	1895	363	330 – 740

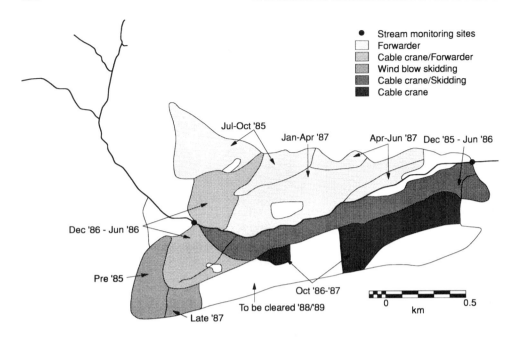

Figure 2 Felling schedule for the Hore sub-catchment

Two Landsat images, one recorded on 27th September 1985, and the other on 6th September 1989, were analysed to determine the extent of clear-felled area. Each image consists of seven bands (three visible, three infrared, and one thermal), with a ground resolution of 28.5 m (apart from the thermal's 120 m). After registering each image to a 1:50 000 base map, an unsupervised classification was used to determine the extent of standing forest, clear-felled area, and grassland within the whole of the Hafren forest, including the upper Severn catchment. In such a classification, a clustering algorithm is used to divide the total population of picture elements (pixels) in the area of interest according to their distribution in multispectral space (Belward *et al.*, 1990), without any user input. Each classification was done using six bands, the maximum allowed, of the Landsat image. The thermal band was not used as it contained the least information. Each initial classification identified sixteen different classes; these were reduced to the three required by the analysis of the statistics of each band within the classes, and from local knowledge. Figure 3 (a) shows the final classification obtained for the upper Severn catchment at each sampling date, and Figure 3 (b) shows the extent of felling within the lower reaches of the Hore. The areal extent of each class at both sampling dates for all the areas of interest are shown in Table 1. This shows that 87 ha of forest was felled in the Hore; this represents 43% of its forested area, or 28% of its total area.

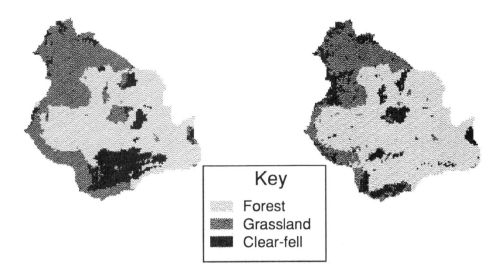

Figure 3   (a) Extent of clear-felling in the Upper Severn catchment

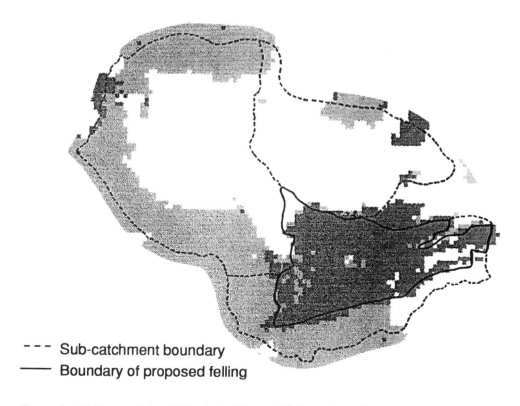

Figure 3   (b) Extent of clear-felling in the Hore and Hafren sub-catchments

CATCHMENT RUNOFF

The upper Severn catchment has been extensively monitored since the early 1970s (Kirby *et al.*, 1990). Streamflow is measured as the stage, or river level, within a trapezoidal flume. In 1975, steep stream structures, specifically designed to measure flow in streams characterised by steep gradients, heavy sediment loads, and high flood/drought flow ratios (Harrison & Owen, 1967) were installed at the outfalls of three sub-catchments (the Hore, Hafren and Tanllwyth) of the upper Severn catchment. More recently, in 1986, a further steep stream structure was installed mid-way up the Hore sub-catchment, above the area to be felled. Details of all of these gauged areas are given in Table 1.

The effects of the clear-felling have been assessed by comparing streamflow losses from the Hore and Hafren sub-catchments prior to and following the felling. These sub-catchments are similar in size, and are forested mainly in their lower reaches, though to different extents (Table 1). Although some felling, most as a result of windthrow problems, did occur in both sub-catchments prior to 1985, and for the Hafren after 1985, the areas involved were much smaller than for the Hore post-1985 felling (Table 1; Figure 3).

Table 2 shows the annual flows of the Hore and Hafren sub-catchments for the years 1976-91 (inclusive). The pre-felling data from 1976-84 have been used to

Table 2   Annual runoff (mm) from the Hore and Hafren sub-catchments

Year	Hafren	Hore
1976	1265	1220
1977	2103	2035
1978	1906	1863
1979	2210	2201
1980	2097	2047
1981	2183	2172
1982	2010	1881
1983	2119	2073
1984	1655	1566
Average 1976-84	1950	1895

Year	Hafren	Hore Measured	Regressed	Increased Flow
1985	1975	1899	1921	−22
1986	2201	2203	2153	+55
1987	1877	1950	1820	+130
1988	2094	2161	2043	+118
1989	1844	1937	1786	+151
1990	2056	2070	2004	+66
1991	1909	1962	1853	+109

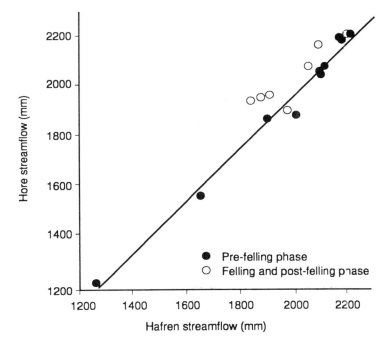

Figure 4   Regression of the annual flow of the Hore catchment on that of the Hafren

formulate a regression against which the effects of the clear-felling can be assessed (Figure 4). As expected, pre-felling runoff from the 46% afforested Hafren are greater than those from the 77% afforested Hore by an average 55 mm per year. This trend is reversed during and following the felling phase, with a maximum increase in the annual flow of the Hore of 151 mm in 1989 (Table 2). An initial comparison of the flows from the Hore and upper Hore agree with these findings.

STREAMWATER CHEMISTRY

Rainfall and streamflow at the outfalls of the Hore and Hafren sub-catchments (Figure 1) have been sampled for chemical analysis at weekly intervals since May 1983. Rainfall is sampled at two sites spanning the range of altitude in the upper Severn catchment. The two samples are bulked for analysis purposes. Stream water is collected by 'grab sampling'. An initial appraisal of the chemistry of the two streams suggested that different mineralogy within the bedrock were producing significant differences in pH values and base cation concentrations (Neal *et al.*, 1992). Therefore, the stream sampling programme was extended in September 1984 to include the outfall of the upper Hore (Figure 1). This provides the 'control' to determine the effects of the felling in the lower Hore. The chemical constituents analysed and the procedures used are described in detail in Neal *et al.*, 1986.

The clear-felling produced an almost immediate effect on the chemistry of the streamflow. These are described in detail in Neal *et al.*, (1992). Figure 5 shows the concentrations of the main affected constituents — nitrate, potassium, dissolved organic carbon, and aluminium — in the streamflow of the Hore and upper Hore before, during, and after the felling. The concentrations of nitrate and potassium were particularly affected, reaching levels four times those in the control, and remaining high four years after the start of felling. A phase shift from a winter to an autumn peak concentration has also occurred. Similar but smaller increases have occurred in the concentration of dissolved organic carbon, but here no phase shift has been observed. Winter concentrations of aluminium increased during the first two years of felling, but after that they have reduced almost to 'normal' levels.

Quantifying these effects in terms of chemical budgets is difficult as the information on stream concentration is not continuous. Chemical inputs in rainfall were simply estimated, for each period, as the product of the amount of rainfall and the volume-weighted mean concentration of the two rainfall samples. Chemical outputs in the streams were calculated by Beale's ratio, an unbiased estimator designed to reduce

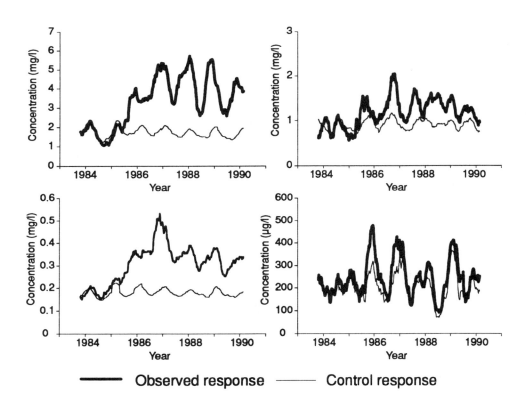

Figure 5 Streamflow chemistry in the Upper and Lower Hore catchment following felling

the risk of a bias introduced by the 'grab sampling' strategy. Details of the technique are given in Durand *et al.*, (in press). Table 3 gives annual hydrochemical budgets for nitrate, potassium, dissolved organic carbon, and aluminium in the Hore and Upper Hore before, during and after felling. This shows, in particular, enhanced losses of nitrate following felling, with net losses greater than 30 kg ha^{-1} a^{-1} five years after the start of felling.

Table 3   Annual hydrochemical budgets (bulk precipitation minus stream output) in the Hore and Upper Hore (kg ha^{-1} a^{-1})

				Hore			
	1984	1985	1986	1987	1988	1989	1990
Nitrate, NO$_3$	-5.3	+2.0	-35.3	-35.0	-44.8	-39.2	-33.6
Potassium, K	+1.0	+0.3	-3.1	-4.1	-2.8	-2.3	-0.9
DOC	-8.6	-17.6	-23.2	-24.3	-22.2	-14.6	-22.6
Aluminium, Al	-5.2	-6.3	-10.8	-7.6	-8.3	-6.0	-9.5
				Upper Hore			
		1985	1986	1987	1988	1989	1990
Nitrate, NO$_3$		+8.5	-13.6	-2.6	-0.3	+2.0	+0.1
Potassium, K		+1.4	-0.8	-0.2	+1.0	+0.6	+1.8
DOC		-20.5	-24.8	-21.3	-20.9	-11.0	-22.9
Aluminium, Al		-5.6	-8.5	-6.2	-7.3	-4.7	-8.9

SEDIMENT LOSSES

Sediment losses from the Plynlimon catchments have been monitored since 1973. This includes bedload, using traps installed upstream of stream gauging stations, and suspended sediment, collected by 'gulp' samplers. In addition, stream turbidity is monitored continuously, and the values are calibrated against actual suspended sediment concentrations.

For the Hore, a major bedload trap was constructed at the lower end of the sub-catchment before the start of clear felling. In addition, a network of minor weirs and traps have been installed on some of the drains and feeder streams to the Hore. This network is designed to provide a range of data, from overall total sediment yields for whole subcatchments using the bedload traps, down to continuous records of suspended load discharges. The network also enables the effects of different timber extraction techniques to be determined.

Prior to felling, the Hore was yielding a mean annual bedload output of 11.8 t km^{-2}. The initial impact of the clearfell was a decline in the bedload trapped at the downstream end of the catchment to 8.3 t km^{-2} (in 1986). This was attributed to containment of sediment behind timber debris within the channel and drains. However, as debris dams broke down or reached capacity, there was a gradual rise in bedload yield to a maximum value of 54.5 t km^{-2} in 1988.

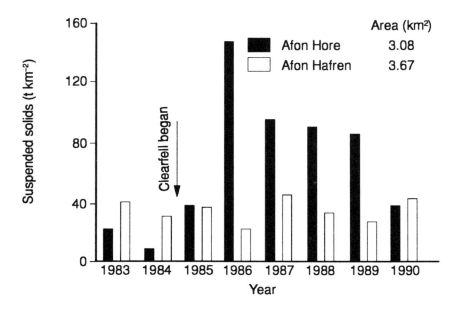

Figure 6 Streamflow chemistry in the Upper and Lower Hore catchment following felling

There was an immediate rise in suspended sediment concentration for any given discharge as soon as felling began. Suspended sediment concentrations from the automatic samplers and turbidity meters at the outfalls of the Hore and the Hafren have been combined with flow measurements to give annual suspended sediment losses from the two sub-catchments (Figure 6). In the first two years following felling, there was an increase in sediment concentration by an order of magnitude for moderate to high flows. This increase was reflected in higher annual yields of suspended sediment from 24.4 t km^{-2} rising to 141.0 t km^{-2} in 1986. Most of the initial increase was associated with road widening, when road material was carried direct to the stream network through road drains. Later, there was considerable ground disruption by machinery, including forwarders and skidders, used during the felling work. This was in contrast to those areas where aerial cable techniques were dominant.

DISCUSSION

The Hafren forest clear-felling study at Plynlimon, mid-Wales is the most detailed and long-running study of the effects of this important phase of the forestry cycle. The initial results show that clear-felling of established forestry is accompanied by:

- an increase in streamflow,
- an increase in concentration of certain chemical species,
- an increase in bedload and suspended sediment concentration.

These results are consistent with what has been observed in the past from other, mainly overseas, studies.

In terms of streamflow losses, the maximum increases of 151 mm in annual flow is remarkably similar to the 160 mm for a 40% reduction in forest cover suggested by Bosch and Hewlett (1982), following their analysis of the results of approximately 60 clear-felling studies. In the UK, Calder and Newson (1979) predict an average reduction in flow of about 10% following increased afforestation of 50% of canopy coverage. Reversing this prediction, the felling of approximately 40% of the forest in the Hore should result in an average 8% increase in flow, or 152 mm (Table 2). However, such generalisations must be treated with some degree of caution, as the rate of felling, and the location of the felled areas relative to the stream network are important considerations. The increase in flow is a result of reduced transpiration and rainfall interception by the forest. Both these processes vary seasonally and with rainfall and other climatic conditions. More detailed analysis of the data is required before this seasonal effect can be resolved.

Many reviews of studies into the effect of forest felling on streamflow chemistry have been given, including Sopper (1975) and Martin et al., 1984. The results vary a great deal, being dependent not only on the areal extent of felling, but also on the techniques employed, climatic conditions, soil types, and topography, and it is extremely difficult to generalize. Nevertheless, the main results presented here (Figure 5; Table 3) are in broad agreement with those given in the reviews, mainly of studies in the USA, and with other recent studies in the UK (Adamson & Hornung, 1990; Stevens & Hornung, 1987). Detailed process studies suggest that the increases in nitrate and potassium concentrations are as a result of reduced nutrient uptake by the vegetation following felling, whilst the enhanced concentrations of aluminium originate from the soil's cation exchange store, following the release of nitrate (Neal et al., 1992).

Although a number of studies into the effects of clear felling of established forestry on sediment yields have been conducted, many have been complicated by the construction of access roads and the burning of vegetation, practices generally not carried out in the UK. In terms of 'felling only' operations, most research has concentrated on the effects of different timber extraction techniques. A review is given in Packer (1965) and another in Soutar (1989). Sediment loss rates vary enormously, for the same reasons as chemical losses, and it is only possible to make the general observation that the cutting of timber in itself does not appear to affect sediment losses adversely. Rather, it is the soil disturbance associated with timber extraction that causes the problems. This was also the case for the clear-felling in the Hore (Leeks, 1992), and at Balquhidder in central Scotland (Johnson, 1993).

The clear-felling schedule in the Hore — one block of 87 ha representing 43% of its forested area or 28% of its total area in five years — proceeded at a faster rate than is normal for UK forestry practice and hence the effects may be exaggerated. Nevertheless, the results do show a deterioration in streamflow quality that may have downstream implications. Nitrate and potassium concentrations, even at their highest levels, are well below the EC maximum permissible for potable water. Moreover, these high values are unlikely to cause increased algal growth in upland reservoirs, since this is largely limited by the concentration of reactive phosphorus. Of greater concern are the enhanced concentrations of dissolved organic carbon: these may cause water quality problems because of the presence of colour and the formation of potentially harmful bichloromethane and chloroform following reaction with chlorine during water treatment. The enhanced concentrations of aluminium may have detrimental effects on the survival of fish (UKAWRG, 1988), whilst increased manganese levels associated with stream bank disturbance, not reported on here, may also affect the stream biota.

Sediment losses from UK uplands are generally small compared with reported losses from other areas of the world. However, losses as a result of ground disturbance can be troublesome as reported following the afforestation of 11.5% of the catchment of the Cray Reservoir in the Brecon Beacons, south Wales (Stretton, 1984). It was found necessary to divert the course of one spectacularly eroding drain, which had contributed approximately 10 800 t of sediment to the reservoir, in order to reduce the sediment pollution within the reservoir back to acceptable levels for water supply. As a result of this and other case studies, new official guidelines have been produced (Forestry Commission, 1993) to minimise the effects of the various forestry practices, including clear felling, on the chemistry and sediment losses within streamflows. Studies such as the one at Plynlimon, will improve our understanding of the processes involved and may contribute to future guidelines.

REFERENCES

Adamson, J.K. & Hornung, M. 1990. The effect of clearfelling a sitka spruce (*Picea sitchensis*) plantation on solute concentrations in drainage waters. *J. Hydrol.*, 116, 287-298.

Belward, A.S., Taylor, J.C., Stuttard, M.J., Biqual, E., Matthews, J. & Curtis, D. 1990. An unsupervised approach to the classification of semi-natural vegetation from Landsat Thematic Mapper-data. A pilot study on Isley. *Int. J. Remote Sensing*, 11, 429-445.

Bosch, J.M. & Hewlett, J.A. 1982. A review of catchment experiments to determine the effect of vegetation changes on water yield and evapotranspiration. *J. Hydrol.*, 55, 3-23.

Calder, I.R. & Newson, M.D. 1979. Land-use and upland water resources in Britain - a strategic look. *Wat. Resour. Bull.*, 15, 1628-1639.

Durand, P., Neal, C. Jeffery, H.A., Ryland, G.P. & Neal, M. (in press). Major, minor and trace element budgets in the Plynlimon afforested catchments (Wales): general trends, effects of felling and of climate variations. *J. Hydrol.*

Forestry Commission, 1993. Forests and Water — Guidelines.

Harrison, A.J.M. & Owen, M.W. 1967. A new type of structure for flow measurement in steep streams. *Proc. Instn Civ. Engrs.*, 36: 273-296.

Johnson, R.C. 1993. Effects of forestry on suspended solids and bedload yields in the Balquhidder catchments. *J. Hydrol.*, 145, 403-419.

Kirby, C., Newson, M.D. & Gilman, K. 1990. Plynlimon research: the first two decades. *IH Report No. 109.*

Leeks, G.J.L. 1992. Impact of plantation forestry on sediment transport processes. In: *Dynamics of Gravel-bed Rivers* (eds: P. Billi, R.D. Hey, C.R. Thorne and P. Tacconi). John Wiley and Sons Ltd., 651-673.

Likens, G.E., Borman, F.H., Johnson, N.M., Fisher, D.W. & Pierce, R.C. 1970. Effects of forest cutting and herbicide treatment on nutrient budgets in the Hubbard Brook watershed-ecosystem. *Ecol. Monogr.*, **40**, 23-47.

Martin, C.W., Noel, D.S. & Federer, C.A. 1984. Effects of forest clear cutting in New England on stream chemistry. *J. Environ. Qual.*, **13**, 204-210.

Neal, C., Fisher, B., Smith, C.J., Hill, S., Neal, M., Conway, T., Ryland, G.P. & Jeffrey, H.A. 1992. The effects of tree harvesting on streamwater quality at an acidic and acid-sensitive spruce forested area: Plynlimon, mid-Wales. *J. Hydrol.*, **135**, 305-319.

Neal, C., Smith, C.J., Walls, J. & Dunn, C.S. 1986. Major, minor and trace element mobility in the acidic upland forested catchment of the Upper River Severn, mid-Wales. *Q. J. Geol. Soc. Lond.*, **143**, 635-648.

Newson, M.D. 1976. The physiography, deposits and vegetation of the Plynlimon catchments. *IH Report No. 30.*

Packer, P.E., 1965. Forest treatment effects on water quality. In: *International Symposium on Forest Hydrology* (eds. W.E. Sopper and H.W. Lull), 29th Aug - 10th Sept 1965, Pennsylvania State University, 687-699.

Sopper, W.E. 1975. The effects of timber harvesting and related practices on water quality in forested watersheds. *J. Environ. Qual.*, **4**, 24-29.

Soutar, R.G. 1989. Afforestation and sediment yields in British fresh waters. *Soil Use Manag.*, **2**, 82-86.

Stevens, P.A. & Hornung, M. 1987. Nitrate leaching from a felled Sitka spruce plantation in Beddgelert Forest, north Wales. *Soil Use Manage.*, **4**, 3-8.

Stretton, C. 1984. Water supply and forestry - a conflict of interests: Cray reservoir, a case study. *J. Instn Wat. Engrs Scient.*, **38**, 323-330.

UKAWRG, 1988. United Kingdom Acid Waters Review Group, Second Report. *Acidity in United Kingdom Fresh Waters*. HMSO, London, 1-61.

19 The impact of land use on nutrient transport into and through three rivers in the north east of Scotland

A M MACDONALD and A C EDWARDS
Macaulay Land Use Research Institute, Aberdeen, UK

K PUGH
North East River Purification Board, Aberdeen, UK

P W BALLS
Marine Laboratory, Aberdeen, UK

INTRODUCTION

Nutrient transport from the terrestrial to the aquatic environment has important environmental and economic implications. Essential nutrients for plant growth such as nitrogen and phosphorus are routinely applied by farmers in inorganic and/or organic forms to supplement the natural soil nutrient supply and consequently increase crop yield. As crops rarely use more than 50% and 20% respectively of applied N and P fertiliser (Nielsen *et al.*, 1987; Holford & Doyle, 1993), the remaining unused nutrients are vulnerable to loss from the soil system to water courses and may ultimately result in eutrophication problems (Sharpley & Menzel, 1987).

It is generally accepted that there is a link between land use, fertilizer application and surface water quality (Heathwaite, 1993). However, because nutrients such as nitrogen and phosphorus, under these circumstances, reach surface waters from a primarily diffuse or non-point source area, this assumed link is difficult to identify clearly (Dermine & Lamberts, 1987). Nutrient transport is largely determined by the interaction of climatic, topographic and edaphic effects which are modified for various land uses (Marston, 1989). Seasonal patterns in nutrient dynamics are apparent as the principal mechanism governing the movement of N and P from the soil to water courses in the interaction between biological processes and drainage. Surface runoff and subsurface flow, both of which are important in the transfer of N and P from the terrestrial to the aquatic environment, show a distinct seasonal pattern. Precipitation duration and intensity are both important climatic factors in determining the extent of soluble *vs* particulate nutrient losses. Soil moisture status also influences the balance between soil processes such as denitrification and mineralistion (McDowell & McGregor, 1984).

Land use in the north-east of Scotland is predominantly low input hill/upland and forestry systems with more intensive, higher-input agriculture mainly restricted to lower lying coastal areas where the topography and climate are more suitable and the soils more fertile (Bibby *et al.*, 1982). Agriculture in these areas encompasses a range of cropping and livestock rotations, each with different requirements in terms of soil cultivation and fertiliser inputs.

This study concentrates on The Dee, Don and Ythan river catchments (see Figure 1), all of which have relatively low population densities and therefore low effluent inputs, making them ideal for studying the impact of land use on concentrations and loadings of nutrients entering their estuaries.

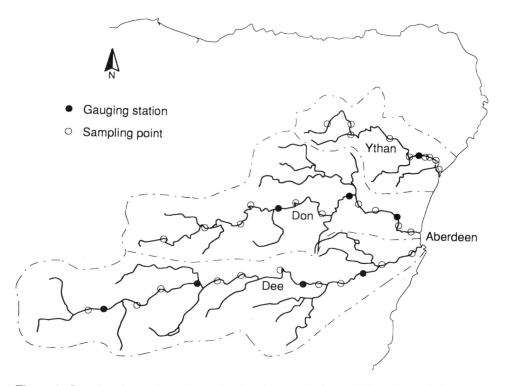

Figure 1 Sample points and gauging station locations on the Rivers Ythan, Don and Dee

MATERIALS AND METHODS

Catchment areas

Situated in the north-east of Scotland, the Dee, Don and Ythan catchments are contrasting both in terms of their physical (Table 1) and land-use (Table 2) attributes. Both the Dee and Don rivers rise in the Grampian mountains and flow in an easterly direction to Aberdeen. Around the sources of these rivers, heather moorland predominates with areas of rough grazing. The topography and climate become more

Table 1   Physical attributes of the three catchments

	Ythan	Don	Dee
Area (km²)	689	1320	2100
River length (km)	60	134	110
Rainfall (mm)	770	810-1060	810-1070
Mean discharge (m³s⁻¹)	7	25	38
Soils*	HIP, BFS, AL	NCG, PG, HIP, A	NCG, PG, HIP, PP

*A = Alpine, sub-alpine; PP = Peaty podzols; HIP = Humus iron podzols; BFS = Brown forest soils; PG = Peaty gley; AL = Alluvial

Table 2   Percentage of land cover features of the Ythan, Don and Dee catchments

Land use summary (%)	Ythan	Don	Dee
Woodlands	2	10	12
Rough grass/moorland mosaic	7	9	22
Moorland	0.4	21	41
Non-agricultural grasslands	0.2	2	2
Agricultural grass	33	24	9
Agricultural crops	56	32	12
Others	1.4	2	2

Based on a classification of Landsat TM satellite (April 1987) reflectance information (*data courtesy of Garry Wright, MLURI*)

favourable towards the east coast and land use becomes more intensive. The smaller Ythan catchment has relatively intensive cereal and livestock production throughout the majority of the catchment and flows into the north sea 20 km north of Aberdeen. The Ythan estuary is currently being investigated as a possible nitrate-vulnerable zone under the EC nitrates directive (91/676/EEC).

The major centres of population in the three catchments are shown in Table 3.

Sampling points
The data used in this study were obtained by the North East River Purification Board (NERPB). The data were from main stream sites only; they were collected either monthly or every second month over the 13-year period 1980-1992. Continuous flow data were collected by the NERPB from gauging stations on the three rivers (Figure 1). The spatial influence of land use on NO_3-N and PO_4-P concentrations was studied by considering data collected from a number of sampling points distributed along the length of each river, together with more detailed assessment from the furthest downstream gauging stations, taken to be representative of each river, prior to

Table 3 Total N and P input from settlements in the Ythan, Don and Dee catchments

Catchment	Settlement	Total N (kg day⁻¹)	Total P (kg day⁻¹)	Population (1991)*
Ythan	Auchterless	0.8	0.18	80
	Fyvie	5.0	1.14	520
	Methlick	2.7	0.62	280
	Ellon	83.0	19.00	8670
	Newburgh	18.0	4.11	1870
	Total	109.5	25.11	
Don	Strathdon	1.3	0.3	125
	Alford	15.1	3.2	1610
	Inverurie	90.1	18.0	9010
	Kintore	36.1	7.2	3610
	Dyce	91.6	18.3	9160
	Total	235.2	47.0	
Dee	Ballater	12.6	2.5	1260
	Dinnet	0.9	0.2	90
	Aboyne	19.5	3.9	1950
	Banchory	60.5	12.1	6050
	Milltimber	59.3	11.9	5930
	Cults	43.5	8.7	4350
	Total	196.3	39.3	

(1991 population estimates as supplied by Grampian Regional Council)

entering the estuary. These gauging station sampling points reflect inputs from the majority of the catchment, are non-tidal, and continuous flow data are available, enabling the calculation of nutrient loadings.

Water analysis

Samples were analysed for NO_2-N and NO_3-N (Henriksen & Selmer-Olsen, 1970) and PO_4-P (Murphy & Riley, 1962) using automated colorimetric procedures. Total oxidised nitrogen (TON), i.e. the sum of NO_3-N and NO_2-N, was determined by the automated cadmium reduction method and expressed in mg l⁻¹. Values for NO_3 concentration were obtained by subtracting NO_2 values, determined separately, from the total oxidised nitrogen values.

Examination of the results showed that relative to NO_3-N, the concentrations of NH_4-N and NO_2-N were generally low (NH_4-N < 0.1 mgl⁻¹, NO_2-N < 0.05 mgl⁻¹); these forms of nitrogen are therefore considered of lesser importance and subsequent attention is focused on NO_3-N. Suspended sediment concentrations were estimated by filtration of 500 ml of the sample through GF/C (1.2 m) filter papers.

Nutrient loadings

NO_3-N, PO_4-P and suspended sediment loadings were calculated by combining concentration values from samples taken at the gauging stations and flow measured at the time of sampling. Exponential relationships between flow and PO_4-P concentration were linearised using log transformations of the data. Linear regression models were then used to estimate daily P concentration using continuous flow measurements for periods in which no samples were taken.

NO_3-N and suspended sediment concentrations were both linearly related to flow although the NO_3-N/flow relationship was relatively poor; consequently, NO_3-N loadings were calculated using mean seasonal concentrations and mean flow taken at time of sampling. Suspended sediment concentrations were closely related to flow, enabling the estimation of daily suspended sediment concentrations which were subsequently used in the calculation of suspended sediment loadings. Nutrient/flow relationships were determined for individual seasons which were defined as spring (March, April, May), summer (June, July, August), autumn (September, October, November) and winter (December, January, February).

Data handling and statistical analysis

All analytical and hydrological data was collated in a database (Oracle Corporation, 1984). Combinations of data were extracted from the database and described using minimum, maximum, means and standard errors. Trends in the data over time were investigated as a time series analysis. Analysis of variance was unsuitable as the dataset was unbalanced (the number of samples collected in each year varied). Trends were taken as significant if the slope of the regression line was significantly greater than 0 or significantly less than 0.

RESULTS AND DISCUSSION

Nitrate in the Rivers Dee, Don and Ythan

A summary of 1980-1992 data for sampling points at the lowest gauging stations (Table 4) shows mean annual NO_3-N concentrations of 6.20, 2.55 and 0.52 mgl^{-1} for the Ythan, Don and Dee rivers respectively, presumably reflecting the differences in land use within each catchment (Table 2). The total quantity of fertiliser N and P applied per catchment was greatest in the Don and least in the Dee. The small intensively managed Ythan catchment however, received the greatest application of N and P fertiliser per km^2 (Table 5) as well as the greatest N and P inputs from sheep and cattle (Table 6) of the three catchments.

Increasing agricultural activity in recent years in the Ythan catchment is thought to have contributed to the high river NO_3-N concentrations and related eutrophication problems in the estuary (Raffaelli *et al.*, 1989; MacDonald *et al.*, in press). Seasonal trends in mean NO_3-N concentrations within and between the three catchments

Table 4 NO$_3$-N (mg l^{-1}) summary (1980-1992)
   (sampled at Ellon, Parkhill and Park Bridge on the Ythan, Don and Dee respectively)

	Ythan	Don	Dee
Mean	6.20	2.55	0.52
Minimum	0.38	1.03	0.13
Maximum	9.68	6.29	1.55
Standard error of the mean	0.10	0.07	0.04

Table 5 Summary of fertiliser N and P (t a^{-1}) applied in the Ythan, Don and Dee catchments
   (survey of fertiliser practice, 1991)

Total fertiliser applied	Ythan		Don		Dee	
	P	N	P	N	P	N
Arable	1164	5360	1284	4745	674	1365
Grass	275	3348	386	5423	208	2574
Total	1439	8707	1670	10168	882	3939
Total applied per km^2	2.1	12.6	1.3	7.7	0.42	1.9

Table 6 N and P inputs (t a^{-1}) from sheep and cattle

N and P	Ythan	Don	Dee
Total N input	3736	5250	2786
N input (km^{-2})	5.4	4.0	1.3
Total P input	402	564	302
P input (km^{-2})	0.6	0.4	0.1

showed a consistent pattern decreasing in the following order: winter, spring, autumn and summer (Figure 2). Similar monthly trends in NO$_3$-N concentrations were observed (Figure 3) between the largely unmanaged upper reaches of the Don — and particularly the Dee catchments — and the predominantly agricultural areas of the lower reaches. This indicates that the principal processes influencing seasonal trends in river NO$_3$-N concentrations, such as plant uptake, mineralisation and denitrification, are similar in managed and unmanaged areas. However, although the seasonal trends were the same between the upper and lower reaches, the concentrations in the intensively managed lower reaches were much greater.

Spring is an important period within the farming calendar with regard to activities directly affecting NO$_3$-N loss, as it is the time of year when ploughing is often carried out and the bulk of fertiliser applied. The increase in NO$_3$-N concentrations in the River Ythan from 1980 to 1992 was most significant (P < 0.001) during the spring months (MacDonald *et al.*, in press) coinciding with an increase in agricultural

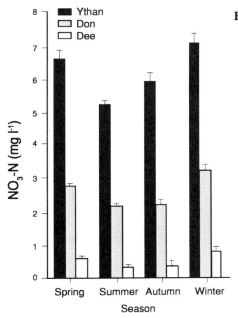

Figure 2   Mean NO_3-N (mg l^{-1}) for the Ythan, Don and Dee (downstream gauging stations) 1980-1992

intensification   reported during this time period. No significant changes were apparent in NO_3-N concentrations during any season in the Dee and the Don over the same time period. The three catchments are located relatively close to one another and as such should have experienced similar climatic conditions from 1980-1992. Climate can be discounted as the main cause of the trend shown by the River Ythan as no such trend was shown by the Don or the Dee rivers. The recent trend towards agricultural intensification is particularly apparent in a catchment such as the Ythan, dominated by agriculture, and would account for the increase in spring NO_3-N concentrations.

The Ythan catchment is relatively uniform throughout in terms of land use and this is reflected in the constant mean spring NO_3-N concentrations along the length of the river (Figure 4). In contrast, land use within the Dee and the Don catchments changes steadily from extensive hill and upland in the upper reaches towards intensive agriculture in the lower reaches. There was a significant increase (P< 0.001) in mean spring NO_3-N concentrations from the upper reaches to the lower reaches in both catchments (Figure 4) indicating a spatial relationship with agricultural intensity. The temporal and spatial nature of these relationships between river NO_3-N concentrations and land use intensity are evidence of the importance of the interaction between edaphic properties and management.

Phosphate in the Rivers Dee, Don and Ythan
Mean PO_4-P concentrations measured at the lowest gauging station sampling points were 0.047, 0.052 and 0.014 mgl^{-1} for the Ythan, Don and Dee respectively (Table 7). The highest mean PO_4-P concentration in the River Don suggests that in contrast

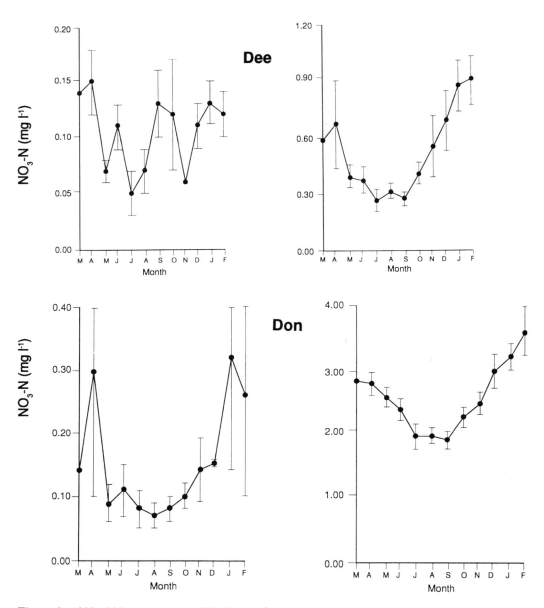

Figure 3 1980-1992 mean monthly NO_3-N (mg l^{-1})

to NO_3-N, the percentage of agricultural land within each catchment is not the principal factor influencing PO_4-P concentration. Wide variation in PO_4-P concentrations along the lengths of each river (Figure 5) suggests the importance of point sources such as effluent inputs from settlements. A comparison of sampling points upstream and downstream of the sewage treatment works for the town of Ellon (Table

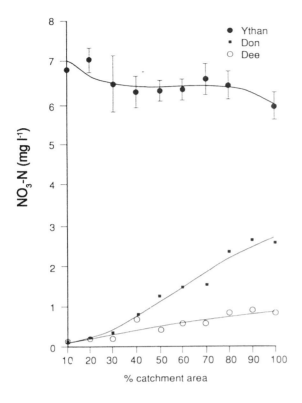

Figure 4   Rivers Ythan, Don and Dee 1980-1992 mean spring NO_3-N (mg l^{-1}) spatial variability

Table 7 PO_4-P (mg l^{-1}) summary (1980-1992)
  (sampled at Ellon, Parkhill and Park Bridge on the Ythan, Don and Dee respectively)

	Ythan	*Don*	*Dee*
Mean	0.047	0.052	0.014
Minimum	0.007	0.005	0.001
Maximum	0.246	0.154	0.093
Standard error of the mean	0.003	0.003	0,002

3) located in the Ythan catchment showed a two-fold increase in mean annual river water PO_4-P concentration (MacDonald *et al.*, in press). A similar comparison including all major settlements in the Ythan, Don and Dee catchments also showed an average two-fold increase in PO_4-P concentration, equating approximately to an increase in concentration of 10% per thousand head of population. Increases in PO_4-P concentration resulting from these effluent inputs are short-lived, concentrations

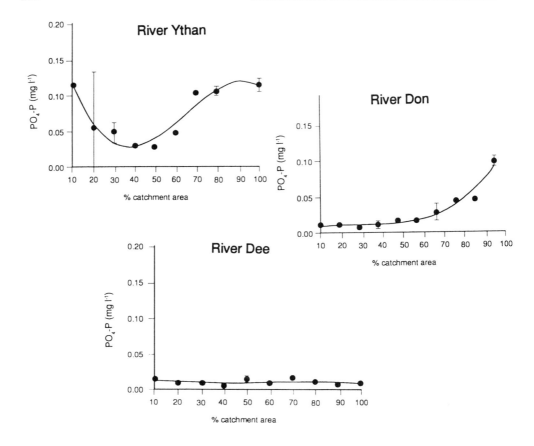

Figure 5 Spatial variability in the mean spring PO$_4$-P (mg l^{-1}) for the Rivers Ythan, Don and Dee, 1980-1992

quickly reverting back to the background concentration. This suggests that rapid biological or physico-chemical nutrient transformation may occur such as P immobilisation or adsorption onto suspended mineral particles (Andrew et al., 1990).

This rapid transformation of P from one form to another makes the calculation of non-point sources such as agriculture difficult and liable to be underestimated. However, by calculating the contribution of point sources to the total P loadings reaching the estuary, the contribution from diffuse sources can be calculated by subtraction.

Suspended sediment in the Rivers Dee, Don and Ythan
Suspended sediment concentrations displayed a similar trend to NO$_3$-N, with the highest mean concentration in the River Ythan and the lowest in the River Dee (Table 8). Phosphorus transport from agricultural land occurs primarily in particulate form (Johnson et al., 1976). Losses of particulate material can be high from heavily grazed permanent pasture, with organic-rich clay and silt sized fractions being preferentially

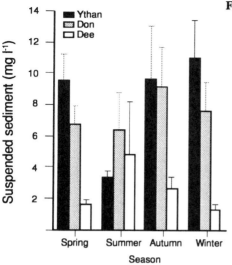

Figure 6 Mean suspended sediment (mg l⁻¹) for the Ythan, Don and Dee (downstream gauging stations) for 1980-1992

Table 8 Suspended sediment (mg l⁻¹) summary 1980-1992 (*sampled at Ellon, Parkhill and Park Bridge on the Ythan, Don and Dee respectively*)

	Ythan	*Don*	*Dee*
Mean	8.11	7.48	2.64
Minimum	1	1	1
Maximum	107	71	47
Standard error of the mean	1.12	0.99	0.92

removed (Heathwaite, 1993). Consequently, soil particles with the maximum concentration of particulate nutrients are transported first. These losses lead to an increase in river-water particulate P, around one-third of which is biologically available (Ryding & Rast, 1989).

It is usually assumed that particulate nitrogen transport from agricultural land is low and most of the nitrogen is transferred in dissolved inorganic form as nitrate (Kissel *et al.*, 1976).

Seasonal trends in suspended sediment concentration differed between catchments (Figure 6). The highest suspended sediment concentrations in the River Ythan occurred during the winter months and the lowest concentration during the summer. The opposite pattern was evident in the River Dee, with the lowest and highest suspended sediment concentrations occurring in the winter and summer months respectively. These opposing patterns are likely to be a reflection of land use and in particular the differences in agricultural management between the two catchments.

Discharge (m³ s⁻¹) and nutrient loadings (t a⁻¹)

Mean annual discharges were calculated as 7, 25 and 38 m³ s⁻¹ for the Ythan, Don and Dee respectively. NO_3-N, PO_4-P and suspended sediment loadings, at the lowest gauging station, are presented in Table 9. The product of intermediate concentrations and flows of the River Don results in the transportation of the highest loadings of the three catchments. The River Ythan and River Dee have similar loadings for all except NO_3-N, despite their different size. Suspended sediment collected at four sampling points on each river and on four separate surveys from September to December 1993 contained a mean particulate P of 15.5 mg g⁻¹ (standard error = 2.28 mg g⁻¹) equal to a mean of 1.5% suspended sediment loading. The largest loss of P associated with suspended sediment was calculated as 82.8 t a⁻¹ in the River Don, with 39.7 and 37.8 t a⁻¹ in the Rivers Ythan and Dee respectively (Table 9). The contribution from sewage treatment works serving the major settlements in each of the three catchments (Table 3) were found to contribute 18, 14 and 26% to the total P loadings (Table 9) of the Ythan, Don and Dee respectively. The difference between effluent inputs and total P loadings gives an indication of the contribution from diffuse sources calculated as 1.20, 0.62 and 0.45 kg P ha⁻¹ a⁻¹ for the Don, Ythan and Dee respectively. The contribution from diffuse sources was highest in the Don catchment, reflecting the large area of agricultural land in the catchment and the high river discharge. The relative contribution of diffuse P sources in the Ythan catchment was lower than the Don catchment suggesting that the total area of agricultural land is more important with regard to P loading than the percentage of agriculture in the catchment. Low fertiliser and animal inputs per km² in the Dee catchment gave rise to the lowest contribution from diffuse sources of the three catchments.

Whilst eutrophication problems have been reported in connection with the River Ythan, no such problems have been encountered in the River Don despite the high nutrient loads. This may be due to the fact that although load is often a useful measure within a mass balance calculation, it not necessarily important in determining vulnerability to eutrophication and in some cases the concentration of nutrient is the critical factor. Eutrophication in the Ythan catchment is unusual in that the estuarine environment is one of the few places where plants respond to NO_3-N, but in this particular situation it is the concentration rather than the load of the nutrient which is important.

Table 9   Total annual loads (t a⁻¹) in the Rivers Ythan, Don and Dee

	Ythan	*Don*	*Dee*
NO_3-N	1369	2010	623
PO_4-P (I)	10.4	40.1	16.8
Suspended sediment	2649	5519	2517
Particulate P (II)	39.7	82.8	37.8
Total P (I = II)	50.1	122.9	54.6

CONCLUSIONS

NO_3-N and suspended sediment concentrations reflect the percentage area of agriculture in the Ythan , Don and Dee catchments. PO_4-P concentrations were closely associated with effluent inputs from settlements. Sewage inputs commonly resulted in a doubling of PO_4-P concentrations, but these peak concentrations quickly decreased downstream. By studying only PO_4-P concentrations, the contribution from agriculture may be underestimated if diffuse sources of P are quickly immobilised or adsorbed onto mineral surfaces, as appears to be the case with point sources. However, by calculating the contribution of sewage works to the total P loadings of each catchment, the remaining proportion of total P unaccounted for by the sewage works input can be taken as an indication of the contribution from diffuse sources. The significant increase in mean spring NO_3-N concentration reported in the River Ythan from 1980-1992 was not observed in the River Don or the River Dee over the same time period. The catchments are all situated in the north-east of Scotland and as such experienced similar climatic conditions over the 13-year period. A trend in rainfall or temperature within this 13-year period which could have caused the significant increase within the River Ythan would also have affected the other rivers in a similar manner. As this was not the case, it is concluded that the increase in mean spring NO_3-N concentration in the River Ythan is due to management practices rather than climate. The fact that the most significant increase occurred in the spring when ploughing is often carried out and fertiliser is commonly applied is further evidence leading to this conclusion.

ACKNOWLEDGEMENTS

Thanks are expressed to the Scottish Office Environment Department for funding this work and to Iain Clinton and Derek Fraser (NERPB) for their help in data retrieval and collation.

REFERENCES

Andrew, N., Sharpley, S. & Smith, S.J. 1990. Phosphorus transport in agricultural runoff: the role of soil erosion. In: Boardman, J., Foster, I.D.L. & Dearing, J.A. (eds), *Soil Erosion on Agricultural Land*. John Wiley & Sons Ltd. 351-366.

Bibby, J.S., Douglas, H.A., Thomasson, A.J. & Robertson, J.S. 1982. *Land Capability Classification for Agriculture*. Soil Survey of Scotland Monograph. The Macaulay Institute for Soil Research, Aberdeen.

Dermine, B. & Lamberts, L. 1987. Nitrate nitrogen in the Belgian course of the Meuse River — fate of the concentrations and origins of the inputs. *J. Hydrol.* **93**, 91-99.

EC Nitrate directive (91/676/EEC).

Heathwaite, A.L. 1993. Nitrogen cycling in surface waters and lakes. In: Burt, T.P., Heathwaite A.L. & Trudgill, S.T., (eds), *Nitrate Processes, Patterns and Management*.

Heathwaite, A.L. 1993. The impact of agriculture on dissolved nitrogen and phosphorus cycling in temperate ecosystems. *Chem. & Ecol.* **8**, 217-231.

Henriksen, A. & Selmer-Olsen, A. 1970. Automated methods for determining nitrate and nitrite in water and soils extracts. *Analyst* **95**, 514-518.

Holford, I.C.R. & Doyle, A.D. 1993. The recovery of fertiliser phosphorus by wheat, its agronomic efficiency and their relationship to soil phosphorus. *Australian J. Agric. Res.* **44**, 1745-1756.

Johnson, A.H., Bouldin, D.R., Gayette, E.A. & Hedges, A.M. 1976. Phosphorus loss by stream transport from a rural watershed: quantities, processes and sources. *J. Environ. Qual.* **5**, 148-157.

Kissel, D.E., Richardson, C.W. & Burnett, E. 1976. Losses of nitrogen in the Blackland Prairie of Texas. *J. Environ. Qual.*, **5**, 288-292.

MacDonald, A.M., Edwards, A.C., Pugh, K. B. & Balls, P.W. (In press). Soluble nitrogen and Phosphorus in the River Ythan system: Annual and seasonal trends. Submitted to Water Research.

Marston, R.A. 1989. Particulate and dissolved losses of nitrogen and phosphorus from forest and agricultural soils. *Progress in Physical Geogr.* **13**, 234-259.

McDowell, L.L., & McGregor, K.C. 1984. Plant nutrition losses from conservation tillage corn. *Soil Tillage Research* **4**, 79-91.

Murphy, J. & Riley, J.P. 1962. A modified single solution method for determination of phosphate in natural waters. *Anal. Chem. Acta* **27**, 31-36.

Nielsen, N.E., Schjørring, J.K. & Jensen, H.E. 1987. Efficiency of fertiliser uptake by spring barley. In: Jenkinson, D.S. & Smith, K.A. (eds), *Nitrogen Efficiency in Agricultural Soils*. Elsevier Applied Science. 62-72.

Oracle Corporation 1984. Interactive Applications Facility: Application Designer's Reference Manual — Version 3.1, Oracle Corporation, California.

Raffaelli, D., Hull, S. & Milne, H. 1989. Long-term changes in nutrients, weed mats and shorebirds in an Estuarine System. *Cah. Biol. Mar.* **30**, 259-270.

Ryding, S.O. & Rast, W. 1989. *The Control of Eutrophication in Lakes and Resevoirs,* Man and the Biosphere series, Vol 1, Parthenon Publishing Group, France.

Sharpley, A.N. & Menzel, R.G. 1987. The impact of soil and fertiliser phosphorus on the environment. *Advances in Agronomy* **41**, 297-323.

20 Effect of deforestation on snowmelt in the Jizera Mountains

SARKA BLAZKOVA
T G Masaryk Water Research Institute, Prague, Czech Republic
CHRISTOPH HOTTELET and LUDWIG N. BRAUN
Swiss Federal Institute of Technology ETH, Zürich, Switzerland
LIBUSE BUBENICKOVA
Czech Hydrometeorological Institute, Prague, Czech Republic

INTRODUCTION

The Jizera Mountains are situated in the Black Triangle, one of the most polluted areas of Europe. As a consequence, on most of the upper part of the mountains, the forest has died and has been harvested.

The effects of deforestation on the hydrological regime under various conditions were described in a large number of studies all over the world. Bosh and Hewlett (1982) presented a review of the results of 94 catchment experiments set up to study the effect of deforestation and afforestation on water yield. Although the direction of change in water yield following forest operations can be predicted, the estimates of magnitude of changes are very approximate due to a large number of factors. The prediction of changes in flood flows is even more difficult (e.g. Flavell, 1982, Hewlett & Helvey, 1970, Swindel & Douglass, 1984). One of the worst examples of total devastation occurred in the Sudbury area in Canada (Winterhalder, 1984). Hydrological and ecological aspects of deforestation were studied in the Hubbard Brook experiment (Bormann *et al.*, 1974). An important contribution to understanding runoff generation processes was presented by Dunne (1978) and a new review of catchment studies has been given by Robinson & Whitehead (1993).

Deforestation was found to speed up the melt process and to result in an increase in flood peaks and in their earlier occurrence. This effect was described e.g. on Fool Creek in Colorado where there is snow cover the whole winter (Troendle & King, 1985). However, an opposite effect has also been reported, i.e. a quicker melt in the forest. In that case snow intercepted by tree crowns was more exposed to the various sources of energy for melt than the precipitation falling as snow in the clearcut watershed (Harr & McCorison, 1979).

Changes in vegetation cover are often studied on paired experimental basins by means of regression relations before and after the cover change on one of the basins. An alternative approach is to use a hydrological model calibrated on the period before the harvesting of forest. The modelled runoff is then used as a reference for the deforested period. An example describing the application of the PULSE model (a modification of HBV) is given in the contribution by Brandt *et al.* (1988). It is stressed that the hydrological modelling approach to the effects of forest management relies heavily on climatological records.

THE UHLIRSKA BASIN

The Jizera Mountains are one of the coldest areas in Bohemia with extreme precipitation totals and intensities. Average annual precipitation total for the period 1901-1950 was 1373 mm. Floods of long return period are caused by summer storms. The snow accumulation, however, is an important part of the water balance of the catchments. The geology is fissured crystalline rocks covered with thin soils.

Snow accumulation is usually observed during December, January and February with possible melting in between (Bubenickova *et al.*, 1985). The meteorological situations bringing the larger quantities of new snow most often are NWc, Wc and Nc (N = north, W = west, c = cyclonic). The most frequent wind direction, which is also connected with the greatest wind velocities at 1500 m a.s.l., is from west and north-west (CHMU, 1994). The absolute maxima of snow depth measured at the station Nova Louka (near the Uhlirska Catchment, 780 m a.s.l.) for the period 1901-1950 were 181 cm in January, 180 cm in February, and 175 cm in March. The values of snow water equivalent decline usually from the second half of March.

An experimental laboratory comprising seven instrumented catchments (2 to 10 km²) was established there by the Czech Hydrometeorological Institute in the early 1980s (Bubenickova *et al.*, 1985). The most pronounced changes which have been found so far are in the stream water chemistry (Bicik, 1993) and in the process of snow

Table 1   Percentages of forest area* in individual age groups** in the Uhlirska Basin (UHUL, 1993)

age gr.	1	2	3	4	5	6	7
1983	14.7	34.9	7.7	1.0	11.7	6.4	3.1
1993	104.6	16.7	49.0	2.5	0.9	12.3	4.0

age gr.	8	9	10	11	12	13	14
1983	6.9	23.7	22.1	10.6	9.3	6.4	3.2
1993	8.8	4.0	0.0	2.3	0.3	0.1	0.0

The numbers give sums of areas of forestry units (i.e. the whole unit is included if more than 50% of it lies within the basin; moreover the boundaries of units were changed in some cases between 1983 and 1993). That is why the totals are not the same and not exactly equal to the basin area.

** *age group 1: 1 to 10 years; 14: 131 to 140 years; etc.*

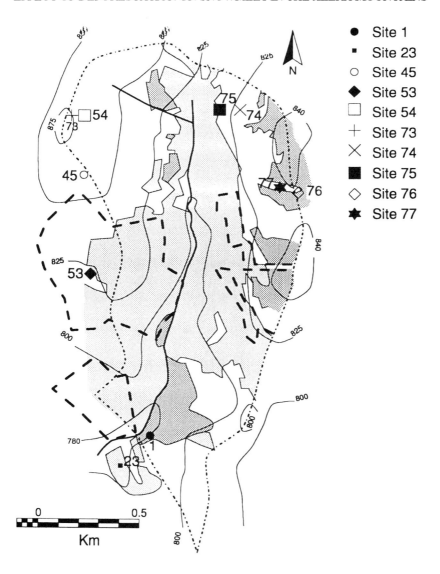

Figure 1 The Uhlirska Basin with contour lines and vegetation cover (simplified from the information from UHUL - unpublished data)

1, 23, 45, etc. snow measurement sites, the notation corresponds to Figure 4

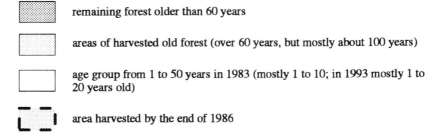

remaining forest older than 60 years

areas of harvested old forest (over 60 years, but mostly about 100 years)

age group from 1 to 50 years in 1983 (mostly 1 to 10; in 1993 mostly 1 to 20 years old)

area harvested by the end of 1986

accumulation and melt.

The Uhlirska Basin on the Cerna Nisa is the smallest one of the experimental catchments (1.87 km^2). Data on vegetation development were obtained from UHUL (Institute for Forest Management). Almost 100 per cent of the basin area is assigned to forest management. Table 1 shows the difference in the age structure of the forest between 1983 and 1993. The trees are conifers with the exception of 0.9 ha in 1983 and 0.8 ha in 1993. It can be seen that during the last ten years the forest cover dropped to under 50 percent. The location of the harvested areas as well as the places with remaining old forest can be seen in Figure 1.

The clearings and the first age group of forest (1 to 10 years) have the character of grassland. The young forest on this area is very sparse and low. The afforestation has a large percentage of failure and has to be repeated for several years in sequence. The interception of both precipitation and fog changes and the conditions of air movement over the harvested areas are also different. The frosty wind near the soil surface prevents the new forest from making more rapid growth.

The second age group (11 to 20 years) is mostly young spruce which, due to the severe climate, is only about 2 m high. However, it covers most of the soil surface and begins to suppress the grass.

Theoretically, a forested area should develop during ten years into the next age group. In the Jizera Mountains this is not the case because of the tough climate and the difficulties with afforestation of the large harvested areas.

RUNOFF MODELLING

Nine years of precipitation and runoff observation (1981-1982 till 1989-1990) covering the period of harvesting, in daily time steps, were available. Precipitation was measured at the climate station Bedrichov (outside the catchment at 777 m a.s.l.) and corrected using data from a number of recording raingauges which have been installed on the mountain ridges in summer.

The HBV-ETH model applied in this study was developed by Jensen (1983) on the basis of the HBV model (Bergstrom, 1976). The more detailed snow- and glaciermelt subroutine is described by Braun and Lang (1986) and Braun and Aellen (1990). To be able to take into account various aspect classes in each elevation belt, Hottelet (1991) developed the version 4 as presented here.

The basin was subdivided into two elevation zones (600-800 and 800-1000 m a.s.l.). The first four years (not yet too much affected by vegetation change) were selected as the calibration period. The calibrated snow routine parameters are the rain correction factor (RCF=1.15), snow correction factor (SCF=1.25), minimum and maximum values of the degree-day factor (CMIN=2.6 and CMAX=2.9), transition temperature snow-rain (T0=-0.8), the parameter controlling the aspect-dependent melt function (REXP=1.5), water holding capacity (CWH=0.04) and refreezing coefficient (CRFR=0.02). The comparison of measured and modelled flows for the snow accumulation and melt parts of the calibration period is shown in Figure 2.

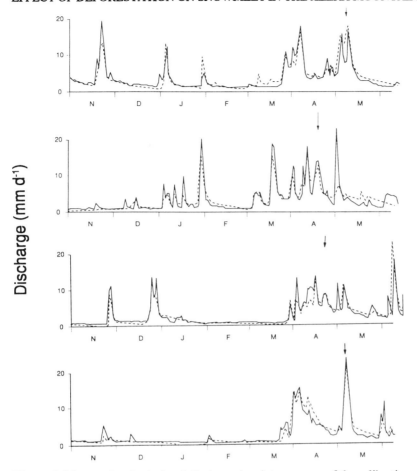

Figure 2 Measured and calculated discharge in winter seasons of the calibration period

The respective values of the Nash-Sutcliffe efficiency criterion R2 are given in Table 2 for the whole years, and the periods with and without snow cover. The flood waves from snow melt are modelled very well, both with respect to magnitude and timing. Summer periods have usually a lower efficiency, the reason being that the daily time step is too long for such a small catchment as far as short events with higher intensity are concerned. The example in Figure 1 is the peak in the beginning of May 1983.

The calibrated set of parameters was then used to model the flows representing the reference (undisturbed conditions) for the period after treatment (timber harvest). The comparison of measured and reference flows is in Figure 3. It is important to bear in mind that in general in the verification period any model (even in unchanged conditions) would be less efficient than in the calibration period. The respective values of R2 dropped both for the whole years and for the periods with and without snow. The one exception is November and December 1987 (a period almost without snow) where two comparatively large peaks are modelled very well.

Table 2 Nash-Sutcliffe efficiency criterion R2

	Whole year R2	Winter without snow cover		Winter with snow cover		Summer		Comments
		N	R2	N	R2	N	R2	
1981/82	.832			212	.825	153	.881	no peaks in summer
1982/83	.830			181	.858	184	.811	
1983/84	.716			197	.804	169	.569	
1984/85	.798			196	.886	169	.419	
1985/86	.655			181	.718	184	.531	
1986/87	.735	30	.342	182	.755	153	.651	
1987/88	.728	76	.906*	121	.831	169	.491	*traces of snow
1988/89	.280			166	.453	199	-.021	
1989/90	.699	15	.350	46	.819			snow w. eq. < 100 mm
		31	.389	43	.748	230	.529	snow w. eq. < 60 mm

N = number of days

On the other hand, the runoff simulation was very poor in 1988-9. During the accumulation period (first half of December) the transition between rainfall and snowfall took place mostly between 700 and 900 m. This fact probably caused large differences in flood peaks in this period. Temperatures during the second half of December and the first half of January kept fluctuating about zero. No other year had such a large difference between snowpacks in the two elevation zones. The failure in summer was probably caused by the occurrence of more short intensive storms than in the other years.

The differences in the affected period which might be interpreted as the consequence of deforestation, i.e. the timeshift (speeding up) of snowmelt, could be seen in December 1985, the end of March and April 1986, April and the beginning of May 1987, to much smaller extent in April 1988, and in March and April 1988-1989.

THE SNOW COVER

On the Uhlirska basin detailed snow measurements have been carried out. The location of sites is shown in Figure 1. In some of the years up to ten sites were observed. No measurements were taken in 1989-90 when there was almost no snow. Snow courses at each site were about 30 m long. The mean snow water equivalent was derived based on snow depth measurements at ten points, and snow water equivalent at three points on each course. Most sites are paired, i.e. measurements were done both in the forest and in clearings. Altitude, aspect and vegetation cover are given in the legend to Figure 4. The snowpack (snow-water equivalent) is an important state variable of the model. Since the measured snowpack was not used in any way in the

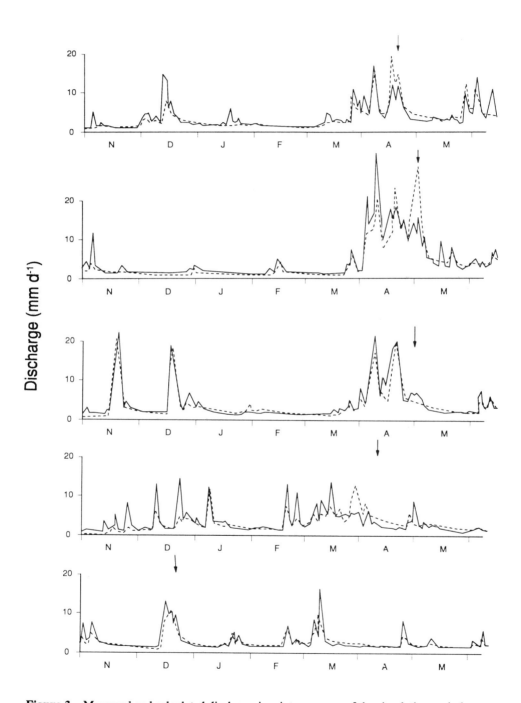

Figure 3   Measured and calculated discharge in winter seasons of the simulation period

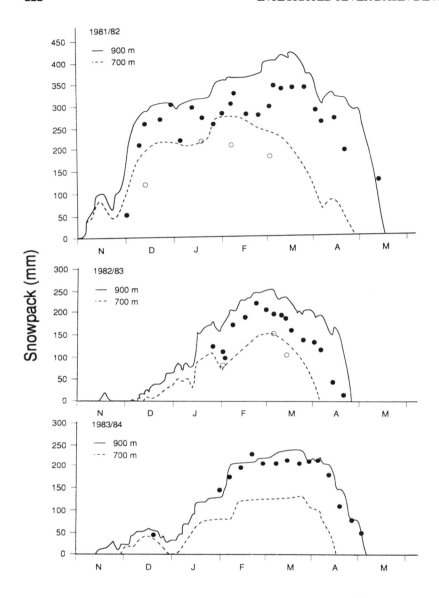

Figure 4 Modelled snow water equivalents for north exposure at elevations of 900 and 700 m
a.s.l. and measured snow water equivalents (sites given in Figure 1; key on page 224)

calibration process it is an excellent means of checking up on the performance of the
model both in the calibration and the simulation periods.

The site No 1 located at the basin outlet in the old forest which has not been harvested
was the one with the most frequent measurements in all the eight years. Its measured
snowpacks lie mostly between the modelled values for 900 and 700 m a.s.l. and the
northern exposition (Figure 4). Since the 1984-5 season a paired clearing (23) has

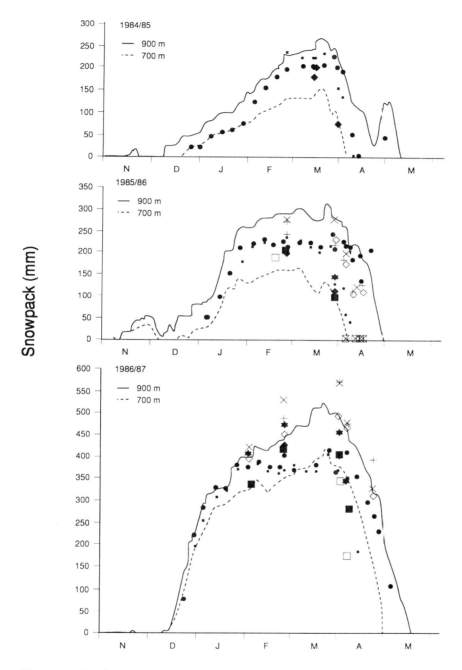

Figure 4 *Contd.*

been observed. On site 45, a clearing facing south, observed in the first two seasons, the snow cover declines very quickly, although the elevation is comparatively high. The combined effect of exposition and vegetation cover can be seen. Since 1984-5

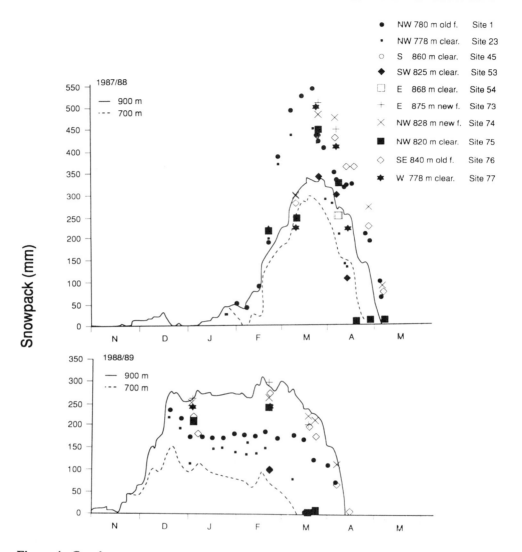

Figure 4 *Contd.*

other clearings 53 and 54 have been observed and since 1985-6 four pairs of forest-clearing sites are available. At the beginning of accumulation the snowpacks on paired sites are usually similar but from the clearings snow melts away about three weeks earlier.

The largest snowpacks mostly occur at the two sites 73 and 74 with new forest (about 20 years old). In the seasons 1986-7 and 1987-8 the measured values at these sites are even larger than modelled values for 900 m a.s.l. and a northern exposure.

For all the ten sites presented in Figure 4, and grouped according to the vegetation cover, the mean deviations from computed snowpack at 900 m a.s.l. and North exposition are shown in Table 3.

Table 3 Mean deviations of the measured snowpack from the computed snowpack at 900 m a.s.l. and North exposition (mm of snow water equivalent)

veget.	NF	NF	OF	OF	CL	CL	CL	CL	CL	CL
expos.	NW	E	NW	SE	NW	NW	E	W	SW	S
elev.	828	875	780	840	820	778	868	835	825	860
site	74	73	1	76	75	23	54	77	53	45
81/82	n	n	-62.	n	n	n	n	n	n	-150.
82/83	n	n	-62.	n	n	n	n	n	n	-103.
83/84	n	n	-13.	n	n	n	n	n	n	n
84/85	n	n	-42.	-40.	n	-49.	-114.	n	-115.	n
85/86	-58.	-74.	-51.	-91.	-198.	-103.	-201.	-212.	-185.	n
86/87	32.	43.	-58.	-7.	-92.	-75.	-134.	-35.	-77.	n
87/88	105.	94.	89	.92.	12.	4.	-13.	-6.	-63.	n
88/89	-20.	-18.	-93.	-42.	-146.	-134.	-134.	-138.	-175.	n

NF - new forest (20 years), OF - old forest, CL - clearing, n - not observed

The year 1987-8 is rather puzzling. It differs from the others in the start of accumulation which did not set in properly till February. The relations both of the measured sites between themselves and the measured and modelled values are different. Up to the half of March the biggest snowpack was observed at the lowest altitude near the outlet. The measured values are mostly larger than modelled ones and the measured snowpack lasted longer. It is however necessary to point out that there is not so much difference between measured and modelled runoff and therefore the modelled snow-water equivalent has not been underestimated.

It seems possible that after harvesting the old forest the snow can be more easily redistributed by wind from the new clearings to the places with the remaining forest. Different snow distribution can, however, be caused also by an unusual meteorological situation. For the year in question both the assumptions might be valid. By the end of 1986 a comparatively large clearcut area was created across the western part of the watershed divide (Figure 1). The most frequent and strongest winds come from

Table 4 Wind direction for the days with snowfall as a percentage (computed from measurements at 7, 14 and 21 h, in the periods November to April) at the station Bedrichov (CHMU, 1994)

wind direction degrees	81/82	82/83	83/84	84/85	85/86	86/87	87/88	88/89
160-210	26.5	25.8	31.1	7.0	27.0	22.3	42.4	33.4
310-360	53.0	57.0	42.2	46.4	43.6	54.1	41.9	44.5

160-210 - wind blowing from the south
310-360 - wind blowing from the west, north-west and north

the western direction. This year, however, was not quite typical as far as the wind direction is concerned. There was about the usual percentage of wind from the west, north-west and north but the proportion of wind coming from the south and south-west was markedly higher than in the other years (Table 4).

From Figure 4 it is clear that not only the vegetation cover but also the meteorological situations were different in the calibration and simulation periods. The accumulation and melt process was rather similar for the four years of the calibration period while in the simulation period the important meteorological variables (amount of precipitation, temperatures and wind) showed greater variation.

CONCLUSIONS

Flows computed by precipitation-runoff model HBV-ETH calibrated during (relatively) undisturbed period were used as a reference for the period covering the timber harvest.

The speeding up of the snowmelt process is quite obvious from the comparison of the snowpack on sites with different vegetation cover. The largest values of snow water equivalent can be seen in places with new forest (20 years). In both old and new forest snowmelt is delayed in comparison with clearings, particularly those exposed to the South.

The effect of deforestation on runoff could not be easily distinguished from the effect of meteorological conditions and the effect of model and calibration errors.

More detailed forest management information, more meteorological variables in the hydrological model and a shorter time step might help to explain the effect of deforestation further. It is essential to understand in detail the impact of forest management on the snow accumulation and melt process. Large mountainous areas of the Czech Republic are affected by deforestation due to air pollution. An intensified melt on a basin where large snowmelt floods occur might indicate the need to reconsider the design values.

ACKNOWLEDGEMENT

The authors are indebted to the crew of the Jizera Experimental Basins of CHMI for providing data. This joint project was financed by the Department of Geography, ETH Zürich and the T.G. Masaryk Water Research Institute, Prague.

REFERENCES

Bergstrom, S. 1976. Development and application of a conceptual runoff model for Scandinavian catchments. Swedish Meteorological and Hydrological Institute, Report RHO No. 7, Norrköping.
Bicik, M. 1993. Hydrological changes in the Jizera Mountains after deforestation caused by

emissions. In: Robinson, M. (ed.), Methods of Hydrologic Basin Comparison, Proc. Fourth Conf. European Network of Experimental and Representative Basins, Oxford, 1992, *IH Report No.120*, Institute of Hydrology, Wallingford, UK. 57-63.

Bormann, F.H., Likens, G.E., Siccama, T.G., Pierce, R.S. & Eaton, J.S. 1974. The export of nutrients and recovery of stable conditions following deforestation at Hubbard Brook. *Ecological Monograph 44*, 255-277.

Bosh, J.M. & Hewlett, J.D. 1982. A Review of catchment experiments to determine the effect of vegetation changes on water yield and evapotranspiration. *J. Hydrol.* **55**, 3-23.

Brandt, M., Bergstrom, S. & Gardelin, M. 1988. Modelling the effects of clearcutting on runoff - examples from Central Sweden. *AMBIO* **17**, 307-313.

Braun, L.N. & Lang, H. 1986. Simulation of snowmelt runoff in lowland and lower Alpine regions of Switzerland. *IAHS Publ. No. 155*, 125-140.

Braun, L.N. & Aellen, M. 1990. Modelling discharge of glacierized basins assisted by direct measurements of glacier mass balance. *IAHS Publ. No. 193*, 99-106.

Bubenickova, L., Chamas, V. & Jirovsky, V. 1985. Experimentalni povodi Jizerske hory (Experimental Catchments in the Jizera Mountains). Report CHMU (Czech Hydrometeorological Institute), Praha.

CHMU 1994. Unpublished data. Czech Hydrometeorological Institute.

Dunne, T. 1978. *Field studies of hillslope flow processes*. In: Kirkby, M.J. (ed.): *Hillslope Hydrology*. John Wiley and Sons, Chichester, UK. 227-294.

Flavell, D.J. 1982. The rational method applied to small rural catchments in the South West of Western Australia. Hydrology and Water Resources Symp., Melbourne, Australia. 49-53.

Harr, R.D. & McCorison, F.M. 1979. Initial effects of clearcut logging on size and timing of peak flows in a small watershed in Western Oregon. *Wat. Resour. Res.* **15**, 90-94.

Hewlett, J.D. & Helvey, J.D. 1970. Effects of forest clear-felling on the storm hydrograph. *Water Resour. Res.* **6**, 768-782.

Hottelet, Ch. 1991. Zur Charakteristik des Wasseräquivalentes der Schneedecke auf der Churfirstennordseite. Albert Ludwigs Universität, Freiburg i. Br., Germany.

Hottelet, Ch., Braun, L.N., Leibundgut, Ch. & Rieg, A. 1993. Simulation of snowpack and discharge in an alpine karst basin (in press).

Jensen, H. 1982. Modell HBV für die Tagesmittel des Abflusses. Kurz- und langfristige Abflussprognosen für den Rhein bei Rheinfelden, Versuchsanstalt für Wasserbau, Hydrologie und Glaziologie, ETH Zürich, Bericht 100.30, 4-14.

Robinson, M. & Whitehead, P.G. 1993. A review of experimental and representative basin studies. In: Robinson, M. (ed.), Methods of Hydrologic Basin Comparison, Proc. Fourth Conf. European Network of Experimental and Representative Basins, Oxford, 1992, *IH Report No.120*, Institute of Hydrology, Wallingford, UK.

Swindel, B.F. & Douglass, J.E. 1984. Describing and testing nonlinear treatment effects in paired watershed experiments. *Forest Sci.* **30**, 305-313.

Troendle, C.A. & King, R.M. 1985. The effect of timber harvest on the Fool Creek watershed, 30 years later. *Wat. Resour. Res.* **21**, 1915-1922.

UHUL 1993. Unpublished data and maps. UHUL (Institute for Forest Mangement).

Winterhalder, K. 1984. Environmental degradation and rehabilitation in the Sudbury Area. *Laurentian University Review*. **16**, 15-47.

21 Aspects of environmental impact evaluation of water resources in flood plains: the case of the Ludueña stream flood plain, Argentina

NORA POUEY, MARGARITA PORTAPILA,
EMMA SCHOELLER and ERIK ZIMMERMANN

Dept. of Hydraulics & Sanitary Engineering, National University of Rosario, Argentina

INTRODUCTION

Human activities are orientated towards the use of natural systems as resources, transforming natural potential into economic values.

Every use of any ecological potential changes the ecological aspect and has a negative impact. Therefore, the interaction between nature and man causes an increase in the negative feedback.

There is a continuous removal and transformation of ecological values into capital assets, leaving an increasingly distorted natural system. Such systems are naturally stable. Stability is only obtained by means of human efforts. In other words, the economic usage is only possible through the instability of the natural system, transforming it into an artificial balance which is guaranteed and maintained via a continuous energy input. Only the unstable natural systems can produce economic values.

The objective of this study has been to show the consequences that uncoordinated human activity produces on hydraulic resources and on the related components of the environment. Taking these impacts into account, one must achieve necessary socioeconomic development without interruptions. The evaluation of human activities in relation to water resources should not only be limited to hydroecological factors, but it should also be considered as a part of an environmental system in which physical, chemical, economic, social and cultural components are assembled.

Researches into the consequences of human activity demonstrate clearly that, up to the present, responsible decision-makers have been overlooking and under-estimating some of the negative impacts which have meant that projects have been less profitable and more expensive than they might have been. Even though the effects taken into account are produced by human activity, for a complete evaluation of the environmental impact, these effects should be considered as an additional burden on situations brought about by natural processes; so that synergy can be allowed for where the concomitant action of different forces makes the total effect larger than the sum of two or more effects of the actions carried out independently.

METHODOLOGY

The method chosen to evaluate the environmental impact has been a cause-effect matrix which is part of the identification methods group. The initial stage of identifying the impacts is very important because once the effects are known, the consequences can be evaluated fairly precisely; in the case where data are not available or the potential damage cannot be strictly evaluated, conservative solutions may be adopted as a precaution against missing information and the lack of actual knowledge. Environmental factors can be also used as check lists.

The components of the matrix are human actions and impact signals situated on two axes. Notes taken in every matrix cell can show only the existence of an interaction (single matrix of interaction) or can assess the interactions in a quantitative or qualitative way (quantified or graduated matrices). Likewise, the matrix can be modified so that it can be adapted to the user's particular needs. Other features of this matrix are its low cost, its multi-subject character with which to evaluate impacts, and its simplicity which makes it easy to understand and to communicate results. However, it should be noted that the time variable is not considered.

Human activities taking place in the south of the Santa Fe Province were considered in order to carry out this evaluation and, in one way or another, all the activities are related to the water resource regarding its relation to flood process as well as considering the variations in water quality.

DESCRIPTION OF THE ECOSYSTEM BEING STUDIED

The region, which constitutes a privileged area for the agricultural exploitation and people and industrial settlement, is in the south of Santa Fe Province, within the Pampas, mainly in the sector known as 'Pampa Ondulada'.

There is a general drainage area, consisting of a chain of hills in a west to east direction, called 'lomas', which are the consequence of the erosion caused by water from surface water courses in the Pampas sediments. This has produced wide terraced fluvial valleys, in which several rivers and streams crossing the subregion from west to east, have shaped their beds so that they could flow into the Paraná River. The most important water courses in the region are the Saladillo, Ludueña, Frías, Seco, Pavón and San Lorenzo streams.

Other features of the area are: flat slopes (0.06 to 0.07%) and a sharp ravine from the North of Puerto San Martín to Riachuelo. The climate is warm, with an average annual temperature of 16.5° C., which has a variation of 13°C. between the average summer and winter temperatures.

Rainfall is between 800 and 1,000 mm, with most rainfall occurring between October and April, and less during winter months, with March characteristically the month when the highest rainfall is recorded. The water balance of the region does not show a deficit; on the contrary, it can show an excess during some months, especially during March.

HUMAN ACTIVITY AND THE LUDUEÑA STREAM BASIN

Human action on an ecosystem implies the alteration of some or several environmental factors, thus interaction among them is modified and so is the system's balance. The alteration rate, and hence the modification of the environment, depends on the type of production that characterizes the socioeconomic system in which man is placed and the stability or complexity of the natural ecosystem. Thus, there is no longer a natural ecosystem but a human system which has new dynamics and functionality ruled not only by natural laws but also by social laws.

In the original ecosystem state, much of the run-off was slow in character and the remaining water, by means of evaporation and infiltration into the ground, could accumulate in the profile. Excess water, delayed by natural vegetation, could flow slowly without eroding, thus decreasing the magnitude and effects of soil degradation.

Nowadays, natural surface run-off characteristics are modified through three main causes:

- infrastructure works (roads, buildings, etc.)
- land usage
- urbanization processes.

Not all the structural factors of an ecosystem have the same speed of response to external pressures. Within natural ecosystems, they cannot be expressed as having a cause-effect relationship; there is a mitigating capacity in the presence of pressures which tend to modify or alter their functionality. Within ecosystems subject to human influences, the move towards an artificial balance tends to increase, imposing a simplification which leads to the lack of a capacity to mitigate the pressures which tend to alter it. Responses then become linearly cause-effect ones. Specifically, this is expressed as the transformation of surface hydrologic dynamics, of the organic substances balance, of the general circulation of nutrients and of the physical degradation of the soil. All this unleashes erosive processes and stresses flood magnitude, which are the most extreme manifestations of environmental modification. These transformations imply the lack of ecosystem stability and the search for a new balance. In an unstable environment the lack of water and soil, and the generalized damage results in an increase in energy consumption to obtain the same production unit. Evidently, this system can neither sustain itself nor increase its production through time.

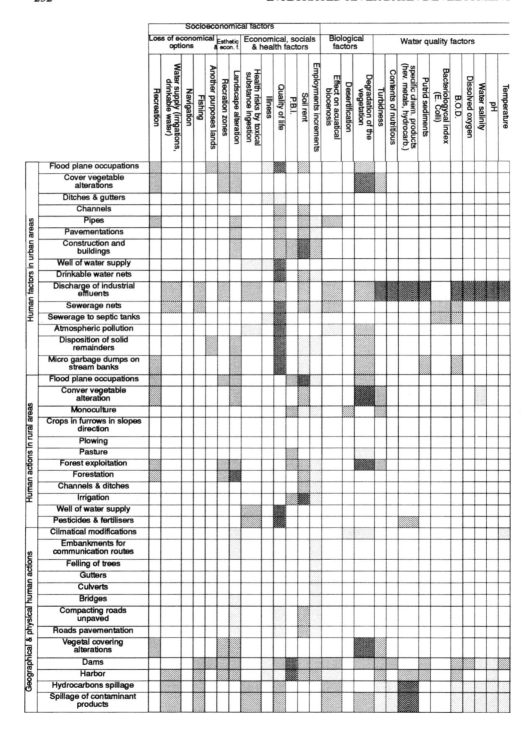

Figure 1 Evaluation matrix of the impact of human actions on water resources

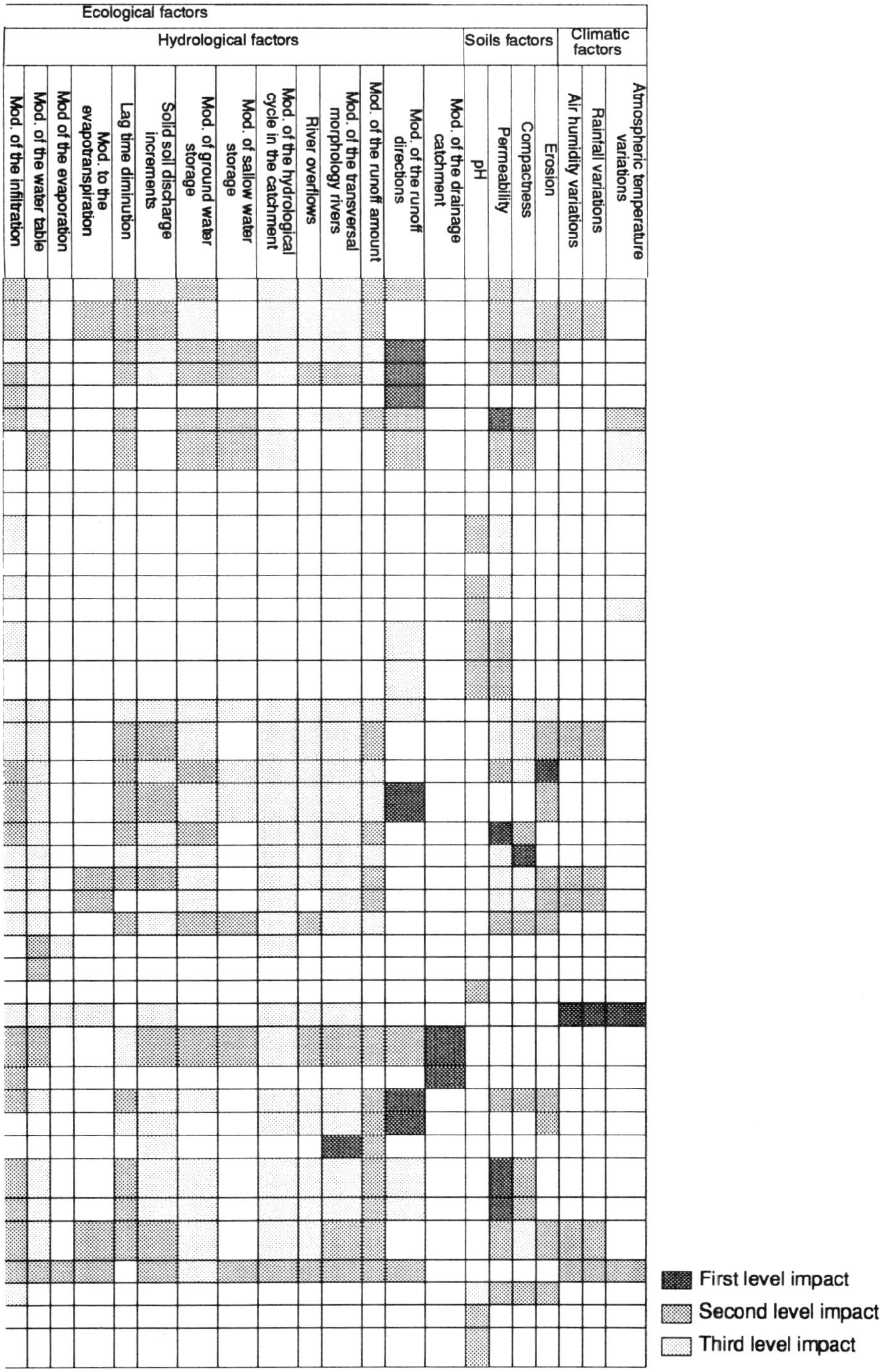

	Ecological factors															Soils factors			Climatic factors		
	Hydrological factors																				
Mod. of the infiltration	Mod. of the water table	Mod. of the evaporation	Mod. to the evapotranspiration	Lag time diminution	Solid soil discharge increments	Mod. of ground water storage	Mod. of sallow water storage	Mod. of the hydrological cycle in the catchment	River overflows	Mod. of the transversal morphology rivers	Mod. of the runoff amount	Mod. of the runoff directions	Mod. of the drainage catchment	pH	Permeability	Compactness	Erosion	Air humidity variations	Rainfall variations	Atmospheric temperature variations

■ First level impact
▨ Second level impact
☐ Third level impact

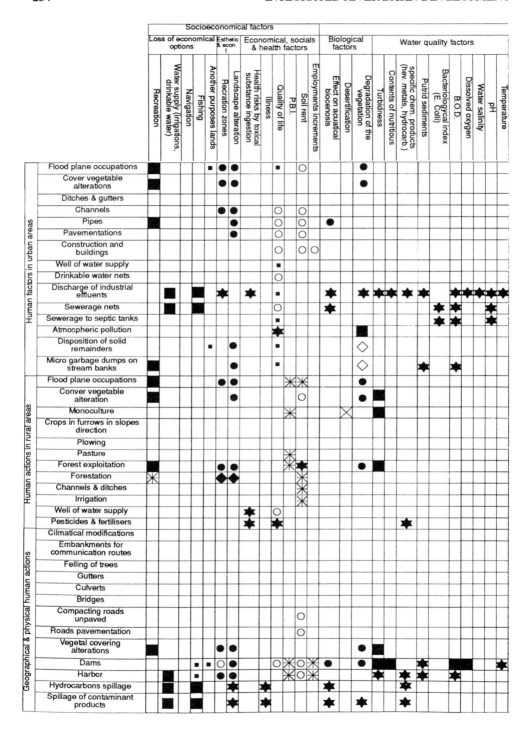

Figure 2 Evaluation matrix of the impact of human actions on water resources

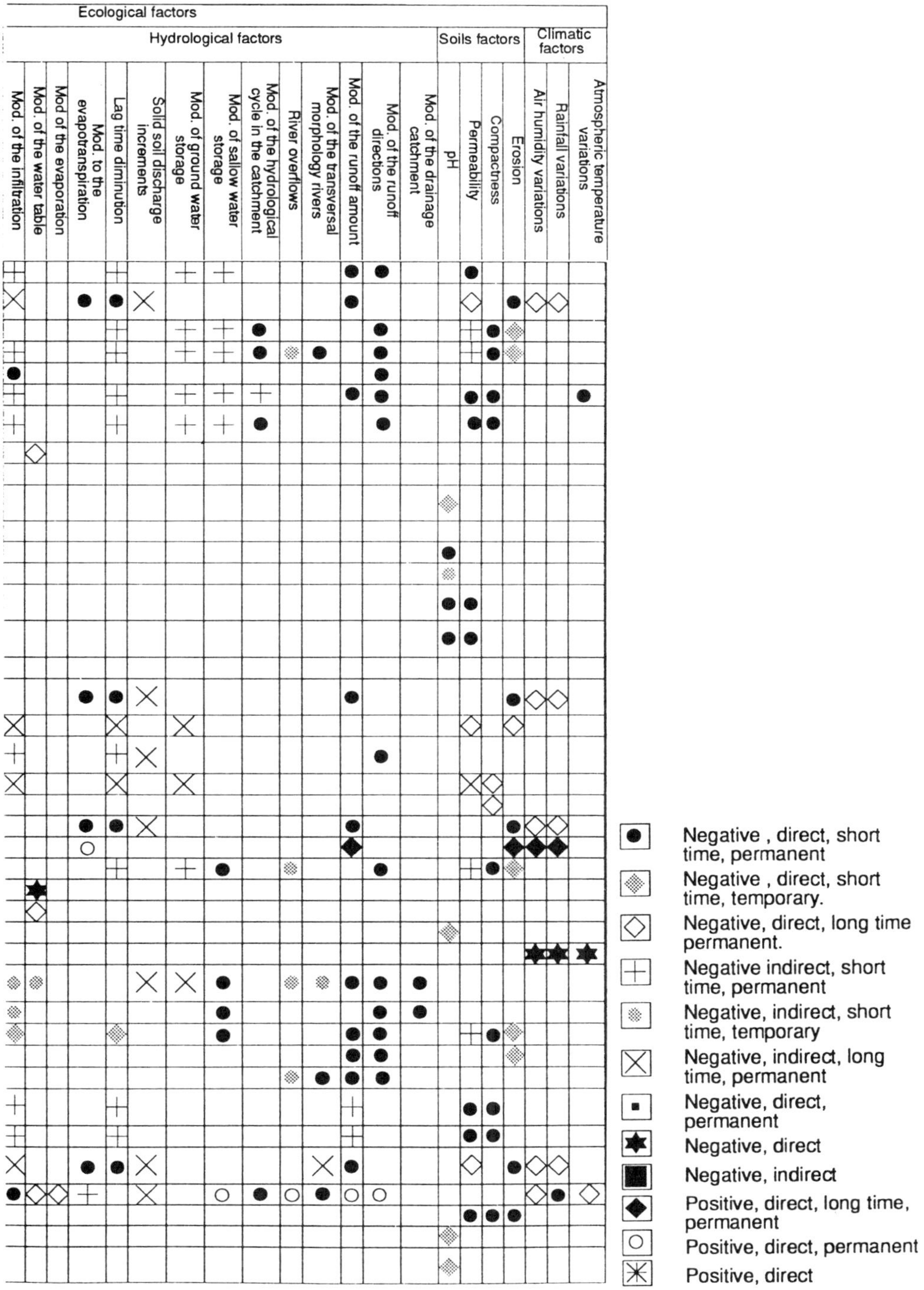

Column headers (left to right):

Ecological factors

Hydrological factors | Soils factors | Climatic factors

- Mod. of the infiltration
- Mod. of the water table
- Mod. of the evaporation
- Mod. of the evapotranspiration
- Mod. to the evapotranspiration
- Lag time diminution
- Solid soil discharge increments
- Mod. of ground water storage
- Mod. of sallow water storage
- Mod. of the hydrological cycle in the catchment
- River overflows
- Mod. of the transversal morphology rivers
- Mod. of the runoff amount
- Mod. of the runoff directions
- Mod. of the drainage catchment
- pH
- Permeability
- Compactness
- Erosion
- Air humidity variations
- Rainfall variations
- Atmospheric temperature variations

Legend:

- ● Negative, direct, short time, permanent
- ▦ Negative, direct, short time, temporary.
- ◇ Negative, direct, long time permanent.
- ⊞ Negative indirect, short time, permanent
- ▨ Negative, indirect, short time, temporary
- ⊠ Negative, indirect, long time, permanent
- ▪ Negative, direct, permanent
- ★ Negative, direct
- ■ Negative, indirect
- ◆ Positive, direct, long time, permanent
- ○ Positive, direct, permanent
- ✳ Positive, direct

VARIATIONS DUE TO LAND USAGE

During the 70s, because of the energy crisis among other factors, there were significant changes in agricultural policy in developed countries. Restrictions imposed on meat and dairy exportation caused a general fall in international prices for cattle production, although the demand for soya bean and forage crops flourished. This made regional mixed systems move towards an agriculture based almost exclusively on double cropping of wheat-soya. Profitability diminished remarkably in the maize area, with cattle raising almost disappearing, and — as a consequence — rotation of semi-permanent pasture.

Two elements relating to land usage can be seen here. On the one hand, we can see a very high proportion — in many cases between 30 and 50 % — of double wheat-soya crop which implies an intensive land usage. If we add to this the large percentage of land subject to continuous agricultural activity for more than 20 years, we can explain partially the damage to the soil resources that we can see now in the region.

One of the most serious and visible consequences of such physical degradation of soil resources is the decrease in the speed of infiltration. This means that less water remains within the soil profile — which is damaging for agriculture — and a greater volume of excess water runs off the cultivated or non-vegetated surface, carrying within it particles of soil. Some of this water discharges through canals, gutters or roads, etc., decreasing periods of concentration, increasing maximum stream volumes, and increasing flooding in rural and urban areas.

This situation is worse when there is a great quantity of furrows, which may be parallel, common or oblique to the contour lines, over and above their influence on run-off, the presence of water lying on the surface, and infiltration.

Another consequence of the decline in cattle raising is the existence of pests and diseases which has increased the use of pesticides, hence increasing environmental pollution which affects animals as well as man.

VARIATIONS DUE TO INFRASTRUCTURE WORKS

Generally, infrastructure works block the laminar runoff. By the end of the last century, this could already be seen with the building of railway embankments. There are substitution of railway embankments where it can be seen that their heights are greater than the ones of the bed; and so they behave as real dams. The consequent effect of this is to dam up water and slow its runoff during periods of intensive rainfall. This leads to a rise in the water table level and a progressive salinization of the affected lands.

It should also be noted that highway planning and road design contributes to the runoff channelisation through gutters with fixed discharge points, i.e. sub-basins are defined in practice. Sub-basins may work with inner drainage ['endorreica'] or with outer drainage ['exorreica'] depending on the rainfall magnitude. It can also happen that some of these sub-basins (as defined by the road embankment design), which formerly belonged to the original catchment, nowadays contribute to a nearby basin.

We should be able to say that bridges over channels and culverts through embankments prevent these damming effects but the fact is that due to poor project design and building, as well as a complete (or almost complete) lack of maintenance, they do not act properly.

URBANIZATION PROCESSES

Rosario's development nearby Ludueña stream basin
In 1888 Rosario increased its urban central area, with the first urban settlement of the future Alberdi district. The Rosario-Sunchales railway had already been built which crossed over the river bed and had the effect mentioned above. This effect has increased substantially due to the multiplication of additional railways and roads with their embankments crossing over the river. In 1905 the city covered the flood channel of the stream, giving rise to the 'Arroyito' and 'Refinería' districts bordering the southern bank. During the following years this inappropriate development continued to occupy the flood plain, including the low-lying areas among the railway embankments. Thus, by 1931, urban usage covered both banks of the stream.

There is no doubt that when the sale of plots and buildings on the flood plain of the stream were authorized, it meant the creation of condemned districts. Because of the characteristics of the land already mentioned - low-lying, prone to flooding and salinized, and with a high water table - the lands were sold to the lower income groups. Thus, the working class district 'Empalme Graneros' took form where the flood waters dammed by the railway embankments would pond. As a consequence of the high costs the provision of social services and the reasons mentioned above, the district was destined to suffer an impoverishment of its way of life.

The high water table has a series of consequences: it prevents or makes difficult the movement of domestic sewage towards the cesspit, a classic individual solution within low density population areas and without any kind of sewerage services. It prevents the pavement construction or makes it difficult, as in this case with clay sub-strata, where concrete pavements are quickly damaged by the pumping effect. The rising water table also caused a serious problem in the 'La Piedad'' cemetery when there was a flood and coffins from the low niches floated on the rising water table.

Before 1931 there had been an attempt to solve the problem of flooding by channelising the stream. After that, in 1941, the section that crossed the city was culverted. In 1966/67 the section between the common limits and Belgrano Railways was also channelised. The situation became more serious because the urban development in this area, as in other parts of the city, continued together with the spread of industrial plants, effluents from which went into the same sewerage system. All this resulted in a high level of biological pollution and of water corrosion of the stream, especially during periods of low flows.

The flood plain is never suitable for urbanisation, unless for green belt development. A good approach to this matter was the imposition of the so-called law of 'non

aedificandi' applied to the Ludueña stream valley; it would also have been advisable
to insist that all means of communication that crossed it allowed free water drainage,
thus keeping the area clear for green spaces. However, this poor choice gave rise to
a depressed urban development and practically turned the Ludueña stream into an
open sewer: what made the situation worse is that it flows into the river next to Alem
Park.

PRESENT ASSESSMENT

Data on man's activity related to hydrological environmental factors and data on
hydrological characteristics of the Ludueña stream basin were gathered together so
as to show the link between the facts and environmental factors of the matrix
suggested earlier, justifying in this way the impact identified. When trying to gather
data several difficulties were encountered:
- scattered data
- incomplete information
- lack of data concerning environmental situations.

The quality of the information obtained does not allow a quantitative evaluation of
the environmental impacts of human activities on the hydraulic resource, leaving us
to identify impacts only through the cause-effect matrix elaborated above.

Figures 1 and 2 show in more detail the matrices and the impacts identified.

MEASURES TO AVOID AND MITIGATE NEGATIVE IMPACTS

Based on the comparison between the natural conditions of the system and its present
state, and on the following evaluation of environmental impacts caused by human
activity, programmes concerning environmental recovery as well as environmental
conservation should be defined. The following list shows the main programmes that
should be implemented as a consequence of the identification of the impacts made in
the suggested matrix:
- land usage control program
- Land occupation control program
- hydrologic impact control program
- water quality control program
- flooding risks management program
- public health and sanitary situation program
- environmental education program

To be able to carry out these programmes, and to obtain a better evaluation of future
projects, it is essential to have the necessary data. We suggest therefore the following
programme of relevant data collection and analysis:
- Parameter selection to evaluate impacts on all the environmental factors under
  consideration

- Selection of methods of data collection and sampling periodicity for all the parameters, according to the different environmental factors.
- Methods to be used in the process of relevant information selection, having the objective of showing the evolution of the identified environmental impacts.

CONCLUSION

To be able to understand and to interpret the dynamics and functionality of a system, the previous analysis shows the importance of the integration of natural resources together with the technological resources used by man, and of the elements that are the result of the interaction between these factors.

Human activity implies resource damage which is directly proportional to the pressure of use man exerts over the ecosystem. This pressure of use and the use of more and more aggressive tools make the problem worse. Present technology adapts itself to that damage but cannot reverse it, introducing an irrational aspect to the use of the ecosystem natural factors. So it is essential to balance the ecosystem, to keep the resource and to rationalize production if there is an expectancy of having a continuous and more efficient production. Note that depressed districts, floods, building plots, and wrongly sited buildings, poor sanitary conditions, water pollution, etc., are not a natural product of urban development, but they are exclusively the unavoidable products of unplanned urban development.

Evidently, there is a close relationship between a certain level of environmental quality and the process of socioeconomic development. An environmental policy should be designed and implemented in order to preserve, restore, or improve environmental capacity with the aim of supporting the development process, especially in the long term. This policy is capable of being elaborated at different levels, covering different features of the environmental aspects of a problem. Thus, it can be simplified to a unique, timely, specific environmental measure of protection destined to solve a special problem of a restricted scope. These problems may also be faced at a global perspective, consolidating an environmental plan which is, by nature, a long-term activity. This environmental plan is the most comprehensive approach to the design and implementation of environmental protection measures, including — apart from the spatial aspects of environmental planning — those elements related to socioeconomics. In short, this approach means the inclusion of an environmental policy within the total process of development planning, stressing particularly long term planning. Thus, it implies the realization that society and the environment constitute two closely linked systems and that environmental quality is a concept beyond pollution problems or resources usage. On the contrary, environmental quality constitutes a complex process of interaction between many factors, some of them being basically socioeconomic and others mainly natural ones.

BIBLIOGRAPHY

Ashby, E. "Reconciliar al hombre con el ambiente"; Ed. Blume; Barcelona, 1981

B.A.H.C. "Biospheric Aspects of the Hydrological Cycle"; Report N° 27; Universitat Berlin; 1993

Beer, T."Environmental Oceanography", Pergamon Press, New York, 1983

Bertalanffy, L.V. "Teoría General de los Sistemas"; Fondo de la Cultura Económica, México, 1990

Bertalanffy, L.V. y otros. "Tendencias en la teoría general de sistemas"; Ed. Alianza; Madrid, 1984

Bolea, M.T. "Evaluación de Impacto Ambiental"; Ed. Mafre; Barcelona, 1984

Bourgoignie, G.E. "Perspectivas en ecología humana"; Colección Nuevo Urbanismo; Madrid, 1976

Brailovsky, A.G. "Memoria verde, Historia ecológica argentina"; Ed. Sudamericana; Argentina, 1991

Caspary, H. "Forest Decline and Soil Acidification as Biospheric Aspects of the Hydrological Cycle"; Proccedings of the Vienna Symposium, IAHS Pub. Nro. 204; 1991

Caubet, C. & Frank, B. "Manejo ambiental en Bacia hidrográfica"; Fund. Agua Viva; Florianópolis, 1993

C.E.P.A.L."Desarrollo Sustentado"; Ed. Naciones Unidas, 1991

Colombo, A.G. "Environmental Impact Assessment"; Commission for the Europeam Communities; Ed. KLUWER ACADEMIC PUBLISHERS; 1991

Comision Mundial Sobre Medio Ambiente y Desarrollo. "Nosso futuro comum"; Ed. Fundaçao Getulio Vargas, R.J.; Brasil, 1988

Contador, R.C. "Avaliaçao social de projetos"; Ed. Atlas; Brasil, 1981

Correa de sa Freire, L.M. "Evoluçao e desenvolvimento de sistemas estuarinos. O caso do estuário do Rio Santa Maria"; Tese de Mestrado; COPPE; RJ; Brasil, 1989

Diegues, A.C. "Desenvolvimento sustentado, gerenciamento geoambiental e o de recursos naturais"; Cadernos FUNDAP; Sao Paulo; Ano 9; Nro. 16; 1989

Dyer, K.R. "Estuaries: A physical introduction"; A Wiley Insterscience Publication, Great Britain, 1973

Freire Motta, V. "Processos sedimentológicos em estuários"; Conferencia efetuada no INPH/PORTOBRAS; 1978

Fundacao Estadual de Engenharia do meio Ambiente. "Perfil Ambiental"; Rio de Janeiro; 1989

Gallopin, G. "El medio ambiente humano"; CIFCA; Argentina, 1977

Gasquet, R. "Medio Ambiente y política"; Documento ECO 92; 1992

GRINOVER, L. "O planejamento fisico-territorial e a dimensao ambiental"; Cadernos FUNDAP, São Paulo; Ano 9; Nro. 16; 1989

Hurtubua, J. "Perspectivas del pensamiento ecológico"; UNAM; México, 1976

IIED - América Latina "Medio Ambiente y Urbanización"; Nro. 41; Argentina, 1992

Ilpies "La gestión del medio ambiente en América Latina. Problemas y posibilidades"; Documento MAM-50; 1987

Ilpes "Los estudios y evaluaciones de impacto ambiental"; Documento MAM-46; 1986

Ilpes "Metodologías para la gestión ambiental: evaluación de impacto ambiental; planificación física integrada; cuentas patrimoniales"; Documento MAM-72; 1991

James, A. "An introduction to water quality modelling"; Ed. Wiley; 1984

Lago A. e Padua, J.A. "O que e a ecologia"; Ed. Brasiliense; Rio de Janeiro, 1984

LEAL, J. "A gestao do meio ambiente na America Latina: Problemas e possibilidades"; Cadernos FUNDAP, Sao Paulo; Ano 9, Nro. 16; 1989

Lwmw Machado, P.A.: "Dereito ambiental brasileiro"; Ed. R. dos Tribunais; Sao Paulo, 1989

Magrini, A. "A evaluaçao de impactos ambientais"; CENDEC/IPEA/SEPLA; Brasilia, 1989

Margalef, R. "Ecología"; Ed. Omega; Barcelona, 1986

Margalef, R. "Perspectivas de la teoría ecológica"; Ed. Blume; Barcelona, 1978

McDowell, D.& O'Connor, B. "Hydraulic behaviour of estuaries"; Cambridge, 1977

Ministerio de Agricultura de la Pcia. de Santa Fe "Programa de investigación aplicada para la planificación del uso, manejo y conservación de los recursos naturales suelo y agua en cuencas de llanura"; Argentina, 1984

Odum, E.P. "Ecología: el vínculo entre las ciencias naturales y las sociales"; Companía Editorial Continental S.A.; México; 1990

Odum, E.P. "Ecología"; Ed. CEC; México, 1990

Odum, E.P. "Ecología"; Ed. Guanabara; Brasil, 1988

O.I.T. "Capacitación en administración del medio ambiente"; Programa OIT/PNUMA; Vol. 1,2,3,4,5; Ginebra, 1990

O.I.T.& PNUMA "Administración de proyectos y medio ambiente"; Ginebra, 1990

Olivier, S.R. "Ecología y Subdesarrollo en América Latina"; Ed. S. XXI; Argentina, 1986

O'Kane, J.P. "Estuarine water-quality management"; Ed. Pitman; London, 1980

Orchard, R. "Tendencias en la teoría general de sistemas"; Alianza; Madrid, 1984

Popolizio, E. "Los sistemas de escurrimiento de las llanuras del NEA como expresión del sistema mórfico"; Proccedings of the Olavarría Synposium; Vol. III; UNESCO; 1983

Rosemberg, R.; Kardnopp, D. "Introduction to physical system dynamics"; DSF A 02002.87; EEUU, 1987

Sachs, I. "Ecodesarrollo: Crecer sin destruir"; Ed. I.M.A.; París, Francia, 1989

Saeijs, H.L.F. "Changing Estuaries"; Governement Publishing Office; The Hague, 1982

S.E.M.A.N: Perfil ambiental municipios de Macaé"; Río de Janeiro, 1989

Sunkel, O. y Gligo, N. "Estilos de desarrollo y medio ambiente en América Latina"; Ed. Fondo de Cultura Económica, México, 1980

Tamames, R. "Ecología y desarrollo"; Ed. Alianza; Madrid, 1985

Tujchneider, O. "Impacto ambiental por manejo inadecuado de los recursos hídricos"; Facultad de Ingeniería y Ciencias Hídricas; UNL; Santa Fe; Argentina, 1991

Vassallo, O.: "Evaluación de impactos ambientales"; FCEIA; Argentina, 1991

Washington D.C., Congress. "Wetlands: Their use and regulation"; Office of Technology Assessment, 1984

William, R. "El medio ambiente y el hombre"; Ed. Limusa; México, 1971

Zimmermann, E.D. "Las acciones del hombre y sus efectos en los procesos de inundaciones"; FCEIA; Argentina, 1991

PART VI

EROSION AND SEDIMENTATION

22 CALSITE: A catchment planning tool for simulating soil erosion and sediment yield

PETER A BRADBURY
HR Wallingford Ltd, Howbery Park, Wallingford, UK

INTRODUCTION

CALSITE is a GIS-based software package for the "CALibrated SImulation of Transported Erosion. The term "transported erosion" is used to describe the eroded soil which reaches the catchment outlet, as distinct from "source erosion" which applies to soil eroded from each grid-cell or "pixel" under study. CALSITE has been designed by HR Wallingford as a catchment planning tool for identifying the main sources of sediment delivered to reservoirs and for predicting the effects of land use change on soil erosion and sediment yield.

Soil erosion and sediment transport within a catchment may be studied using a range of different techniques. Complex physically-based models are being developed to help understand the nature of soil erosion and sedimentation processes, such as the SHESED (Wicks *et al.*, 1992). Such models require powerful computers and detailed input data, particularly when applied to large river basins. Semi-empirical equations have been used with some success to predict soil erosion on a catchment scale using GIS technology (see for example Hession & Shanholtz, 1988). Although such techniques provide little insight into the nature of processes and may have some theoretical limitations, they can serve as useful tools for the planning of soil conservation programmes. Furthermore, such approaches are generally less demanding than physically-based models in terms of computer power and input data.

As CALSITE is intended as a catchment planning tool for use in the developing world, where it is likely that catchment data are scarce or lacking, the package has been developed using semi-empirical relationships and is designed to run on an IBM PC/AT.

CALSITE makes use of the low-cost IDRISI GIS package for the preparation of input map data and for map display. Map information on rainfall, soils, land use and slope is digitized within IDRISI then used within CALSITE to determine soil erosion for each pixel. Typically, a pixel may represent a grid-cell 150 × 150 metres. Digital elevation data are used to predict overland flow routes and to estimate a sediment delivery ratio. The product of the soil erosion and delivery ratio provides an estimate of sediment yield of each pixel contributing to the sediment yield of the catchment. Estimates of sediment yield are calibrated using measurements of sediment load from sub-catchments within the river basin.

The initial development of CALSITE was based upon the Magat catchment in The Philippines, the results of which are reviewed here. The testing of Version 2.0 of CALSITE is currently being undertaken based upon the Upper Mahaweli catchment in Sri Lanka and catchments in The Philippines.

AN OUTLINE OF THE METHODOLOGY

CALSITE requires the use of spatial data in a grid-cell or "raster" format. In the case of the Magat river basin in The Philippines, a grid-cell or "pixel" size of 150 m was used to cover the catchment area of 4180 km².

The four main stages in setting up a CALSITE simulation are described below.
- Determination of source erosion
- Determination of delivery index
- Calibration of delivery
- Determination of sediment yield.

DETERMINATION OF SOURCE EROSION

The CALSITE package currently makes use of the Universal Soil Loss Equation (USLE) of Wischmeier & Smith (1978) to determine source erosion. Although USLE was developed for use in the USA it has been adapted for use in many tropical countries, including The Philippines (David, 1982). Some tests were undertaken using SLEMSA (Elwell, 1980) and it was concluded that USLE was preferable for use in the humid tropics, for which there are currently few alternative formulae. It is recognised that the USLE has some weaknesses when applied to tropical conditions and alternative techniques are currently being investigated.

The USLE is applied to each grid cell within the river basin and consists of the following equation:

$$E_a = R \, K \, CP \, SL \qquad\qquad (1)$$

where E_a = annual soil loss (t ha^{-1}); R = rainfall erosivity; K = soil erodibility; CP = combined cropping management and conservation practice factor ; SL = slope length and steepness factor.

Digital maps of each of the above four factors are produced by digitizing existing maps or by processing existing digital data sets, either using the IDRISI GIS or other GIS packages. Figure 1 shows a map of land cover for the upper Magat catchment in The Philippines which was used to provide the CP factor of the USLE.

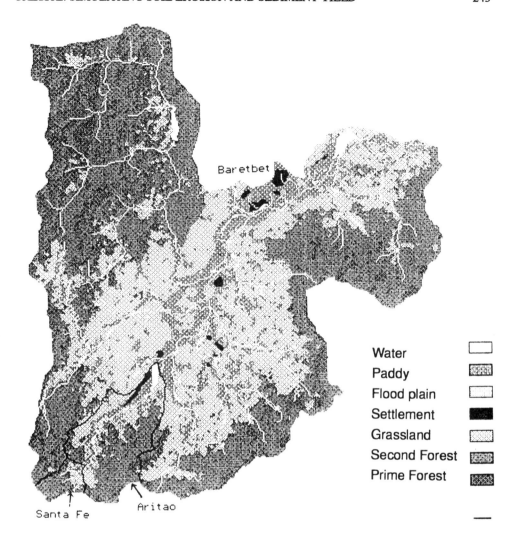

Figure 1   Land cover of the Upper Magat catchment, The Philippines

In some countries empirical relationships have been developed to estimate some USLE factors for conditions different from the USA. The methods used for the Magat basin are based on those currently used in The Philippines (David, 1982) and are described below.

The rainfall erosivity factor (R) of the USLE is normally estimated using the EI_{30} expression where E = kinetic energy of storm rainfall; I_{30} = maximum 30-minute intensity during a storm.

Reliable information on rainfall intensity is rarely available and it is often necessary to estimate rainfall erosivity from total annual rainfall using empirical relationships.

In The Philippines the following equation has been used (NIA/ECI, 1978) based upon research carried out in Indonesia by Bols (1978):

$$R = \frac{2.5\ P_a^2}{0.073\ P_a + 0.73} \tag{2}$$

where R is rainfall erosivity and P_a is the annual rainfall in mm.

A rainfall map or image was created based upon a rainfall-elevation model developed by Blyth & White (1990) for the Magat basin. The values in the image were then converted to rainfall erosivity using Equation 2.

A map of soil erodibility was produced by digitising a reconnaissance soil survey map at 1:50 000 scale. Soil erodibility in the Magat basin varies from 0.26 for clay loam to 0.5 for silty and sandy soils in the valley floors, based upon data provided by the Department of Soils, The Philippines.

Information on land use and vegetation is used to determine the C and P factors of the ULSE. For areas where no accurate and up-to-date land cover map is available, as was the case in the Magat basin, this information can be derived from satellite data. A supervised classification of Landsat Multispectral Scanner imagery provided the seven class land cover map of the Magat basin shown in Figure 1. CP factors were chosen for each class based upon local empirical work (David, 1982).

A digital map of slope was created within CALSITE from digital elevation data then converted to the slope length (LS) factor using the following equation of Smith & Whitt (1948):

$$LS = 0.2\ S^{1.33} + 0.1 \tag{3}$$

Figure 2 shows a digital map of source erosion derived from the product of digital maps of rainfall erosivity, soil erodibility, land cover and slope length. The worst soil erosion corresponds to areas of degraded grassland on steep slopes bordering the forest area as shown in Figure 1. Most of the highland areas of the catchment are protected from soil erosion by primary and secondary forest as can be seen from Figure 1. The lowland parts of the catchment in the east have gentle slopes and are mainly under rice cultivation and have relatively low levels of soil erosion.

DETERMINATION OF DELIVERY INDEX

The "delivery ratio" represents the proportion of source erosion which reaches an outlet. Thus a delivery ratio of 0.5 means that only 50 per cent of source erosion from a pixel reaches a reservoir or other outlet; the rest of the soil erosion is redeposited along the flow path.

Although the physical processes of overland flow routing are complex they are treated in a simple manner in CALSITE. The term "delivery index" is used to refer to an arbitrary value from 0 to 255, which is an estimate of the limiting transporting capacity down a flow path, and is related to delivery ratio. The delivery index value

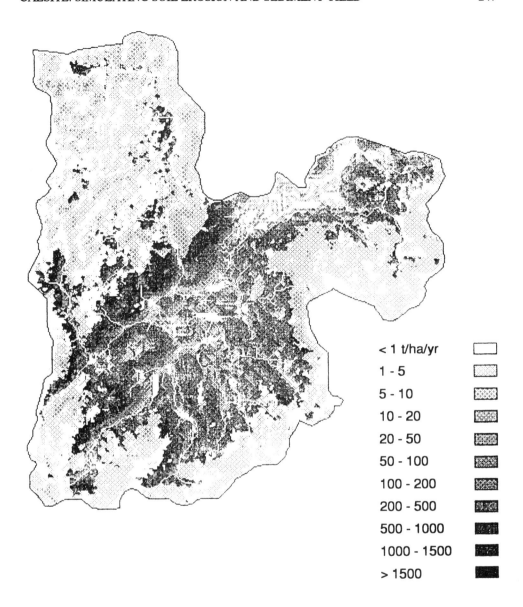

Figure 2 Digital map of source erosion for 1986

is determined for each pixel, using Equation 10 below, then converted to delivery ratio as a part of the calibration procedure.

The determination of delivery index is based on theoretical considerations on the transporting capacity of overland flow, supported by the empirical results of Govers (1990).

The transporting capacity of overland flow (T) can be related to liquid discharge

(Q) and channel slope (S) by the equation:

$$T = c \, Q^a \, S^b \tag{4}$$

where c, a and b are constants which vary according to the particle size of the transported sediment.

Alternatively, it is more useful — as shown below — to present the relationship in terms of concentration of the transported sediment (X):

$$X = T/Q = c \, Q^{a-1} \, S^b \tag{5}$$

The choice of appropriate coefficients for Equation 5 is a matter of debate. The application of empirical relationships derived from sediment transport and channel regime in alluvial rivers and channels, suggest that the appropriate values for the coefficients are a = 1.35, b = 1.66 (For further explanation see Bradbury *et al.*, 1993).

Govers (1990) undertook extensive experimental studies of sediment transport under conditions of sheet flow in a sloping laboratory channel containing sediment of different sizes. For the purposes of CALSITE it is assumed that, for long flow paths crossing several pixels in a relatively steep terrain, the flow is turbulent rather than laminar. Govers' results for turbulent flow suggest that the coefficient 'a' should take a value between 1.80 and 1.04 (larger particle sizes giving lower values of 'a') and the coefficient 'b' should take a value between 1.44 and 1.96. On the basis of Govers' work and taking into account work by Atkinson (1992) who concluded that the D_{50} grain size of eroded material is around 30 μm, it was decided to use the coefficients a = 1.5, b = 1.67 for the Magat basin.

Hence the equation for the limiting concentration of overland flow becomes

$$X = c \, Q^{0.5} \, S^{1.67} \tag{6}$$

Along a particular flow path the sediment delivery from a source pixel will be limited by the pixel with the minimum sediment concentration. Thus the delivery index value of a pixel is related to the minimum transporting capacity along the flow path from that pixel.

In order to calculate Q, the flow discharge for each pixel, two surrogate variables are used: F, a flow parameter, and P_a, the annual rainfall. F is determined by calculating the number of flow paths from each pixel within the image which pass over the pixel under study. Flow paths are determined by the aspect-driven routing algorithm of Lea (1992). These calculations are performed using a digital map of aspect which is created within CALSITE and used to build up a flow paths image as shown in Figure 3.

The relationship between rainfall and runoff is not linear. For part of the Magat catchment Amphlett & Dickinson (1989) developed the relationship:

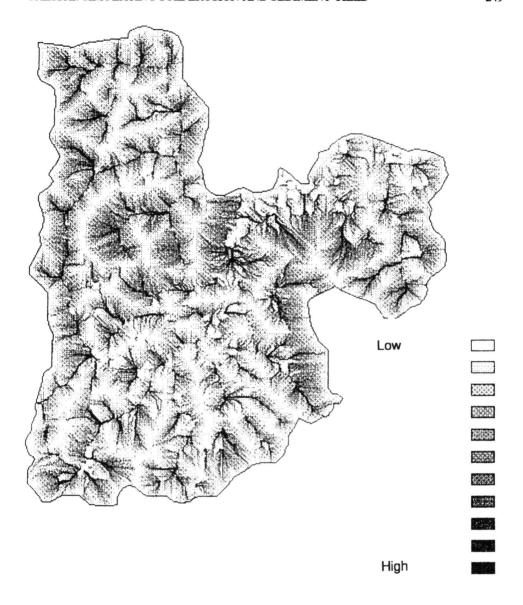

Figure 3   Convergent flows

$$V_a = c_2\, P_a^{1.66} \tag{7}$$

where P_a = total annual rainfall; V_a = total annual discharge, and c_2 is a constant.

This formula was developed for a gently sloping part of the catchment with a relatively low rainfall. To take into account conditions with steeper slopes and higher rainfall, the exponent has been arbitrarily reduced to 1.4.

The Q value in Equation 6 has thus been replaced by

$$V_a = F P_a^{1.4} \tag{8}$$

Combining Equation 6 with Equation 8, and taking into account the minimum transporting capacity down the flow path, the equation for delivery index of each pixel (DI_p) becomes:

$$DI_p = \min [F^{0.5} P_a^{0.7} S^{1.67}] \tag{9}$$

As the annual rainfall value used is that for the pixel under study, rather than the minimum value down a flow path, the equation may be rewritten

$$DI_p = P_a^{0.7} \min [F^{0.5} S^{1.67}] \tag{10}$$

In order to develop the above equation which is suitable for application on a GIS it has been necessary to simplify equations governing the transporting capacity of overland flow. A sensitivity analysis will be undertaken on the variables and exponents of this equation.

Figure 4 shows the final delivery index image for the Baretbet sub-catchment of the Magat basin using the above equation.

CALIBRATION OF DELIVERY

The calculation of sediment yield (SY) involves the basic equation:

$$SY = SE \times f(DI) \times k \tag{11}$$

where SE = source erosion; f(DI) = the calibrated delivery index or delivery ratio value and k is a scaling constant.

The bold assumption is made that if the predicted and measured values of sediment delivery do not match each other then it is the delivery ratio value that is incorrect. The calibration is therefore performed on the "look up table" which converts the delivery index values from 0 to 255 into delivery ratio values from 0 to 1. In reality the variance between observed and measured values may be mostly due to errors in predicting source erosion, however in the absence of source erosion measurements it is assumed that source erosion estimates are correct.

Figure 5 shows a look up table for converting delivery index to delivery ratio. The shape of this graph is modified following an analysis of annual sediment load measurements for sub-catchments within the main basin. A "threshold" (t) value of delivery index is assumed below which all source erosion is re-deposited; a "saturation" or upper threshold value (s) is also assumed above which all source erosion is transported and none is deposited.

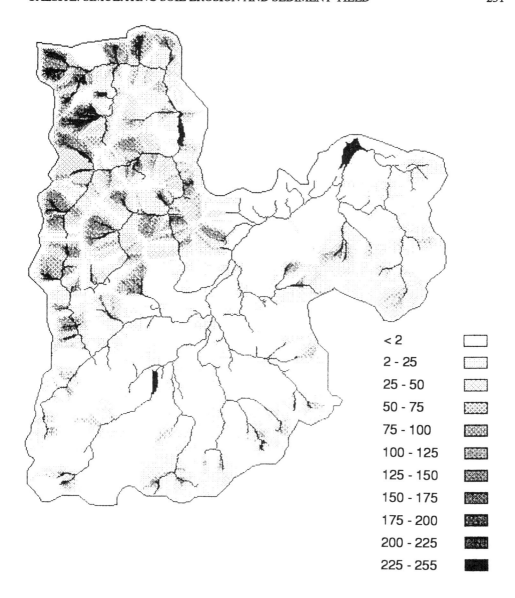

Figure 4   Delivery index

In the Magat basin an upper threshold value is achieved for the main river system where on an annual basis a negligible amount of sediment is redeposited (Atkinson, 1992). Thus the CALSITE model assumes that once eroded sediment reaches a river system, 100 per cent of that sediment reaches the reservoir or catchment outlet.

The calibration procedure examines the variance between observed and measured sedimentation rates for t values between 0 and 50 and s values between 1 and 100,

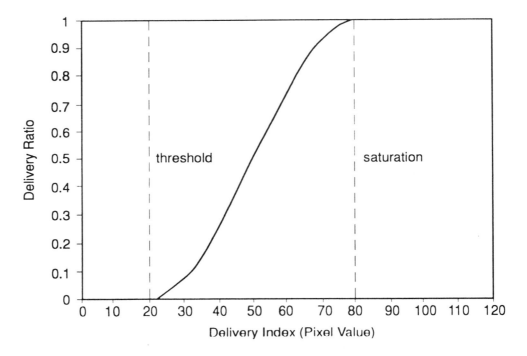

Figure 5   Delivery ratio 'look-up' table

where s is greater than t. A graph is displayed by CALSITE showing variance in colour for a plot of t against s. The user may accept the default values of t and s for the lowest variance or select alternative values which give a good agreement between observed and simulated annual sediment loads for the calibration sub-catchments under study.

CALCULATION OF SEDIMENT YIELD

Once the source erosion and the delivery index images have been produced and the calibration has been performed, the transported erosion of each pixel is performed using Equation 11. Figure 6 shows the resulting image for the Baretbet area for the year 1986. High areas of transported erosion indicate locations which are the prime causes of reservoir sedimentation. It is these locations where soil conservation efforts should be focused.

In the case of the Magat area reservoir, sedimentation sources appear to be the degraded grassland slopes which have high amounts of source erosion and high delivery index values associated with convergent overland flow paths.

CALSITE has been tested using six years of calibration data for three nested sub-catchments of the Magat basin, as reported in Bradbury *et al.* (1993). Regular

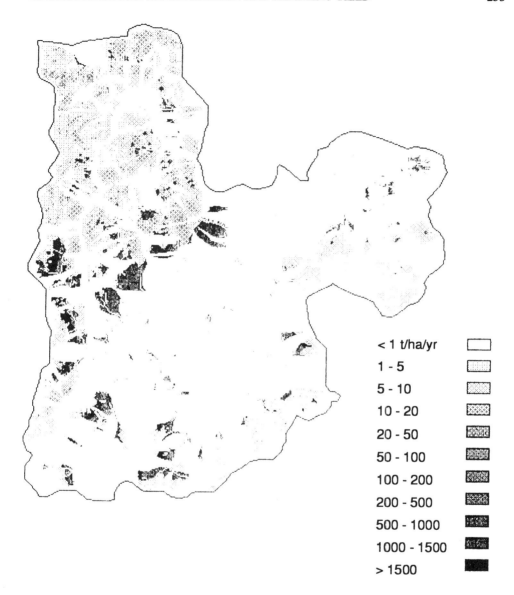

Figure 6   Transported erosion for 1986

sediment measurements were taken as described by Dickinson & Reid (1993) for the Baretbet, Aritao and Sante Fe catchments.

Table 1 shows the results of applying the model for the year 1986 for the three sub-catchments under study. The boundaries of the Santa Fe and Aritao sub-catchments were shown in Figure 1.

Table 1 provides the average source erosion estimates for the three sub-catchments

Table 1 1 986 Simulated and observed erosion rates (t^{-1} ha^{-1} a^{-1})

Catchment	Area (km^2)	Source erosion predicted	Transported erosion predicted	observed
Baretbet	2149	19.05	19.93	25.25
Aritao	144	26.66	23.17	22.00
Santa Fe	15	24.83	41.87	39.70
Calibration coefficients		k	t	s
		3.35	0	10

based upon the USLE technique. For the Baretbet, Aritao and Santa Fe sub-catchments the best fit between observed and predicted is obtained by repeatedly applying different values for t and s. In this case t and s values of 0 and 10 were used, resulting in pixels with delivery index values of greater than 10 contributing 100 per cent of their source erosion to sediment yield.

DISCUSSION

The use of CALSITE in The Philippines has shown that it is a useful method for graphically illustrating predicted soil erosion and for identifying the possible sources of reservoir sedimentation. The use of observed sediment data for sub-catchments enables realistic rates of transported erosion to be mapped. In the absence of reliable sediment estimates for a year under study or for the prediction of erosion, it is possible to use a standard delivery function to convert delivery index values to delivery ratios, which may be based upon years for which sediment data are available. Thus for the accurate prediction of future sediment yields, it may be necessary to make use of calibration data for several previous years.

CALSITE enables the effects of changes in land use on sediment yield to be evaluated. For the Magat catchment as a whole the CALSITE results suggest that if primary forest is degraded to secondary forest and secondary forest is reduced to scrub then sedimentation will be doubled (Bradbury *et al.*, 1993). In a similar manner the effect on source and transported erosion of afforesting the steeply sloping grasslands may be predicted.

FUTURE DEVELOPMENTS

Version 2.0 of CALSITE is planned for release during 1994 and is currently being tested in Sri Lanka, The Philippines and Thailand. It requires the use of the IDRISI

GIS software package for the display and pre-processing of digital map data.

Version 2.0 includes an option to calculate transported erosion for multiple reservoir systems where a high elevation reservoir drains into a lower elevation reservoir. This is based upon the identification of sub-catchments for the higher reservoirs and modifying delivery index values for these areas based upon the known trapping of the dams.

The techniques used to predict source erosion and delivery index are being periodically reviewed in the light of recent empirical and theoretical studies. A simple technique for more accurately estimating rainfall erosivity using a measure which takes some account of intensity, such as the annual rainfall above a daily threshold, is currently being investigated. A modification of the technique of determining delivery index is being considered that will give index values which are more closely related to actual delivery ratios.

Future developments of CALSITE my include the options to take account of gully and channel erosion and to take account of rivers having depositional reaches. A simple hydrological component is under consideration which takes account of the hydrological effects of changing land use.

ACKNOWLEDGEMENTS

The author is grateful to the Overseas Development Administration of the British Government for funding the development of CALSITE and to the National Irrigation Administration of the Government of The Philippines for collaborating with this work.

REFERENCES

Amphlett, M. B. & Dickinson, 1989. Dallao Soil Erosion Study, Magat, Philippines. Summary Report (1984-1987). Report No. OD 111, HR Wallingford, in collaboration with the National Irrigation Administration, Government of the Philippines.

Atkinson, E. 1992. Sediment Delivery in River Systems. OD/TN 48, Overseas Development Unit, HR Wallingford, UK.

Blyth, E. M. & White, S. M. 1990. Statistical analysis of rainfall in the Magat catchment, Luzon, The Philippines. Overseas Development Unit, OD/TN 54, November 1990, HR Wallingford, UK.

Bols, P. L. 1978. The Iso-Erodent map of Java and Madura. Belgian Technical Assistance Project ATA 105, Soil Research Institute, Bogor, Indonesia.

Bradbury, P. A., Lea N. J. & Bolton, P. 1993. Estimating Catchment Sediment Yield: Development of the GIS-based CALSITE model. Report No. OD 125, Overseas Development Unit, HR Wallingford, UK.

David W. P. 1982. Erosion and Sediment Transport. Seminar Paper. College of Engineering and Agro-Industrial Technology. UPLB College, Laguna, The Philippines.

Dickinson, A. & Reid, C. M. 1993. Sediment measurements (1986-1991) in the Magat Catchment, Luzon. The Philippines OD/TN 63. Overseas Development Unit, HR Wallingford, UK.

Elwell, H.A. 1980. A soil loss estimation technique for Southern Africa. Proc. "Conservation 80". Silsoe College, Silsoe, UK.

Govers, G. 1990. Empirical relationships for the transport capacity of overland flow. In: Erosion Transport and Depositional Processes. Proc. Jerusalem Workshop, 1987. *IAHS Publ. No. 189.*,

Hession W. C. & Shanholtz V. O. 1988. A geographical information system for targeting non-point-source agricultural pollution. *J. Soil Wat. Conserv.* **43**, .

Lea, N. J. 1992. An aspect driven kinematic routing algorithm. In: Parson, A. J. & Abrahams (Eds) *Overland Flow — Hydraulics and Erosion Mechanics*. UCL Press, London.

NIA/ECI 1978. Draft feasibility report — Magat watershed management project. Volumes I – III. National Irrigation Administration & Engineering Consultants Inc., Manila, Philippines.

Smith, D. D. & Whitt, D. M. 1948). Evaluating soil losses from field areas. *Agric. Engng.* **29**, 349-396.

Wicks, J. M., Bathurst, J. C. & Johnson, C. W. 1992. Calibrating SHE soil-erosion model for different land covers. *J. Irrig. Drain. Engng*, **118**, 708-723.

Wischmeier, W. H. & Smith, D. D. 1978. Predicting rainfall erosion losses: a guide to conservation planning. *USDA-SEA Agric. Handbook 537*, Agricultural Research Service, USDA, Washington DC, USA.

23 The use of GIS techniques to evaluate sedimentation patterns in a bedrock controlled channel in a semi-arid region

A. W. VAN NIEKERK AND G. L. HERITAGE
Centre for Water in the Environment, University of the Witwatersrand, South Africa

INTRODUCTION

The Sabie River rises on the eastern slopes of the Mauch Berg in the Transvaal Drakensberg in South Africa at an altitude of about 2200 m and flows eastwards for some 210 km to its confluence with the Incomati River in Mozambique. The catchment area is about 7096 km². It is a perennial, mixed bedrock/alluvial channel which is incised into the African 1 and 2 plantation surfaces (Partridge and Maud, 1987). A large proportion of the Sabie River catchment drains the so-called homeland areas of Gazankulu, Lebowa and KaNgwane (Figure 1) which were established during the apartheid era and have large, impoverished rural populations. There is an urgent need to upgrade existing water resources in these areas for industrial and agricultural development and for domestic use. The lower reaches of the Sabie River are in the Kruger National Park (Figure 1) which is internationally recognised as a major conservation area, and there is a need to provide sufficient water for maintenance of the biota which depend on the river. There is thus severe pressure on the limited water resources in this semi-arid environment and water resource managers are investigating options to provide water for human consumption while minimising ecological impacts.

The organisation, structure and development of stream communities are largely determined by the organization, structure and dynamics of the physical stream habitat, together with the pool of species available for colonization (Wevers and Warren, 1986). It therefore becomes necessary to understand these physical patterns across time and space (Frissel *et al.*, 1986) and relate them to biotic patterns in order to establish the ecological effects of changes to the physical system. The geomorphology and nature of changes in a fluvial system result from complex interactions of climate, geology, water discharge and sediment influx (both in the main channel and from tributaries), the bed and bank characteristics, the development of the vegetation along the rivers and the effects of human interference (river

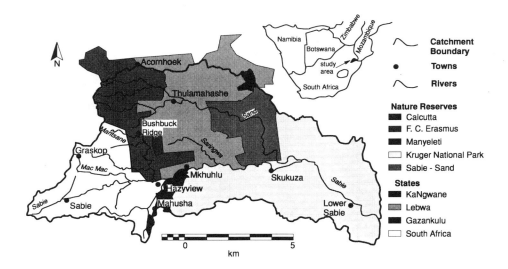

Figure 1 Political boundaries and nature reserves in the Sabie River catchment

structures and water abstraction) (Morisawa, 1985). Increased demands on limited water resources and changes in land usage lead to modification in the flow regime and consequently, the morphology of river systems. This leads to changes in available habitat for different faunal and floral species dependent on the river and an associated adjustment in species distributions.

The flow regime of the Sabie River is characteristic of semi arid systems. Extremes of discharge occur with low winter baseflows (daily average) of approximately 1 m^3 s^{-1} and occasional high summer flows (daily average) in excess of 100 m^3 s^{-1}. Two major tributaries in the upper, wetter areas of the catchment, the Maritsane and Mac Rivers are perennial. The major tributary in the Eastern Transvaal Lowveld, downstream of the Maritsane confluence, is the Sand River which has become a seasonal river as a result of increased abstraction in recent years. There are a number of small, ephemeral tributaries along the length of the Sabie River in the Lowveld. Sediment inputs into the Sabie River are episodic and subsequent reworking redistributes the sediments according to patterns dictated by the underlying geology (van Niekerk and Heritage, 1993). The high variability in flow regime and channel type has resulted in a large diversity in riparian biota. In order to develop models to predict geomorphic changes in response to changes in water and land use, it has become necessary to analyse the location of sediment inputs and quantify them as well as to investigate the competence of different river sections to redistribute them.

LAND USE AND SOILS

A detailed study of development potential and management of water resources in the Sabie River catchment was commissioned by the South African Department of Water

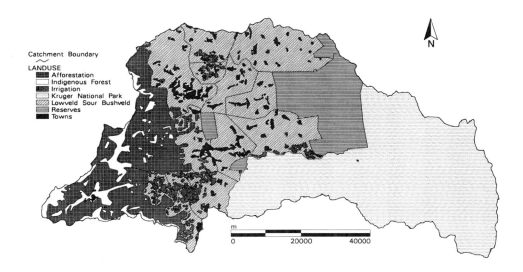

Figure 2 Land use in the Sabie river catchment

Affairs (Chunnett, Fourie and Partners, 1990). In this study, a soil/slope class map was produced and the detailed data on land use in the catchment were used to produce a land use map (figure 2). These data have been captured in the ArcInfo GIS.

The portion of the catchment downstream of Sabie town is underlain by granites and gneisses of the Basement Complex and sediments in the river are characteristic of the soils generated on these rock types. The soil/slope class map along with estimates of soil erodibility from Chunnet, Fourie and Partners (1990) and P. Cheshire (*pers. comm.*) were used to define six slope/erodibility classes (Figure 3) according to broad slope (Table 1) and soil type (Table 2) criteria as follows:

    Class 1 : Moderate slopes and low erodibility
    Class 2 : Steep slopes and below average erodibility
    Class 3 : Moderate slopes and below average erodibility
    Class 4 : Moderate to flat slopes and moderate to above average erodibility
    Class 5 : Moderate to steep slopes and above average erodibility
    Class 6 : Moderate slopes and above average erodibility

Monthly runoff volumes for different subcatchments in the Sabie River under a variety of development scenarios (e.g. natural undisturbed conditions, existing conditions, maximum afforestaiton) for the period 1922 to 1986 were simulated using

Table 1 Slope criteria

Slope description	Slope criteria
Steep	>15%
Moderate	15% to 5%
Flat	<5%

Table 2   Soil erodibility criteria

Soil erodibility description	Soil type
Above average	Very fine sand
	Loamy very fine sand
	Fine sandy loam
	Very fine sandy loam
	Loam
	Silt loam
	Silt
Average	Loamy fine sand
	Sandy loam
	Sandy clay loam
	Clay loam
	Silty clay
Below average	Fine sand
	Sandy clay
	Colluvial cover
	No soil

the Pitman (1972) model (Chunnett, fourie and Partners, 1990 and 1990a). The Pitman model uses meteorological data (monthly rainfall and mean monthly evaporation) and catchment parameters to simulate monthly rainfall and mean monthly runoff volumes at specified points in a catchment and is calibrated using existing flow data. The mean annual runoff (MAR) for the different development conditions is calculated from these data.

Afforestation in the upper catchment has led to a significant reduction in the natural runoff of approximately 17% (Chunnett, Fourie and Partners, 1990). Soil erosion and its associated problems in the homeland areas (Figure 1) is a consequence of severe overgrazing, the removal of tree cover for firewood and in some areas, poor agricultural practices. Drought years are common, thus exacerbating the problem. During drought years, overgrazing occurs in the Nature Reserves since natural game migration routes have been severed. The burgeoning rural population and past restrictions on population movement have resulted in the development of massive rural settlements with increased pressure on the land and water resources. Increased water abstraction for agriculture and extensive afforestation in the wetter upper catchment (Figure 2) have resulted in a decrease in the natural runoff in the Sabie River (Chunnett Fourie and Partners, 1990).

SEDIMENT YIELD AND SEDIMENTATION INDICES

The application of the American Universal Soil Loss Equation (USLE) (Wischmeir and Smith, 1978) for South African conditions is problematic in that the American

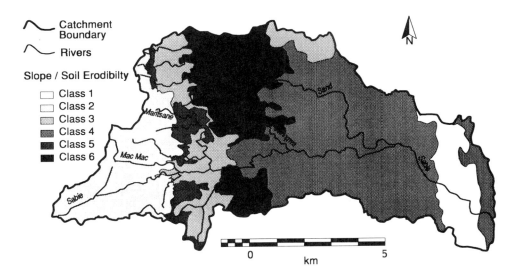

Figure 3 Slope/soil erodibility classes in the Sabie River catchment

factor values are not necessarily applicable to local conditions and the development of correct local factors on a meagre budget with a lack of skilled manpower may prove to be prohibitive (Elwell, 1984). Furthermore, river sediment sampling and reservoir sedimentation data, necessary for calibration and verification of the model, are limited in South Africa and are almost non-existent in the Sabie River catchment.

Degrees of relative land degradation have been estimated using the classification of Viljoen *et. al.* (1993) and a study of Landsat TM imagery in conjunction with ground truthing. Three classes are clearly distinguishable:

- forested areas
- game reserves showing an intermediate degree of vegetation denudation
- homeland areas showing a high degree of vegetation denudation.

Chunnett, Fourie and Partners (1990) recommend the delineation of zones with equal sediment yield potential based on dominant combinations of slope, soil erodibility and land use in the Sabie river catchment. They obtained local sediment yields by adapting values derived from a regionalised sediment yield map, established from data collected throughout South Africa (Rooseboom, 1975), to local conditions. The adaptation of the regionalised data was done in collaboration with Rooseboom and discussions with personnel from the South Africa Department of Agriculture. The Chunnett, Fourie and Partners (1990) local sediment yields and a consideration of degrees of land degradation were used to estimate local sediment yields for different combinations of the controlling variables (Table 3). The land use, slope/soil erodibility and political boundary maps were combined using the ArcInfo GIS to delineate zones with different slope/soil erodibility, land use and vegetation degradation characteristics. This map and the estimates of sediment yields (Table 3) were then used to produce the maximum potential sediment yield map (Figure 4).

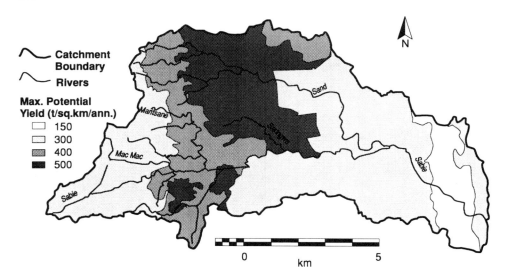

Figure 4   Maximum potential sediment yield map for the Sabie River catchment

Table 3   Estimated sediment yields for different dominant combinations of slope, soil
erodibility   and land use in the Sabie River catchment

Description	Estimated yield $t\ km^{-2}\ annum^{-1}$
Forests on steep slopes; soils with above average erodibility; low degree of vegetation denudation	300
Forests on steep and moderate slopes; soils with above average erodibility; low degree of vegetation denudation	400
Other land use activities (settlements, subsistence farming and cattle) on moderate slopes, soils with above average erodibility, high degree of vegetation denudation	400
Other land use activities on moderate slopes; soils with above average erodibility, high degree of vegetation denudation	500
Irrigated crops on moderate to steep slopes; soils with above average erodibility; intermediate degree of land degradation	500
Game reserves on moderate to flat slopes; soils with moderate to above average erodibility; intermediate degree of vegetation denudation	300
Game reserves on moderate to flat slopes; soils with low yield potential; intermediate degree of land degradation	150

Using the above recommendations (Table 3), it is reasonable to assume a maximum sediment yield potential of 300 t km^{-2} annum^{-1} for the whole catchment under undisturbed conditions except for the area in the Kruger National Park which has a low yield potential of 150 km^{-2} annum^{-1}.

Rooseboom *et al.* (1992) collated all available information relevant to sediment yields in South Africa and developed a yield map for the region. Their early attempts to quantify maximum sediment yield potential revealed that sediment availability is the determining factor in sediment yield processes throughout Southern Africa. Statistical analysis on a regional basis to overcome the wide variability observed in sediment yields was used to develop a new method for estimating sediment yields from ungauged catchments. Regionalised sediment yield potentials are weighted to different areas according to soil erodibility and the size of the area.

This approach results in a value of 155 t km^{-2} annum^{-1} for the Sabie River catchment which forms a small portion of region one of nine regions in South Africa. The game reserves do not fall into any of the regions since no data are available in these areas and a value of 155 t km^{-2} annum^{-1} was assumed for these areas also. This regional approach does not quantify the impact of specific variables such as land use, rainfall intensity, slopes and vegetation in a meaningful way and since there were extremely limited data from the Sabie River catchment for use in the statistical analysis, the first analysis provides the most reliable estimate of relative sediment yield potentials. The values are, however, necessarily conservative since they were used to estimate sedimentation rates in proposed future dams.

The GIS was used to overlay the subcatchment and maximum sediment yield potential maps and calculate the total area within each subcatchment with a specific yield potential. The products of the areas and yields were summed in the GIS to calculate annual sediment tonnages generated in each of the important subcatchments (Table 4 and Figure 5) for existing land use (present) and undisturbed (natural) conditions. The tonnage yielded by the different subcatchments under undisturbed conditions has increased by between 0% and 26% for existing conditions (Table 4) with the major increases occurring in the rural areas where there has been a significant increase in population and major land use change with increased land degradation. Replacing indigenous forest with exotic plantations has not had a major effect in the upper catchment. Associated with change in land use is a reduction of MAR simulated by Chunnett, Fourie and Partners (1990 and 1990a) ranging from 13% to approximately 30% at different key points on the rivers (Figure 5).

The availability of sediment for transport plays a dominant role in determining sediment load in rivers (Rooseboom *et al.*, 1992) and it is therefore reasonable to assume that all of the sediment will ultimately be delivered into the river channels. Simulated existing and natural MAR (Chunnett, Fourie and Partners, 1990 and 1990a) (Tables 5 and 6) show that the bulk of the sediment (80%) is produced in the dry subcatchments (35 % of natural MAR) downstream of the Maritsane confluence (Tables 4, 5, 6 and Figure 5). Large sediment loads will thus be introduced into the Sabie River episodically (during seasonal and ephemeral floods) and subsequently reworked. Since sediment availability is the dominant factor in determining sediment

Table 4 Calculated maximum sediment yields and simulated Mean Annual Runoff (MAR)

River catchment	Area (km²)	Present annual yield¹ (t)	Natural annual yield¹ (t)	Rooseboom annual yield² (t)	Present % annual yield¹	Natural % annual yield¹	Rooseboom % annual yield²	Difference (%) (1)	(2)
Upper Sabie	770	252209	231000	119271	12	13	12	-8	-53
Maritsane	472	162929	141300	73084	7	8	8	-13	-55
Noord Sand	254	102359	75600	39430	5	4	4	-26	-62
Sand	1910	766943	573300	296073	35	32	31	-25	-61
Middle Sabie	1466	554873	438900	227151	25	24	23	-21	-59
Lower Sabie	1389	342633	342633	215315	16	19	22	0	-37

Table 5 Simulated natural mean annual runoff (MAR) — evapotranspiration losses are not included

River catchment	Natural MAR³ (hm³)	% Natural MAR
Upper Sabie	320	41
Maritsane	192	24
Noord Sand	50	6
Sand	169	21
Middle Sabie	49	6
Lower Sabie	13	2

[1] Calculated using the approach established by Chunnett, Fourie and Partners (1990).
[2] Calculated using the approach established by Rooseboom et al. (1992).
[3] Values derived from simulated natural runoff (Chunnett, Fourie and Partners, 1990).

(1) Difference between present annual yield and natural annual yield.
(2) Difference between present annual yield and Rooseboom annual yield.

Table 6 Transport and reworking indices — evapotranspiration losses are included in MAR values

Location (as in Figure 5)	Natural MAR[4] (hm³)	Present MAR[5] (hm³)	Present Annual Yield (t)	Natural Annual Yield (t)	Rooseboom Annual Yield (t)	Transport Index[6] (t hm⁻³)	Transport Index[7] (t hm⁻³)	Transport Index[8] (t hm⁻³)
A	192	-	162929	141300	73005	736	-	-
B	50	-	102359	75600	39060	151	-	-
C	158	137	766943	573300	296205	3629	5598	2162
D	320	-	252209	231000	231000	722	-	-
E	513	-	415138	372300	192355	726	-	-
F	563	404	517497	447900	231415	796	1281	573
G	603	437	1072370	886800	458180	1471	2454	1048
H	761	574	1839313	1460100	754385	1919	3204	1314
I	764	577	2181946	1802733	969525	2360	3782	1680

[4] Values derived from simulated natural runoff (Chunnett, Fourie and Partners, 1990a).
[5] Values derived from simulated present runoff (Chunnett, Fourie and Partners, 1990a).
[6] Ratio of Natural Annual Yield[1] to Natural MAR.
[7] Ratio of Present Annual Yield[1] to Natural MAR.
[8] Ratio of Rooseboom Annual Yield[1] to Present MAR.

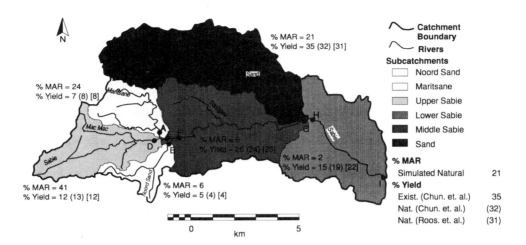

Figure 5 Sabie River subcatchments showing simulated MAR and maximum existing and
natural () potential sediment yield percentages determined using Chunnett, Fourie and
Partners (1990) and Rooseboom et al. (1992) []

load in South African rivers, a storm of one magnitude may transport large volumes
of sediment as suspended and bedload and a subsequent storm of the same magnitude
would transport less material as most of the available sediment has been removed by
the first.

Sedimentation indices (Table 6) are calculated as the ratio of annual sediment
tonnage generated upstream of the considered river section to the MAR at that point.
This provides a relative measure of the amount of water available for transporting
sediments to specific points and subsequently reworking them.

DISCUSSION

There is an increase in the sedimentation indices throughout the catchment from
natural, undisturbed conditions to the present day (Table 6) implying a progressive
increase in sedimentation from the 1920s, when the catchment was essentially
undisturbed. A study of time sequences of aerial photographs between 1940 and 1985
of the Sabie River in the Kruger National Park (Vogt, 1992) has revealed that there
was an overall increase in sedimentation and vegetation in the river during this period.
Increased short-term sedimentation in the Sabie River is also reported by Kruger
National Park personnel (F. Venter, *pers. comm.*). Chunnett, Fourie and Partners
(1990), however, from a comparison of photographs of the Sabie River during 1944
and 1886, found that of fifty current hippo pools, seven did not exist in 1944, twelve
were smaller during 1944, five were larger in 1944 and six were essentially unaltered.
A study of a single hippo pool on the Sand River during 1944, 1965, 1974 and 1986

(Chunnett, Fourie and Partners, 1990), found that the pool was virtually non-existent during 1944 but that it had cleared by 1965. Subsequent to 1965 the hippo pool started to trap sediments, which resulted in a substantially reduced size of pool in 1986 compared to that which existed in 1965 and 1974, but still larger than that which existed in 1944. The studies of hippo pools on the Sabie and Sand Rivers illustrate the dynamic nature of localised sedimentation processes. No conclusions can be drawn concerning overall sedimentation since firstly, the studies do not consider hippo pools which may have existed in 1944 and had shrunk or disappeared by 1986 and secondly, a single morphological type only was considered.

The high sediment yield and low runoff from the Sand River catchment (Table 4) which is predominately occupied by the territories of Gazankulu and Lebowa, is clearly reflected in the high sedimentation indices at location C (Figure 5). The indices in the reach A-E-F (Table 6) are similar, indicating that increased sedimentation is not to be expected downstream of the Maritsane and Noord Sand River confluences. There is a large increase in indices from points F and G, indicating that increased sedimentation is to be expected in this reach. The dramatic increase in indices from location G to H and H to I indicate that a further increase in sedimentation is expected in the Sabie River reaches downstream of the Sabie/Sand River confluence.

Close agreement is found when comparing the picture emerging from the sedimentation indices with actual patterns of sediment accumulation derived from recent (1986) 1:10000 aerial photographs and extensive ground truthing. The Sabie River and its tributaries upstream of location F display predominately bedrock features with localised sediment accumulations upstream of major bedrock controls. There is a progressive increase in valley fill and alluvial features between locations F and G although bedrock features are present in the form of rapids, bedrock anastomosing channel sections and bedrock core bars (van Niekerk and Heritage, 1993). The Sand River is contributing large quantities of sediment to the Sabie River and there is a massive accumulation of sediment at the confluence of the two rivers (van Niekerk and Heritage, 1993). There are large accumulations of alluvium in the lower reaches of the sand river with a progressive increase in bedrock features in an upstream direction. The Sabie River widens downstream of the Sand River confluence and there is a dramatic increase in valley infill in both extent and thickness between locations G and I. Bedrock features observed upstream are largely buried although rapids do occur in the active channel where it has cut down to unweathered bedrock (van Niekerk and Heritage, 1993). The Sabie River steepens for 5km upstream of location I, where it cuts through the Lebombo mountains to the coastal plains in Mozambique. Sediments are flushed through this well defined valley and the valley floor consists predominately of unweathered bedrock.

CONCLUSIONS

Although the competence of the river to transport sediments is a function of the local stream power and available sediment sizes, the approach described here provides a useful tool for developing a qualitative intuition into the river dynamics as well as a feeling for the effects of different land and water management policies. There has been and increase in sedimentation in the Sabie River as the catchment progressed from an undisturbed state to present day conditions. Increased sedimentation occurs in areas of preferential deposition in reaches having relatively high sedimentation indices. There is a high correlation when comparing the temporal and spatial patterns of sedimentation with changes in values of the calculated sedimentation indices. The sedimentation indices provide a clear indication of the temporal and spatial effects of land use and water management policies and clearly highlight areas of concern. In order to prevent extensive geomorphic change and loss of geomorphic variability through excessive sedimentation which will reduce habitat and hence, biological diversity, water resource management must be balanced so as not to reduce channel competence further by uncontrolled damming and abstraction in the headwaters. Afforestation and agricultural development in the catchment needs to be managed so as ensure that sufficient water is available for the predicted population without excessive impact on the riparian biota. Similarly, the effects of land use practices on sediment production must be monitored and erosion control measures implemented.

The technique described will be greatly enhanced through the development of improved predictive sediment yield models with increased reliability and greater transferability.

ACKNOWLEDGEMENTS

The funding of this project by the South African Water Research Commission is gratefully acknowledged. The assistance of the National Parks Board research personnel in the Kruger National Park has been invaluable in ensuring the smooth running of the project. We thank the GIS laboratory at the University of Pretoria for providing the basic ArcInfo covers which were used for the modelling and to produce the maps.

REFERENCES

Chunnet, Fourie and Partners. 1990. Water resources planning of the Sabie River catchment. South Africa Department of Water Affairs Report No. P.X.300/00/0490, Vols 1 to 10.

Chunnet, Fourie and Partners. 1990a. Kruger National Park Rivers Research Programme, water for nature, hydrology, Sabie River catchment. South Africa Department of Water Affairs Report No. P.X.300/00/0390.

Elwell, H. A. 1984. Soil loss estimation: a modelling technique. In: Hadley, R. F. and Walling, D. E. (eds), *Erosion and sediment yield, some methods of measurement and modelling*. Geo Books, Norwich, England.

Frissell, C. A., Liss, W. J., Warren, C. E. & Hurley, M. D. 1986. A hierarchical framework for stream habitat classification: viewing streams in a watershed context. *Environmental Management*, **10**, 199-214.

Morisawa, M. 1985. *Rivers*. Longman, London.

Partridge T. C. & Maud, R. R. 1987. Geomorphic evolution of Southern Africa since the Mesozoic, *South African Journal of Geology*, **90**, 179-208.

Rooseboom, A., Verster, E., Zietsman, H. L. & Lotriet, H. H. 1992. The development of the new sediment yield map of Southern Africa. South Africa Water Research Commission Report No. 297/2/92.

van Niekerk, A. W. & Heritage, G. L. 1993. Geomorphology of the Sabie River: overview and classification. University of the Witwatersrand, Centre for Water in the Environment, Report No. 2/92., Johannesburg, South Africa.

Viljoen, M. J., Franey, N. J., Coward, D & Wedepohl, C. 1993. Classification of vegetation degradation in the North-Eastern Transvaal using SPOT imagery, *South African Journal of Science*, **89**, September, 429 - 431.

Vogt, I. 1992. Short-Term geomorphological changes in the Sabie and Letaba Rivers in the Kruger National Park, Unpublished M.Sc. dissertation, University of the Witwatersrand, Johannesburg.

Wevers, M. J., & Warren, C. E. 1986. A perspective on stream community organization, structure and hydrobiology. *Archive fur Hydrogiologie*, **108**, 213-233.

Wischmeier, W. H., & Smith, D. D. 1978. Predicting Rainfall Erosion Losses - Guide to Conservation Planning, US Department of Agriculture, Agriculture Handbook No. 537.

24 Sediment transport in the arid zone ephemeral channels in India

K D SHARMA
Central Arid Zone Research Institute, Jodhpur, India

INTRODUCTION

The need for water is becoming more critical each year in the arid regions as increasing urban and expanding industrial and agricultural demands put more pressure on the existing water supplies. In addition, the sedimentation problem adds to the need to develop practical approaches to its monitoring since the arid zones produce record concentrations of suspended sediment (Jones, 1981) which not only deteriorate the quality of the water but also affects the physical and biological conditions of reservoirs (Sharma *et al.*, 1984). A compilation of river suspended sediment yields for moderate size drainage basins suggests that arid basins produce 36 times more sediment than humid temperate basins and 21 times more than the humid tropical equivalents (Reid & Frostick, 1987). Because the sediment transport is so large, and because the infrequency of flash floods makes gauging uncertain, there is a considerable need for predictive models of sediment transport in arid regions to aid in the development of water resources. In the present study, depending upon the dominance of hydrological processes, the arid zone drainage basin is conceptualized into an upland phase — where the runoff is related only with rainfall at the drainage basin surface (i.e. in hilly/mountainous regions), and a channel phase — where the transmission losses play an important role. Separate sediment transport models are developed for these phases and coupled to predict the sediment transport at the drainage basin outlet.

GOVERNING EQUATIONS

UPLAND PHASE

Many models of sediment transport by water in upland basins route sediment dynamically by solving the continuity equation for sediment transport (Bennett, 1974). The solution of this equation is generally accomplished using numerical methods which are not only unstable, but also uncertain due to variable friction losses. A closed form solution of the governing differential equation under steady state conditions not only reduces the number of computations but also reduces the instabilities associated with the numerical solutions.

Sediment movement downslope obeys the principle of continuity of mass expressed as follows (Nearing et al., 1989):

$$\delta Q_s/\delta X = D_f + D_i \tag{1}$$

where Q_s (kg s^{-1} m^{-1}) is mass transport rate per unit width; X (m) is downslope distance; D_f (kg s^{-1} m^{-2}) is net flow detachment rate, and D_i (kg s^{-1} m^{-2}) is net rainfall detachment rate. The assumption of quasi-steady state allows the deletion of the time terms from Equation 1. D_i is negligible since the transport capacity of rainsplash is very low (Lu et al., 1989). The value of D_f in arid zones, where the initial potential sediment load is in excess of the transport capacity (Foster et al., 1980; Sharma, 1992), has been estimated by a first order reaction model of the type:

$$D_f = G\,(T_c - Q_s) \tag{2}$$

where G (m^{-1}) is a first order reaction coefficient and T_c (kg s^{-1} m^{-1}) is the transport capacity of flow.

The hydrological variables required to drive the upland sediment transport model are the flow depth, h (m), and the effective runoff duration, t_r (s). The flow depth can be estimated from Manning's equation:

$$h = (q\,n/S^{0.5})^{0.6} \tag{3}$$

where q (m^z s^{-1}) is flow discharge, n is Manning's roughness coefficient and S is mean slope. The effective duration of runoff can be calculated as:

$$t_r = V_{up}/P_r \tag{4}$$

where V_{up} (m^3) is total runoff volume and P_r (m^3 s^{-1}) is peak runoff rate.

The shear stress acting on the soil, T_s (kg m^{-1} s^{-2}), is calculated as:

$$T_s = \tau\,h\,S \tag{5}$$

where τ (kg m^{-2} s^{-2}) is the specific weight of water.

Several generalised formulae have been developed for computing T_c. However, Alonso et al. (1981) concluded that the Yalin equation (Yalin, 1963) provided reliable estimates of T_c for shallow overland flow associated with upland erosion. The Yalin equation is defined as:

$$T_c/(SG)\,d\,p_w^{0.5}\,T_s^{0.5} = 0.635\,\delta\,[\,1 - (1/\beta)\ln(1+\beta)\,] \tag{6}$$

$$\beta = 2.45\,(SG)^{-0.4}\,(Y_{cr})^{0.5}\,\delta \tag{7}$$

$$\delta = (Y/Y_{cr}) - 1 \qquad\qquad \text{(when } Y < Y_{cr} ; \beta = 0) \tag{8}$$

$$Y = (T_s/p_w)/(SG - 1) \, g \, d \tag{9}$$

where SG is the specific gravity of the particle (2.65 for fine sand and silt), d (m) is the diameter of the particle, p_w (kg m^{-3}) is the mass density of water, Y is dimensionless shear stress, Y_{cr} is dimensionless critical shear stress from the Shield's diagram, g (m s^{-2}) is the acceleration due to gravity, and β and δ are parameters as defined by Equations 7 and 8. The modified Yalin equation (Foster *et al.*, 1980) which is capable of dealing with mixtures of particles of varying diameter and density was used in the present analysis.

Combining Equations 1 and 2, the upland soil erosion model is derived as:

$$(\delta Q_s/\delta X) + G \, Q_s - G \, T_c = 0 \tag{10}$$

Equation 9 is a linear inhomogeneous ordinary differential equation having a closed form solution as:

$$\ln (T_c - Q_s) = - G \, X + \ln C \tag{11}$$

where C (kg s^{-1} m^{-1}) is a constant of integration and is equal to $T_c - Q_s$ at X = 0.

CHANNEL PHASE

The channel storage of sediment can have a great effect on sediment transport and therefore the sediment supply has to be taken into account for sediment transport modelling in arid environments (Hadley, 1977; Reid & Frostick, 1987). As the flood flows traverse coarse unsaturated sediments in the ephemeral channels, the sediment transport capacity is progressively reduced by transmission losses of runoff and this results in the deposition of sediments (Walters, 1990; Sharma, 1992).

The suspended sediment dynamics can be represented by a spatially lumped continuity equation and a linear storage law. For a time interval Δt (s) these relationships can be written as:

$$I_s(t) = Q_s'(t) + dS_s(t)/dt \tag{12}$$

and $\quad S_s = K_s \, Q_s' \tag{13}$

where $I_s(t)$ (kg s^{-1}) is sediment input, $Q_s'(t)$ (kg s^{-1}) is sediment discharge, $S_s(t)$ (kg) is sediment storage, K_s (s) is a sediment storage coefficient and t(s) is time since the beginning of sediment discharge. For an instantaneous sediment inflow to the channel, the outflow from the first reservoir may be expressed as:

$$Q_s'(t) = (V_s/K_s) \exp(-t/K_s) \tag{14}$$

where V_s (kg) is the weight of mobilized sediment. By successively routing through n identical reservoirs, the sediment outflow from the nth reservoir is:

$$Q_s'(t) = [V_s/K_s \, !(ns - 1)] \, (t/K_s)^{ns-1} \exp(-t/K_s) \tag{15}$$

where ns is a dimensionless shape parameter. Differentiating Equation 15 with respect to time and using the condition that $dQ_s'/dt = 0$ at $t = t_p$ (t_p(s) is time to peak sediment discharge) gives:

$$t_p = (ns - 1) \tag{16}$$

On substituting the value of K_s in Equation 15, the sediment impulse response becomes:

$$U_s(0,t) = Q_s'(t)/V_s$$

$$= [(ns-1)^{ns}/t_p \, !(ns-1)] \, [(t/t_p) \exp(-t/t_p)]^{ns-1} \tag{17}$$

where $U_s(0,t)$ (s^{-1}) is an ordinate of instantaneous unit sediment graph (IUSG) at time t. The IUSG convoluted with V_s generates the sediment graph at the drainage basin outlet.

LINKAGE BETWEEN UPLAND AND CHANNEL PHASES

The mobilized sediment V_s can be calculated by a regression model of the type:

$$V_s = a + b \, V_i + c \, [V_{up}(X,W) - V(X,W)) \tag{18}$$

where V_i (kg) is the inflow sediment calculated from the area integration of the upland sediment graph (Equation 11); V_{up} (m^3) is the inflow runoff volume, V (m^3) is the outflow runoff — both in a channel reach of length X (m) and average width W (m) — and a, b, and c are relationship parameters.

HYDROLOGICAL DATA

Sediment transport data were collected for nine years (1979–87) from 15 channel reaches in the Luni basin situated in arid northwest India (Sharma *et al.*, 1993). Each channel reach was sampled at a minimum of two stations, one in the upland region and the other in the down channel section. In the upland region the mean depth, width and gradient of the channels were 1.2 m, 158 m and 0.00245, respectively, whereas in the channel phase these were 3.6 m, 1958 m and 0.0012, respectively.

The drainage basin areas varied from 104 to 950 km^2 in the upland region and 1449 to 5492 km^2 in the downstream valley.

Hourly sediment concentrations were determined from samples collected simultaneously using three to five US DH-48 depth integrating suspended sediment wading type hand samplers, employing the equal transit rate method as recommended by Jones (1981) for arid regions. Discharge measurements were by current meter and velocity area method, according to standard United States Geological Survey practice. The resulting data allowed a reasonably accurate representation of the variation in sediment concentration during each flow event, as well as the computation of suspended sediment discharge.

RESULTS AND DISCUSSION

Absolute values of sediment concentrations in the upland basins receiving runoff from different terrains were as follows:

Limestone	0.2 to 13.0 g l^{-1}
Phyllite, schist and shale/slate	0.4 to 29.0 g l^{-1}
Gneiss and granite	0.2 to 18.0 g l^{-1}
Rhyolite	5.7 to 28.9 g l^{-1}

Downstream in the channel phase the sediment concentrations rose sharply to between 1.0 and 453.0 g l^{-1}. Nearly 90 % of the suspended sediments by weight have particle sizes ranging between 0.002 and 0.2 mm. Increases in the sediment concentrations in the channel phase were attributed to the cessation of smaller flows as a result of high transmission losses in the channel alluvium, leaving a large amount of loose material to be picked up by subsequent flows of greater magnitude (Hadley, 1977; Sharma et al.,1984). This is the contrary of the corresponding process in humid regions, where the sediment concentration is further diluted, with downchannel increase in the discharge.

The upland sediment transport model (Equation 11) was tested for ten arid upland basins located within the Luni basin in the Indian arid zone (Sharma et al., 1993) and with areas ranging between 104 and 1520 km^2. The basin complexity was accounted for by dividing the basin into three zones: upper, middle and lower, according to degree of steepness and the stream order (Sharma et al., 1992). The calibration options for T_c were (a) reference slope, (b) dual slope, and (c) average shear. Values of G and C were determined by the least squares techniques at each stage of the flow hydrograph, i.e. rising, peak and recession, for 90, 68 and 76 events, respectively. Independent events at each stage of the hydrography were used to evaluate G and C for the best fit calibration option.

The model parameters G and C are associated with the soil erosion process. According to Equation (11) the sediment transport is proportional to the difference between the transport capacity and the actual transport. The parameter G was found to be in the range of 0.0034 to 0.0069 m^{-1} for the best fit calibration method. For G values greater than unity most of the resulting sediment concentrations were

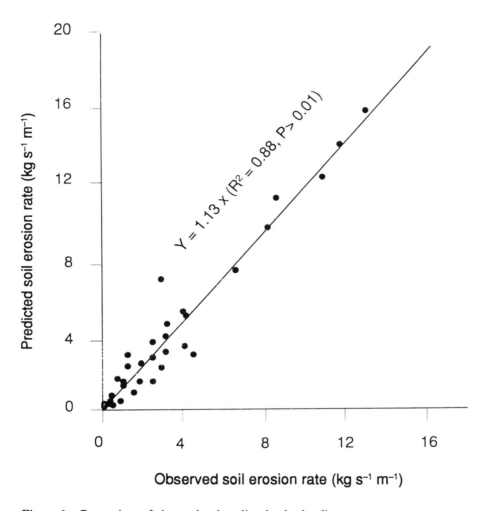

Figure 1   Comparison of observed and predicted upland sediment transport rates

negligible (Laguna & Giraldez, 1993). Singh & Regl (1983) suggested a reasonable value of G as 0.0030 m^{-1}. The parameter C varied from 2.4 to 61.3 kg s^{-1} m^{-1}. A comparison of observed and predicted sediment transport rates (Figure 1) shows good agreement. Furthermore, when using the optimum calibration method, the maximum deviation between the observed and predicted sediment transport rates was less than 6.4 per cent.

As a test to verify and validate the IUSG concept of sediment transport in the channel phase, observed sediment graphs for four representative channels covering the dominant lithologies in the region were compared with the corresponding sediment graphs obtained from the model. For these drainage basins, the peak runoff rate and the maximum sediment transport rate coincides for the discrete storms. To determine the IUSG parameters t_p was taken from a regression model between the peak flow and time-to-peak at the drainage basin outlet and K_s was considered as approximately equivalent to the travel time of the runoff crests in the trunk stream

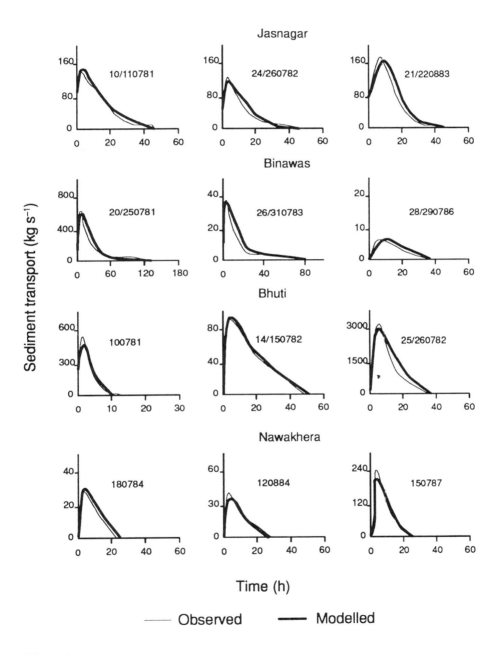

Figure 2 Comparison of observed and predicted sediment graphs at the drainage basin outlets

(Sharma *et al.*, 1993). Parameter values for the sediment mobilization model (Equation 18) are given in Table 1.

Figure 2 depicts the comparison of the observed and predicted sediment graphs at the drainage basin outlet. The sediment graphs generated by the IUSG technique are noted to be good approximations to the actual storm sediment graphs (coefficients of determination for the hourly sediment transport rates between the observed and modelled curves vary from 0.90 to 0.96, $P > 0.01$). This implies that in arid regions the sediment transport is dependent on the availability of erodible material within the channel bed which is hydraulically controlled.

Table 1 Parameter values of sediment mobilization model

Channel reach	Length (km)	a	b	c	R^2
Alniawas – Jasnagar	23	1119	0.61	−0.0001	0.99**
Pipar - Binawas	24	259	0.10	−20.5840	0.98**
Sanderao - Bhuti*	22	−4998	1.29	−0.0011	0.96**
Posalia - Nawakhera	64	4294	−0.03	0.0001	0.95**

* Disorganised, losing stream
** $P > 0.01$

CONCLUSIONS

Depending upon the dominance of hydrological processes, the arid zone drainage basins are divided into an upland phase — where the runoff is directly related to the rainfall, and a channel phase — where the transmission losses dominate. In the upland phase a physically-based model of sediment transport utilising the Mannings' turbulent flow and Yalins' sediment transport capacity equations predicts the sediment transport to within 10 per cent. In the channel phase the sediment transport is governed by the availability of erodible material within the channel bed and is accurately predicted by a conceptual instantaneous unit sediment graph model. A coupling of these two models through a regression model, interlinking the sediment inflow in the channel phase and the transmission losses of runoff in the trunk stream, shows a close agreement between the computed and observed sediment transport rates at the outlet to the drainage basins. By means of this simple approach, in which a limited number of parameters are involved, a reasonable accuracy is attained that satisfies the practical requirements in water quality modelling and in designing the efficient sediment control structures for maximum trap efficiency in the arid regions.

ACKNOWLEDGEMENTS

Dr J. Venkateswarlu and Mr A. S. Kolarkar provided facilities, encouragement and permission to conduct this study. Dr R. P. Dhir and Mr N. S. Vangani helped during the course of this work.

REFERENCES

Alonso, C.V., Neibling, W.H. & Foster, G.R. 1981. Estimating sediment transport capacity in watershed modelling. *Trans. Amer. Soc. Agric. Engineers,* **24**, 1211-1220, 1226.

Bennett, J.P. 1974. Concepts of mathematical modelling of sediment yield. *Wat. Resour. Res.,* **10**, 485-492.

Foster, G.R., Lane, L.J., Nowlin, J.D., Laflen, J.M. & Young, R.A. 1980. A model to estimate sediment yield from field sized areas. In: W. Knisel (Ed.) *CREAMS — A field scale model for Chemicals, Runoff and Erosion from Agricultural Management Systems,* USDA Conservation Research Report No. 26, Beltsville, 34-64.

Hadley, R.F. 1977. Some concepts of erosional processes and sediment yield in a semi arid environment. In: T.J. Toy (Ed.) *Erosion: Research Techniques, Erodibility and Sediment Delivery,* Geo Books, Norwich, 73-81.

Jones, K.R. 1981. *Arid Zone Hydrology,* FAO, Rome.

Laguna, A. & Giraldez, J.V. 1993. The description of soil erosion through a kinematic wave model. *J. Hydrol.,* **145**, 65-82.

Lu, J.Y., Cassol, E.A. & Moldenhauer, W.C. 1989. Sediment transport relationships for sand and silt loam soils. *Trans. Amer. Soc. Agric. Engineers,* **32**, 1923-1931.

Nearing, M.A., Foster, G.R., Lane, L.J. & Finkner, S.C. 1989. A process based soil erosion model for USDA - water erosion prediction project technology. *Trans. Amer. Soc. Agric. Engineers,* **32**, 1587-1593.

Reid, I. & Frostick, L.E. 1987. Flow dynamics and suspended sediment properties in arid zone flash floods. *Hydrol. Processes,* **1**, 239-253.

Sharma, K.D. 1992. *Runoff and Sediment Transport in an Arid Zone Drainage Basin.* PhD Thesis. Indian Institute of Technology, Bombay.

Sharma, K.D., Vangani, N.S. & Choudhari, J.S. 1984. Sediment transport characteristics of the desert streams in India. *J. Hydrol.,* **67**, 261-272.

Sharma, K.D., Dhir, R.P. & Murthy, J.S.R. 1992. Modelling sediment transport in arid upland basins in India. In: D.E. Walling, T.R. Davies & B. Hasholt (Eds), Erosion, debris flows and environment in mountain regions, *IAHS Publ. No. 209,* Wallingford, 169-176.

Sharma, K.D., Murthy, J.S.R. & Dhir, R.P. 1993. Streamflow routing in the Indian arid zone. *Hydrological Processes,* **7**,

Singh, V.P. & Regl, R.R. 1983. Analytical solutions of kinematic equations for erosion on a plane. I: rainfall of infinite duration. *Adv. Wat. Resour.,* **6**, 2-10.

Walters, M.O. 1990. Transmission losses in arid regions. *J. Hydraul. Div. ASCE,* **116**, 129-138.

Yalin M.S. 1963. An expression for bedload transportation. *J. Hydraul. Div. ASCE,* **89**, 221-250.

25 Sedimentation issues of reservoirs in the northern region of China

CHENG WEIJUN
Ministry of Water Resources, Beijing, China

GENERAL GEOGRAPHIC AND SOCIO-ECONOMIC CHARACTERISTICS

The northern region of China, as defined in this paper, is the area which is backed by the Inner Mongolia Plateau in the north, faces the Bohai Sea in the east, and is bounded by the Yellow River in the south and the west. Its physical location ranges from 110° to 122° longitude east and from 34° to 42° latitude north. The total area is 430 10^3 km². As far as the administrative region is concerned, it includes the municipalities of both Beijing and Tianjin, the whole of the Shanxi province, the majority of the Hebei province, the northern part (along Yellow River) of the Sandong and Henan provinces, as well as a small part of the Inner Mongolia Autonomous Region and Liaoning province — see Figure 1.

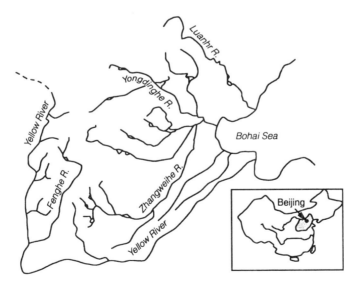

Figure 1 Map of the northern region of China

This region consists of a mountainous area of 290×10³ km² in the north west, and a plain area of 140×10³ km² stretching away to the south east. The mean height above sea level of the mountainous area ranges from 500 m to 2,000 m. The western part is bounded by the western edge of the Loess Plateau of the Yellow River Basin where the mean height above sea level is 800 to 1400 m. The middle reach of the Yellow River is well known for its specific morphology with loose soil and serious erosion.

The northern region of China involves the whole Hai Luanhe basin, and branches on the eastern and northern banks in the middle section of the Yellow River Basin. The area of the Hai Luanhe basin accounts for three-quarters of the total area, with the Yellow River basin for the remaining quarter. Hydrometric networks (precipitation, evaporation, river discharge, etc.) have been set up in both basins. Based on 34 years of data from the 1,400 raingauges in the region from 1956 to 1989, it is estimated that the average annual rainfall is 540 mm — see Figure 1. The maximum yearly rainfall was 790 mm in 1964 and the minimum was 386 mm in 1972.

Table 1   Frequency analysis of average annual rainfall

Average annual rainfall		For various frequencies (in 10⁹m³)			
mm	10⁹m³	20%	50%	75%	95%
540	232.2	271.5	231.8	201.3	163.8

In this region, only a few small tributaries of the Yellow River in the west have retained their natural conditions, and development and utilization of water resources in the other rivers are at a high level. Especially since the 1980s, the natural hydrological conditions of the rivers have changed. Discharges at the gauging stations cannot provide information on the natural runoff and the quantity of water for irrigation and industrial use; the storage changes of the reservoirs and diversions of water across catchments should all be taken into account in the calculation of the restorable runoff data. According to the results of a comprehensive evaluation of water resources made for similar river systems and areas in the region, the average annual surface runoff is 33,800 million m³, which is equivalent to 79.1 mm of rainfall. Of this, the annual runoff for the mountainous area is 26,300 million m³ (depth 91.4 mm) and 7,500 m³ 53.6 mm) for the area of the plains — see Table 2.

Table 2 Frequency analysis for average annual runoff

Area	Average annual runoff		For various frequencies (in 10⁹m³)			
	mm	10⁹m³	20%	50%	75%	95%
Whole	79.1	33.8	44.3	30.8	23.3	16.6
Mountain	91.4	26.3	33.7	24.0	18.7	14.0
Plain	53.6	7.5	10.8	6.2	3.9	2.2

The sediment contents of the rivers in the region may vary considerably. The largest average annual sediment concentration may reach about 500 kg m³, while the smallest is less than 1 kg m³. The small tributaries of Yellow River in the western part of the region flow through the eastern edge of the Loess Plateau, where the land surface is generally covered by 10-100 m of soil. In addition, a semi-arid climate with uneven and concentrated rainfall in summer causes serious erosion during the flood season and wind erosion at other times of year. The average annual sediment content can reach 100-500 kg m³ and the annual sediment transportation rates are 5,000-10,000 t km². The sediments in the river come mainly from the mountainous area. According to the analysis of the data on sediment measured in the hydrometric networks, the average annual sediment yield in the region is 523 million tonnes. The average erosion modulus is 1,810 t km² (see Table 3), estimated separately for the Yellow River and Hai Luanhe River Basins (Lu X.R, 1992).

Table 3 Sediment yields

Mountainous area	Area (10³km²)	Yearly average yield (10⁶t)	Average erosion rate (t/km²)
Northern Plain of China	288	523	1,810
Yellow River Catchment	99	341	3,440
Hai Luanhe River Basin	189	182	960

PROPOSED STRATEGY FOR THE RESERVOIR SEDIMENTATION PROBLEM IN THE REGION

The northern region is one of the most populous areas in China. The cultivated land per capita in this region is a little higher than the average for the country as a whole; water resources, however, are short. The available water resources per capita and per mu (15mu=1 hectare) are 442m³ and 249³ respectively. This is not only much lower than the national average but also the lowest among the large basins in China. On the other hand, annual precipitation varies greatly from year to year and the rainstorms are usually concentrated within a short time during the flood season under the influence of the monsoon. Consequently, changes in annual runoff and the distribution of discharges in the rivers are extremely uneven. More difficulty emerges in flood protection due to the frequent switch between rising and falling limbs on the flood hydrograph. For decades many reservoirs of large, middle and small size have been constructed to regulate the surface runoff in this region, with the original purpose of some reservoirs being mainly for flood alleviation only. Nowadays, however, most of the reservoirs are for agricultural irrigation, electricity generation and domestic water supply. According to the latest data, 2,300 reservoirs of various scales and with a total capacity of 28.2×10⁹ m³ have been constructed in the northern region of China.

There are 32 large reservoirs with storage capacity above $23.4 \times 10^9 m^3$, see Table 4.

Table 4 Reservoirs in the Hailuanhe River and the Yellow River basins

Region	Total Number	Storage Capacity	Large Scale Number	Large Scale Storage	Middle Scale Number	Middle Scale Storage	Small Scale Number	Small Scale Storage
Northern Region of China	(2300)	28.2	32	23.4	121	3.2	2147)	(1.6)
Hail-Luanhe C'ment	(1900)	26.5	30	22.6	96	2.5	(1774)	(1.4)
Yellow R. C'ment	(400)	1.7	2	0.8	25	0.7	(373)	(0.2)

Note: ( ) is the approximate value

The large reservoirs in the Hai Luanhe River basin cover 83% of the mountainous area. The soil erosion on the mountainous area of $189 \times 10^3 km^2$ is very serious because of the low amount of forest cover on a large area of clay and rock soil and human activities. The soil erosion over about one-third of the area is worse than the average level. It is estimated that about 44% of the sediment yield is deposited in the mountainous reservoirs every year, 44% in the plain area, and the remaining 12% flows into the Bohai Sea. Yongdinghe River — also called the small Yellow River — is the most serious soil erosion area in the Hai Luanhe River basin. The mean annual sediment transportation is about 95×10^6 tons, and the sediment yield modulus is 2,090 t km^2. These values indicate that the sediment yield of the 23% mountainous area exceeds 52% of total sediment. The cumulative deposited sediment volume in the Guanting Reservoir (the first flood prevention reservoir for Beijing, located in the lower stretch of the river and operated since 1954) has reached $0.6 \times 10^9 m^3$. This value accounts for more than a quarter of the reservoir design storage of $2.27 \times 10^9 m^3$. In other words, the storage capacity of the reservoir is reduced by 1% per year on average. Owing to climate fluctuations or less rainfall and the human effects on the basin during the last two decades, the input flow to the reservoir has been reducing sharply. This, together with the existing serious sediment deposit, is reducing the capability to supply water to the capital city of Beijing (Liu X.R, 1992).

The whole region has varying soils and the vegetation cover differs from place to place but with overall less forest in the mountainous area. The wide south-eastern plain is mainly agricultural. The climate of the region belongs to the continental monsoon climate of the temperate zone. Winds from the north west dominate in winter and from the south east in summer. It is cold in winter with a little rainfall or snow; springs are dry with sandy and dusty weather. It is warm in summer with a lot of rainfall and is clear in the autumn. Temperature differences are large throughout the whole region. The annual temperature is between 1.5-14°C and increases from north west to south east. The lowest temperature occurs in July which may be up to 40°C. The sunshine is between 2,400 to 3,100 hours.

The Jing-jin-tan (Beijing, Tianjin and Tangshan). Economic Zone and the Shanxi Energy Base lie in this region which also includes one of the main cereal and cotton production zones in China. As the statistics in 1987 showed, the population in this region was 130 million, the cultivated land was 209 million mu (15mu=1 hectare), of which the area under irrigation was 89 mu, and the domestic gross product (DGP) was 231.8 billion yuan ($63 billion).

BASIC HYDROLOGY OF RIVER BASINS IN THE NORTHERN REGION OF CHINA

Regarding the mountainous area of 99×10^3km² in the Yellow River basin of the region, the soil is easily eroded because it is located in the eastern edge of the Loess Plateau with scarce forest cover and few human cultivation activities. For example, the Fenhe catchment (total area about 4×10^4km²) had 234 reservoirs of various sizes with a total storage capacity of 1.55×10^9m³ by the end of the 1980s. There are two large reservoirs: one is Fen He Reservoir on the main stream of Fen He river with a storage capacity of 7.32×10^8m³. It has total silt deposits of 3.3×10^8m³ from the time it commenced operating in 1959 up to the end of 80s. The other one is Wenyuhe Reservoir located in the middle stretch of this river and with a storage capacity of 1.1×10^8m³. Similarly, it has total silt deposits of 0.22×10^3 from 1960 to 1989. It should be stressed that since the importance of water and soil conservation was realized early in Fen He catchment, considerable hydraulic and ecological measures such as constructing sediment deposit barriers, developing terraced fields and planting water conservation trees and grass, have been undertaken. These works play an important role in storing water and preventing sediment. By the end of the 80s the real measured runoff in the outlet section of Fenhe river had been reduced by more than 50%, and the sediment had been basically controlled. The reduced flow and sediment are estimated by comparative studies on the outputs of the conceptual models or relationships between rainfall and runoff and sediment yield over the basin at different periods with changed conditions. The reduction of water and sediment in this river is the most outstanding among those in the branches of Yellow River in the Northern region of China, as shown in Figure 2 (Gu Wenshu, 1993).

The ageing problem of reservoir functions due to sediment siltation is very common in the small-sized reservoirs in the northern region of China, in particular. In other words, the reduction of reservoir regulation function is very evident because of the sediment yield, and sediment transportation in river systems is very large. On the other hand, since the sediment input to the reservoir is constant, the regulation of reservoir sediment cannot be done once for all time. From the analysis of 60 reservoirs of varying sizes in the Shanxi province, the deposit loss rate of reservoirs (in the ratio of reservoir deposit value to the reservoir storage capacity) is 30%. This influences the efficiency and benefit of irrigation and water supply to the same extent.

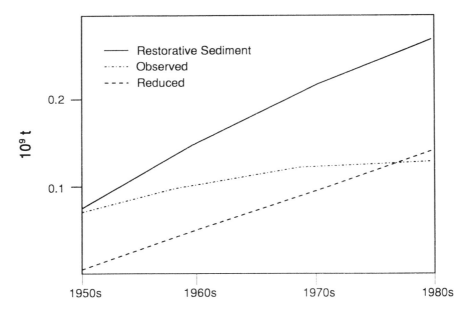

Figure 2 Cumulative curves of sediment yield for the Fenhe catchment (from Gu, W.S.)

Furthermore, for some small-sized reservoirs not only was the role of original flood protection badly affected but also dam failures were caused when the flood overtopped the dam due to the huge sediment siltation in the reservoir. Statistically, Shanxi province has the highest accident rate of small-sized reservoirs in the country, increasing by 35.9% from 1954 to 1990. About 40% of the accidents are due to the problem of sediment siltation (Xia M.D., 1992).

Regarding the problems of reservoir sediment in the region described here, one of the most important strategies is to reinforce the work of soil and water conservation of the catchment, i.e. to plant trees and grass, to build terraced fields and to construct dams for sediment deposit as well as to adopt reasonably scientific agricultural cultivation. At present, the work of small-sized catchment harness in a comprehensive way has been popularized in the country. It has been proved that the method of management, i.e. farmers or groups of farmers taking charge of the work of the land to harness and own the related benefit, is very useful for water and soil conservation. There is no doubt that water and soil conservation takes a long time and usually needs several decades or more to show its remarkable results. This work, however, is the basic way to retard sediment deposits in the reservoirs and to improve the ecological environment of the catchment. The results of reduced sediment deposit in the Fenhe catchment in 1980s is a notable example.

Second, more attention must be paid to the regulation of sediment reservoirs operation in order to retard the worsening of sediment deposit problems in existing reservoirs. China has undertaken some tests and research work on the practical needs and gained some useful experience in recent years. Up to now, the types of regulation

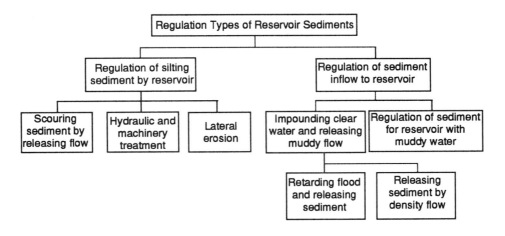

reservoir sediments can be generally summarized as given in Xia, M.D., 1992.

The lateral erosion of sediment shown above is a new approach developed by the Institute of Water Resources Science in the north-west region. The method consists of breaking down the flood plain deposits and flushing them out by the combined actions of scouring and gravitational erosion by the great transverse gradient of the flood plain formed by sedimentation. Experience with this method showed that it consumed limited water but had a high efficiency along with low costs. This method seems to be suitable to the northern region of China where water resources are scarce. The small Hongqi Reservoir in Shanxi province, for example, had a very serious problem with sediment deposition. It adopted a hydraulic absorbing device for scouring sediment in 1977. This device was stopped five years later because of the high cost and low efficiency. Less than three months after the lateral erosion method had been adjusted for the same reservoir, it regained a storage capacity of $180 \times 10^3 m^3$. The efficiency has been increased by 2.8 times, and the cost is only one-sixth of that consumed before (Xia M.D., 1992).

CONCLUSION

Being located in the northern region of China with specific climate and land features, completed reservoirs of various scales have severe sedimentation problems. Some small-sized reservoirs have been silted up and have lost efficiency quickly, and with eventual dam failure, are likely to become a top priority for the undertaking of appropriate counter measures in coping with sedimentation issues of reservoirs in the region. Improvement in soil-water conservation over the basin area and active regulation of reservoirs as regards both flow and sediment are considered to be the main avenues to slowing down the siltation of reservoirs which can then lead to reasonable or sustainable utilization of water resources in the region.

REFERENCES

Lu Xueren. 1992. Research on water resources of northern region of China. Science-Tech Press. Beijing (in Chinese).

Gu Wenshu. 1993. The analysis report on water and sedimentation changes and their impacts on the Yellow River (in Chinese).

Xia Maiding. 1992. On the strategies and management of reservoirs in northern part of China. Northwest Water Resources and Water Engineering (Quarterly) 14 (in Chinese).

26 Erosion control of the Meghna River, Bangladesh

BERT TE SLAA, FORTUNATO CARVAJAL MONAR and
HARRY LABOYRIE
HASKONING Royal Dutch Consulting Engineers & Architects, Nijmegen, The Netherlands

INTRODUCTION

This paper presents the approach which was used to plan and design river bank protection and training works required along some of the main rivers in Bangladesh. In view of the river bank erosion problems, changes in bed levels and the influence of various future projects on the behaviour of the rivers, it is important to be aware of morphological processes when designing protective works. Proposed measures against erosion should fit into a long-term strategic plan.

River training works at strategic locations are generally seen to be the most suitable means of fixing the courses of rivers. This paper includes the primary technical issues in the engineering development of river training alternatives, type of river bank protection, data required and design criteria including the needs for monitoring and maintenance. The paper also pays attention to techniques for predicting future behaviour of rivers, including the use of satellite imagery.

Background

The Meghna, one of Bangladesh's major rivers, flows through the eastern part of Bangladesh and discharges into the Bay of Bengal (see Figure 1). The Meghna River drains an area of 77 000 km², of which about 47 000 km² is located in Bangladesh. The Ganges joins the Brahmaputra near Aricha and thereafter takes the name of the Padma. The Padma joins the Meghna not far from Chandpur. The Lower Meghna River conveys the melt and rain water from the Ganges and Jamuna (local name for the Brahmaputra river) basins, combined in the Padma River, and from the Upper Meghna basin to the sea. The total catchment area is about 1 600 000 km². Maximum discharges in the Lower Meghna can be as high as 160 000 m³ s⁻¹. The major contribution of the discharge originates from the Jamuna River (annual average 20 000 m³ s⁻¹) and the Ganges River (annual average 11 000 m³ s⁻¹).

Like other rivers in Bangladesh, the Meghna (both Upper and Lower) erodes its banks at many points. A number of locations require prompt attention to prevent further damage or other severe nuisance, such as destruction of commercial and

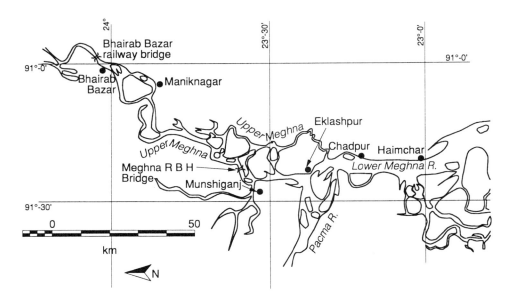

Figure 1 The Meghna and Padma Rivers

industrial properties, disruption of rail and road transport and power supply and displacement of people.

Following the disastrous floods of 1987 and 1988, Bangladesh attracted considerable international interest in helping to find a long-term solution to its flood problems. Over the past three years, the Flood Action Plan (FAP) has provided a framework for a wide range of investigations into the technical, economic, social and environmental issues to be faced in formulating plans for the control and management of Bangladesh's main rivers. At the same time, there has also been a considerable progress in developing the tools and techniques needed to pursue a multi-disciplinary approach in the search for ways to alleviate flood, drainage and river bank erosion problems in Bangladesh. Recent studies of river behaviour in relation to bank erosion and river training works, including studies of a bridge across the Brahmaputra (Jamuna) Bridge, have greatly enhanced the understanding of this problem.

Long-term strategic planning is necessary to include the changes in social, economic and environmental conditions in this time frame and to predict how they might affect the future needs for flood protection, river bank protection, river training works, inland navigation, fisheries, irrigation and drainage.

Recent projections indicate that Bangladesh's population will grow to 150-170 million people by 2010. Since agriculture seems unlikely to absorb more than about 30% of the incremental labour force, the remainder will need to find employment in commerce, industry and services. Urban housing and infrastructure will become much more important than in the past, adding significantly to the value of property and services needing protection from river bank erosion. The Meghna River Bank

Protection Study (FAP-9B) aimed to protect those places selected for urgent bank protection works. They are situated along both banks of the Meghna River.

STRATEGIC PLANNING

Year after year the course of the Lower Meghna has shifted towards the east despite all efforts made to protect the river banks. Much attention has been paid in the past to protecting Chandpur and the irrigation projects upstream and downstream of this town, but no comprehensive plan has yet been presented for long-term protection. The study area for the Lower Meghna extends from the confluence of the Meghna and Padma just north of Eklashpur to Haimchar some 10 km south of Chandpur.

The studies included aspects such as data collection, surveys, studies and river engineering. Some of them were interrelated, as shown in Figure 2. The long-term study should provide:

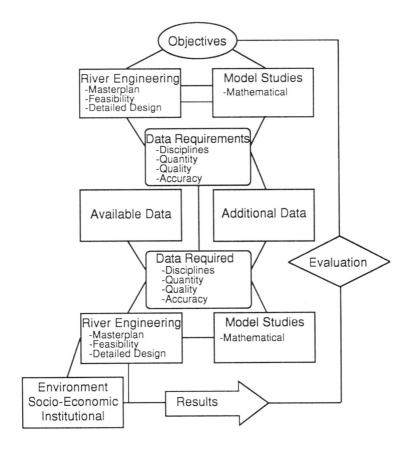

Figure 2   Integral approach

- insight into the long-term natural changes in the planform of rivers;
- insight into the influence of the various FAP and other projects on the morphological behaviour of the rivers;
- determination of the influence of river training works on the natural planform development;
- alternatives for river training works/measures, taking into account that the selection shall be based on a technical and an economic analysis and on social and environmental considerations;
- an implementation schedule for the measures proposed: the schedule shall be flexible and shall allow a phased implementation of long-term river works;
- identification of priority works for early implementation;
- design and cost estimates of river training works, including an investment schedule for phased implementation, monitoring and maintenance.

Inclusion of a short-term plan in a long-term strategic development plan of the Lower Meghna is also a requirement. Therefore, a preliminary assessment was made of a possible river training scheme aiming at a complete regulation of the Lower Meghna, incorporating the measures proposed for Eklashpur, Chandpur and Haimchar. These short-term measures must be sustainable by themselves. In addition to these works, preliminary designs have been prepared for structures in the intermediate zones, covering the stretches between Eklashpur and Chandpur and from Chandpur to Haimchar, aiming at fixing the left bank of the Lower Meghna in its present position.

In dealing with these issues, it was realised that a course of the Lower Meghna forced upon the river by protection works at Eklashpur, Chandpur, etc., is not necessarily the best course for the river or from a geomorphological point of view. However, the town of Chandpur and the land (and all the corresponding infrastructure) upstream and downstream of Chandpur would sooner or later have to be given up to the river if the existing waterfront at Chandpur was not to be included in a protection scheme. Therefore a starting point for the studies was that the existing infrastructure should be protected as well as reasonably possible.

Data collection

Generally, there is a scarcity of reliable and comprehensive data about the rivers of Bangladesh. This especially concerns information on planforms, geomorphology, river morphology and environmental data. Satellite images are very useful instruments to gain insight into the historic changes in channel patterns, and in the average yearly erosion and sedimentation rate of the Padma and Upper and Lower Meghna, but they have only been available since the mid-1970s.

Apart from collection of easily accessible data in the initial phases of the studies carried out in the FAP framework, a great deal of effort has been made to collect additional data by means of site visits, surveys, satellite images and model testing. The site visits provided a good insight into the scope of the erosion problems at the various sites. An important feature of the surveys was the timing of the bathymetric and hydrometric surveys. This timing is of interest because of seasonal variations in river discharge and bed level. Satellite images were used mainly for the geomorphological studies.

Studies

Comprehensive studies were undertaken during a twelve-month period of the protection works along the Meghna River. These included a feasibility study, detailed designs for various alternatives and tender documents for bank protection works for the Bhairab Bazaar Township and Railway Bridge, Munshiganj Town (located on the Dhaleswari River) and Chandpur Town and a feasibility study for bank protection works in Eklashpur and Haimchar. Pre-feasibility studies for bank protection works were carried out for the Meghna Roads & Highways Bridge and Maniknagar, forming a part of the Gumti Irrigation Project.

The geotechnical and geomorphological studies revealed the causes of bank erosion of the Meghna at eight locations. They also demonstrated the importance of the morphological processes for bank erosion in the Padma rivers and scour processes in the Lower Meghna River.

FIXING THE COURSE OF THE MEGHNA

Autonomous morphological development

On the basis of a reconstruction of the planform of the Upper and Lower Meghna (with the help of historical maps, satellite images, cross sections and soundings) a prediction was made for the autonomous development of the future planform.

The erosion will not be limited to Eklashpur but to a reach of some 10 to 20 km south of Eklashpur. The bank erosion will proceed as a kind of cyclic process with a periodicity of some 15 years. The overall trend of the Lower Meghna River over the coming decades will be a continued movement eastwards, with intermittent periods of reduced erosion or possibly even some accretion at times. Without an adequate protection scheme large parts of Chandpur town will gradually be eroded.

Planform characteristics

Planform characteristics of the Upper Meghna have remained essentially the same over the last decades. Special mention should be made of the confluence of the Dhaleswari and Upper Meghna, where the Dhaleswari shows a tendency towards gradual widening during the last 15 years.

In the Lower Meghna the situation is more complex: this river at the confluence of the Upper Meghna and the Padma shows a cyclic pattern (some 15 years) of two major channels which each in turn grow and deteriorate. Two processes are going on: the Padma is moving to the east and the channels in the flood plain migrate and change in number, causing local erosion and accretion.

Local scour

Local scour is the scour caused by man-made structures or natural obstructions. In river bank protection works local scour near groynes, guide bunds, protrusions and along revetments is very important for design purposes. It should be realised that the local scour has to be superimposed on other forms of scour, such as outer bend scour or constriction scour.

A distinction can be made between general scour and scour that occurs more locally. General scour is the reaction of the river on changes in its boundary conditions, like aggradation or degradation owing to accelerated soil erosion, sea level rise, cut-offs of bends etc. More localized scour can be distinguished in a number of different types, notably: constriction scour, caused by a local constriction of the width of the river, confluence scour, occurring in the reach downstream of the junction of two rivers, protrusion scour, occurring when the bank of the river protrudes into the channel, bedform scour, related to the possibility of occurrence of deep troughs downstream of dune crests and local scour, occurring near bank protection works like groynes and revetments.

The conditions related to scour on the Upper Meghna can best be described by the present conditions at Bhairab Bazaar which are characterized by a fairly straight flow downstream of the bridge. The thalweg of the river is approximately in the middle of the river. Some characteristics at Bhairab Bazaar are given hereafter. As far as the maximum scour during the passage of a 22 000 $m^3 s^{-1}$ flood is concerned, the total scour will consist of constriction scour, bend scour and local scour. The total scour depth becomes approximately 40 m in relation to the water level of 7.80 m +PWD; thus the scour depth below PWD to be expected is some 32 m +PWD.

The scour pattern in the Lower Meghna is of another magnitude and is clearly shown at Chandpur (Figure 3). The development of scour here could be drawn from the BIWTA sounding maps over the period 1964 until 1991. It can be observed that initially the scour depths at Puran Bazaar were larger than in Nutan Bazaar, but that especially in recent years the latter scour has increased substantially; in particular, It appeared that the effect of the protrusion scour has increased recently. Furthermore, it was observed that the present scour depths are of the same order of magnitude as in the mid-seventies. A design scour depth more than 66 m was used for Chandpur.

Protection works required in the short term

After studying various alternatives, a selection was made of designs to be studied in greater detail. The selection was based on various factors including the impact of the protection works on the flow pattern, the area required to construct and maintain the protection works, and cost.

Bank protection works to be constructed along the Upper Meghna are in principle not different from works constructed elsewhere in the world along meandering rivers of comparable magnitude. Bank protection works to be constructed along the Lower Meghna - and more specifically at Chandpur - are unique in view of the expected depth of scour holes (> 60m) and current velocities (up to 4 m s^{-1}) during the flood season in this large river. River bank protection works will in principle consist of a revetment which has to be placed on a geotechnically stable slope. Any

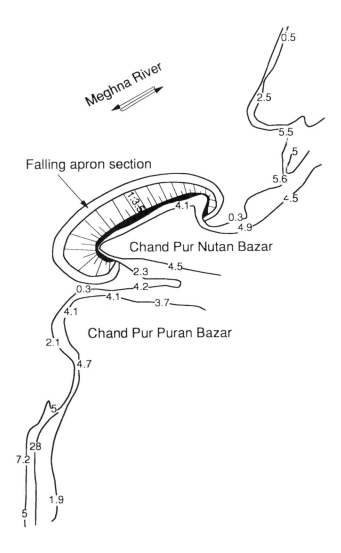

Figure 3 The scour pattern in the Lower Meghna is clearly shown at Chandpur

other type of structure, like massive rock bunds, are too costly, mainly in view of the absence of sufficient (accessible or exploitable) quantities of rock or similar materials in Bangladesh.

The slopes will be formed by placing (hydraulic) fill followed by slope trimming by dredgers. The revetment will consist of above water of open stone asphalt on a geotextile, under water of a fascine mattress ballasted with rock or boulders depending on the site and at the toe of a falling apron of rock or boulders without a geotextile underneath. A typical cross-section is presented in Figure 4.

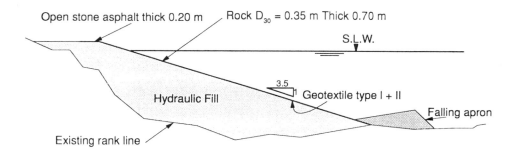

Figure 4   Typical cross-section of the Chandpur Town advanced protection works

Long-term protection scheme

The existing protection at Chandpur has provided some shelter to the Chandpur Irrigation Project and, to a lesser extent, has also favourably influenced river bank erosion north of Chandpur. Therefore, Chandpur should not be considered as an isolated threatened location, but rather as a first fixed (hard) point along a more or less fixed alignment of the Lower Padma and Lower Meghna. Chandpur has so far been defended against river erosion by BWDB. The defence works carried out are constantly being undermined and require much maintenance.

The long-term protection works for the left bank of the Lower Meghna comprise a series of bank protection and river training structures. The most promising of the schemes considered includes the following works:

- Protection of the existing bank at Eklashpur in 1993, 1998 and 2005.
- Protection works between Eklashpur and Chandpur, envisaged for 2002.
- Town protection at Chandpur, first phase in 1993 and additional works in 2003 and 2018.
- Protection between Chandpur and Haimchar, planned for the year 2002.
- Guide bund protection works at Haimchar in 1993, 1998 and 2008.

IMPACT OF RIVER TRAINING WORKS

Geomorphology and planform changes

Along the Lower Meghna River the bank erosion proceeds at a much larger rate than along the Upper Meghna River. Bank protection works are being considered for Chandpur and at Eklashpur. The question raised was: how will the Lower Meghna River react upon these fixed points? It may be expected that the bank erosion between Eklashpur and Chandpur will continue, but the effect will be the development of a curved alignment of the bank. The distance between Eklashpur and Chandpur is about 20 km, of the same order of magnitude as the length of the sand bars in the Lower Meghna River. The implication is that these river training works may impose a meandering pattern upon the Lower Meghna River, which may in time result in accretion on the river's left bank downstream of Chandpur. Considering the celerity of the bank erosion at Eklashpur it may take several decades for this development to start taking place.

In addition it may be questioned whether a distance of some 20 km is not too long. Considering the length of the sand bars, it may be necessary to create an intermediate point between Eklashpur and Chandpur. Therefore some additional fixed points for the guidance of the flow are proposed between the two main fixed points.

It was suggested that the erosion conditions along the left bank may be enlarging one of the channels by dredging. However, more detailed studies and a substantial improvement of the understanding of the morphological processes of these large river systems is needed before final conclusions can be drawn about the effect of the protection works (or any other scheme) on the Padma and Lower Meghna River.

Environmental impact

For each phase of the project, all possible negative environmental impacts have been identified for the various project activities. Most activities associated with river-bank protection have only temporary, short-term or small-scale impacts. This is because a relatively limited area is concerned. Furthermore, most impacts are associated with construction activities and not with the constructed works. The effects are therefore also limited in time. River-bank protection does not have a very intrusive nature: it strives to maintain a steady state at a certain location but does not significantly alter the riverine and estuarine ecosystems.

Socio-economic impact

The Lower Meghna has one of the highest rural population densities in the world, and nearly all cultivable land in Bangladesh is farmed. Agriculture is the backbone of the economy and its practices are adapted to flooding: as the flood water recedes, submerged areas become productive land, especially for growing rice.

It has been recommended that a long-term strategic plan be prepared for future development in the protected areas. Ideally, a monitoring programme should study the changes in water and soil quality, including the effect on aquatic organisms and fish, before the proposed works are carried out (assessment of existing situation), directly after project implementation (assessment of short-term impact), and after one or more years (assessment of long-term impact).

Risk assessment and sensitivity analysis

Until recently, civil and hydraulic engineering projects were designed on a purely deterministic basis. This meant that a certain safety factor was applied, a design load selected without regard to the economic consequences of these choices or to the possibilities of failure which existed. In many cases this meant that a structure was over-designed in one aspect and was therefore too costly but also that it had a good chance of collapsing because another aspect (e.g. lack of maintenance) was neglected.

The procedures and calculation methods developed in the Netherlands for risk analysis and probabilistic design of large hydraulic engineering projects (Delta Project) were applied for the first time on a large scale in Bangladesh for the design of the river training works for the Jamuna Multipurpose Bridge. A risk analysis was also carried out for the bank protection works at three priority sites along the Meghna River.

Probabilistic design implies that, for instance, for calculating resistance against current attack, input parameters were: discharge, Chezy coefficient, Shields constant, stone diameter, specific density of protective elements, slope of the protection, angle of internal friction of protection layer, water depth. The mean values and standard deviation of each of the parameters were used in a probabilistic calculation to determine the risk of failure (which risk of course shall be sufficiently small).

An important side effect of this risk analysis is the saving in the overall construction cost obtained by using probabilistic design methods. This in turn makes it easier to demonstrate the economic viability of the works. From the failure probability for the top event, failure probabilities for the individual elements of the works were determined.

CONCLUSION

At the time of writing this paper, detailed designs and tender documents for the protection works at Bhairab Bazaar, Munshiganj and Chandpur have been prepared. The Government of Bangladesh is now exploring avenues for financing these projects. While the protection works at Bhairab Bazaar and Munshiganj can be seen as stand-alone works, the protection of Chandpur would be the first in a series of hard points aimed at fixing the course of the Lower Meghna.

27 Developing a standard geomorphological approach for the appraisal of river projects

PETER W. DOWNS
Department of Geography, University of Nottingham, UK
ANDREW BROOKES
National Rivers Authority Thames Region, Reading, UK

PURPOSE AND BACKGROUND

The aim of this paper is to review a range of approaches which have been developed for a geomorphological appraisal of river projects and to recommend a more standard approach for the purposes of solving river management problems.

An integrated approach to river basin development devolves upon a recognition of the hydrological and geomorphological integrity of the fluvial system at the catchment scale (Downs *et al.*, 1991). This mutual interaction between hydrology and geomorphology determines the natural routes for surface waters and permits the redistribution of sediments through erosion, transportation and deposition and, in so doing, provides the basis for a variety of plant and animal habitats. Therefore, the dynamics of river channels provide an interesting dilemma to those developing integrated river basin management schemes. At one extreme, channel adjustments are undesirable because they may threaten the functioning of engineering structures and thus serve to confound the best efforts of humans to gain a mastery over nature. However, as this dynamism is natural, its retention should ensure the healthy functioning of the fluvial system and may reduce the need for corrective engineering when original designs malfunction. Furthermore, the desirability of conserving and enhancing natural river features is now enshrined in legislation in many countries (e.g. the UK, Denmark, Germany) and is also appreciated by the public (House & Sangster, 1991; Gregory & Davis, 1993). Given these viewpoints, the 'design with nature' approach to the planning of river channel management is clearly a worthwhile objective.

There has been increasing attention paid to the importance of river morphology in river management ever since the discovery that characteristic relationships exist between flow and the channel cross-sectional and planform morphology, and that river channels react to changes in hydrological regime in a complex but qualitatively understandable manner. As a consequence of these studies, which have included the recognition that human actions can indirectly alter river morphologies, attention has sequentially focused on means of reducing the impact of traditional construction and maintenance measures, techniques for mitigation and enhancement of river environments and, most recently, methods of river channel restoration (Brookes, 1988). In the UK interest in this latter challenge has recently resulted in the inception of a nationwide River Restoration Project (RRP, 1992).

APPROACH

One consequence of these developments has been the increasing need for the appraisal of river channel morphology and, as a consequence of this, numerous methods have been devised; many of which are partially-derived from the pioneering scheme of Kellerhals *et al.* (1976). Some of these have been documented in the published literature, whilst many other methods were derived by organisations or by individuals in academia for particular applications, and are not widely available. In summary, the requirement of geomorphological appraisals for river projects is that recommendations are based upon the systematic interpretation of channel conditions at individual locations.

Given the integrity of the fluvial system, this appraisal must also consider the potential for drainage basin controls or human actions in providing influences from upstream or downstream which may affect the project site. However, temporal and spatial scale considerations mean that river channel dynamics cannot be routinely monitored on an extensive basis and, as it is unlikely that historical records will exist from which information in three-dimensions can be derived, geomorphological evaluation requires interpretive field techniques which infer processes from aspects of the channel morphology (Downs, in press).

A number of schemes have thus been developed, including two by the current authors: one of these evolved as an input to applied scientific research (Downs, 1992), the other out of necessity for rapid baseline surveys to aid management decision-making (NRA, 1990a; 1990b; 1992a; 1993a; 1993b). These approaches enable a rapid, primarily qualitative, continuous assessment of rivers across an entire catchment. Conversely, schemes such as Thorne's (Thorne, 1992; or see Thorne & Easton, 1994) entail detailed, partially geotechnical, evaluations of specific site locations. Clearly, although each of the schemes possesses many common elements, the methods are devised, at least in part, for particular purposes or for specific river environments. However, as geomorphologists in the UK are being increasingly employed by water organizations such as the NRA or as contracted consultants (Brookes, 1994), so cost-effective appraisal techniques become increasingly essential in order to fill the data deficit, and there is clearly an outstanding requirement for a standard approach to *individual project appraisals* which fall in-between catchment or detailed site appraisals. Such an approach must balance the creation of an inventory of channel conditions, i.e. the recording of pools, riffles, point-bars and other features indicative of the channel condition and its dynamics, against an inventory of management interpretations which includes placing the catchment in a spatial and temporal context. This paper reviews the progress made towards such an approach by the authors and is based on appraisals from the Thames basin and the north-west, although the proposed methodology is flexible enough to allow application to a wide range of British rivers.

PROPOSED METHOD

In the NRA Thames Region, geomorphological project appraisals are produced as structured reports which involve a reconnaissance survey as a core element in the decision-making process. The reports proceed from a problem/purpose statement and utilise a desk survey of map information to provide a geological and topographical background to the project. Specific information for the project appraisal is then obtained from a reconnaissance scheme and Figure 1 outlines the framework of such a scheme which is designed to lead the operator through a logical sequence of observations. The schemes need minimal instrumentation, in keeping with a rapid and cost-effective survey: this is a requirement of modern approaches to river management analysis, including the NRA's Conservation and Landscape survey methodologies (NRA, 1992b; 1993c).

Figure 1 shows that, after recording information relating to the site and the purpose of the evaluation, the scheme prompts for regional characteristics of the near-channel environment at various lateral distances away from the channel edge. At the broadest extent these include recognising the geomorphological characteristics of the valley for the purposes of interpreting the Holocene evolution of the site. Secondly, details of current floodplain land uses are required in order to highlight uses which potentially influence the channel morphology, and finally, notes concerning the nature of the riparian zone will establish the potential for channel restoration or enhancement works.

The scheme then focuses on two aspects of the river channel geomorphology. The first concern is with the 'naturalness' of the river channel, judged against stereotypical representations of an undisturbed river channel for that particular geological type, or from adjacent natural reaches. By studying the location of the channel on its floodplain, characteristics of the channel planform, its banks and its bed, the nature of flow and the overall channel dimensions, it is possible to assess the degree of disruption caused by previous channel works and thus begin to formulate the protection status and potential susceptibility of the channel to further disturbance.

The second aspect involves evaluating the style and intensity of current channel adjustments and forms the basis against which it is recommended that decisions must be made. Using evidence obtained from the sedimentary characteristics of the morphology, the type, position and degree of cover of the bank vegetation, and the location of in- or near-channel structures in relation to their intended siting, interpretations are made concerning the degree to which erosion and deposition are currently active in the project reach. Figure 2, for example, illustrates the recovery

OUTLINE
    1) Scope of survey and proposed management plan
        a) Details of survey
        b) Management proposal
        c) General comments

REGION
    2) Valley characteristics
        a) valley description
        b) floodplain land uses
        c) riparian/corridor land uses

SITE
    3) Channel characteristics
        a) position on floodplain
        b) planform characteristics
        c) bank characteristics
        d) substrate characteristics
    4) Channel dynamics — as indicated by:
        a) morphology
        b) vegetation
        c) structures

CATCHMENT
    5) Catchment characteristics with a bearing on the outcome
        a) general information (input from desk study)
        b) upstream
            i) channel character
            ii) potential influences (natural/land use/channel management)
            iii) evidence of previous/recent adjustments
        c) downstream
            i) channel character
            ii) potential influences (natural/land use/channel management)
            iii) evidence of previous/recent adjustments
        d) sediment budget

EVALUATION
    6) Management Interpretation
        a) previous works
        b) evidence of recovery
        c) Project appraisal
            i) conservation value of channel characteristics - reasons reference above
            ii) conservation value of channel dynamics - reasons reference above
            iii) summary conservation value (evaluation of potential loss)
        d) Project recommendations
            i) strength and grounds for opposition to scheme (if any)
            ii) minimum mitigation measures required
            iii) opportunities for local enhancement
            iv) potential for full restoration

Figure 1   Skeletal framework reconnaissance evaluation sheet for individual project appraisals

Figure 2 Channel artificially widened to about three times its natural width, now recovered in the absence of maintenance by deposition of sediment as berms on either side of the low-flow channel

by deposition of a more natural low-flow width in the River Stort at Manuden. Adjustment information can be classified (e.g. Brice, 1981; Brookes, 1987a; Downs, 1992) so that comparisons can be made against chronological models of channel recovery following disturbance (e.g. Simon, 1989). Work is also progressing towards utilising the relation of the style of adjustment to the catchment characteristics in order to predict site-specific responses to changes in channel management and land use (Downs, in press).

The existence of upstream and downstream links between geomorphological processes and the hydrological cycle means that geomorphological project appraisal requires more than just site assessment. Therefore, having become familiar with the nature and dynamics of the study reach it is necessary to incorporate a rapid assessment of those catchment characteristics which may affect the future performance of the project. A common, and valuable approach to this requirement is to consult geological, soil survey and topographic maps (possibly comparing a historical sequence of topographical maps) for information which may provide important background to the 'catchment history' and these are incorporated as an introduction to the project report. With this background, and the site evaluation, it is then possible to use sites selected upstream and downstream of the project reach in order to identify contrasting channel characteristics, and land uses and other channel management operations which may influence the proposed management option. Evidence of recent or previous adjustments of the surrounding river channels should also be noted. With the compilation of these various forms of evidence it should be possible to identify a qualitative sediment budget for the project reach in terms of, for example, upstream or downstream progressing erosion or deposition, and the quantity and calibre of the sediment throughflow.

The final section given in Figure 1 concerns summary interpretation of the foregoing sections in relation to the management proposal. Therefore, using evidence of previous works and the degree of channel recovery (an indication of the stream power and dynamics of the reach), summary conclusions can be made in relation to the conservation value of the river from either its morphological features or its geomorphological dynamics, or both. Both of these categories have formed the basis for recommending River Conservation Sites to the Joint Nature Conservancy Council (Downs & Gregory, in press). They can also be used to justify the designation of an English Nature Regionally Important Geological/Geomorphological Site (RIGS).

On the basis of all this information, the scheme concludes with outline project recommendations which may either approve the original design proposal, suggest modifications to it, oppose it totally, or provide compromise design notes which will conform with the NRA's statutory obligation to further and enhance the conservation of the river environment under the 1991 Water Resources Act. These recommendations are written more fully in the project report and, in consultation with project engineers, recommendations can be located on large-scale maps or plans of the project reach to guide the implementation of the agreed scheme.

TESTING OF METHOD

The method has been tested successfully on a number of sites throughout England and Wales by the NRA. Developing the approach has been as iterative process, updated and modified following each of these initial applications. The following discussion concentrates on two case studies — the Kitswell Brook at Radlett to the north of London and the River Bollin close to Manchester Airport. The Kitswell Brook is an erosion/flood control problem proposed for treatment by the NRA; the Bollin is indicative of a corridor development proposal requiring NRA land drainage consent.

Reconnaissance of the lower Kitswell Brook identified two causes for the river channel change. The first is due to a change in hydrologic regime caused by extensive urbanisation in the catchment, and is evidenced by considerable amounts of bank erosion (see Figure 3). The result is a lack of recognizable bedforms and the non-segregated sand and gravel bed material suggests the continuing mobility of the entire channel bed during high flows. The second cause of change is due to lowering the bed of the mainstream Radlett Brook of which the Kitswell Brook is a tributary. This has led to incision of the lower Kitswell Brook by means of a series of distinct knickpoints which has already caused the partial collapse of one bridge and threaten

Figure 3 Erosion along the Kitswell Brook at Radlett, Hertfordshire

increased bank erosion of riparian properties as they progress upstream. The increased sediment throughput from the Kitswell Brook has led to a discernable plume of sediments at the confluence with the Radlett Brook, the alignment of which clearly suggests the Kitswell as the source. In consultation with the flood defence project manager, proposals have been forwarded to place six low stone-built and dished weirs in suitable locations to arrest the bed degradation and capture some sediment. As gardens abut the stream channel there is no scope for restoration of the channel banks and, according to locations fixed by the reconnaissance survey, larch-pole piling has been proposed as a low-impact structural solution to the bank erosion.

In the case of the River Bollin, a channelization scheme is proposed in the middle reaches and it is the statutory duty of the NRA to indicate the extent of loss of river environment that the proposals may cause. Reconnaissance surveys are being used, in conjunction with previous geomorphological studies (particularly Mosley, 1975), to ascertain the potential impact. The site-specific survey confirmed the existence of a highly mobile channel, which is adjusting to a combination of knickpoint incision and sequential adjustments (and which may be a natural response to meander cut-offs which increase slope). In addition, large inputs of sediment are being received from upstream, which may be due to agricultural practices, other meander activity or the increased extent of the urban area; these are

causing further lateral adjustments of the channel and are instrumental in determining
the nature of the ecological habitat. A new floodplain level is emerging in the form
of bar development in the reaches which have over-widened due to bank failures
(Figure 4). In summary, preliminary findings of the project appraisal are that the
complex suite of geomorphological adjustments surrounding the site may be of
great enough significance to warrant the proposal of a Regionally Important
Geomorphological Site; in this case any channel modifications may cause potentially
significant environmental loss.

Figure 4 Emergence of a new floodplain level above the River Bollin, near Morley Green,
Cheshire. Distinct linear sequences of tree growth across the bar show that its age
decreases from right to left as the river has incised and migrated.

DISCUSSION

The approach outlined in the previous sections can provide the geomorphological
basis against which recommendations can be made at several stages in the
assessment of any proposals intended to modify the channel bed or banks, and
particularly for capital projects, operations works and restoration measures. The
method can be applied to all stages of the project management process from pre-
planning to post-project monitoring and evaluation and is estimated to have costs
ranging from about £35 to £125 per km, including site appraisal and report writing.

One input is a prelude to option evaluation, whereby potential management alternatives are assessed against their regard for the environment and for system integrity. If the reconnaissance evaluation is at an early stage of the project planning process, the information gathered enables management options to be compared. This proactive approach to the incorporation of geomorphological assessment has the advantage that schemes are less likely to produce detailed designs which run counter to geomorphological principles and, therefore, may meet significantly fewer environmental objections from a range of disciplines.

A second input involves providing measures which aid the mitigation of unforeseen or environmentally undesirable consequences of a chosen management solution. This input formed the initial undertaking for geomorphology in river management in the UK, and in the United States (Keller, 1975) and Denmark where large sums of money have been spent in restoring previously straightened river channels (Brookes, 1987b). Where wholesale restoration has been impossible, a growing selection of mitigation techniques have been developed which exist alongside the undesirable scheme, and these often aim to improve habitats by increasing morphological diversity in the managed reach (Brookes, 1988, 1991, 1992; Hey, 1990; Hey & Winterbottom, 1990). Figure 5 shows measures used to mitigate the loss of diversity caused by realignment of the Redhill Brook, Surrey. However, these measures only offer limited environmental benefits when compared to full restoration projects. The rapidity of reconnaissance-based evaluations makes them ideal for appraising the options available for mitigation or restoration.

The third input is in regard to post-project appraisal, which is a vitally important undertaking in relation to improving the operational and environmental performance of future river channel management practices. Given that routine geomorphological involvement in river channel management has only recently gained momentum, this option has very rarely been utilised but, allowing suitable periods between survey, it would provide an invaluable database with which to increase knowledge concerning the complex nature of river channel adjustment.

FUTURE PROPOSALS

This paper has attempted to show how both objective and qualitative elements of geomorphological assessment can be combined to provide directions for taking management decisions in relation to river project appraisal. It is difficult to see how the scientific basis of this could be improved much further given the need for rapid, cost-effective assessment. The results aid the project manager in deciding on a sustainable channel design and it follows that a design which works with nature rather than against will probably be more acceptable from an ecological (biological, fisheries and conservation) viewpoint. However, there is a need to ensure that the recommendations from a geomorphological assessment are carefully balanced against recommendations from other environmental surveys. Clearly the method can be applied to all projects involving works to, or impacts on, the bed and banks

Figure 5 New course of the Redhill Brook in Surrey prior to diversion of flow, showing creation of 'natural' pools, riffles and point bars as well as varied channel cross-sections

of channels. The method should clearly be applied by an experienced geomorphologist since it requires skilled intuitive interpretation and, given this requirement for intuitive judgement, there is a need to ensure consistency by training of relevant staff. This will help to overcome the problems encountered with the NCC (English Nature) River Corridor Survey which did not involve training of surveyors and led to a wide variation of results as a consequence of surveyor error.

Given the simplicity of the approach, its basic strength of flagging up geomorphological concerns at any stage of project appraisal, it is recommended that it should be more widely applied to river management problems. There are many proposals for flood control, riparian works to rivers, development proposals and navigation works potentially affecting the bed and banks of rivers and, within the wider context of integrated river basin development, addressing these issues during development rather than when the problems arise, can provide a cost-effective means of ensuring the sustainability of the development.

REFERENCES

Brice, J.C. 1981. Stability of relocated stream channels, Technical Report no. FHWA/RD-80/158, Federal Highways Administration, US Dept. of Transportation, Washington D.C.

Brookes, A. 1987a. Restoring the sinuosity of artificially straightened stream channels, *Environ. Geol. and Wat. Sci.* **10**, 33-41.

Brookes, A. 1987b. The distribution and management of channelized streams in Denmark, *Regulated Rivers: Research and Management*, 1, 3-16.

Brookes, A. 1988. *Channelized Rivers: Perspectives for Environmental Management*, John Wiley & Sons, Chichester.

Brookes, A. 1991. Design practices for channel receiving urban runoff: examples from the River Thames catchment, UK, paper presented at American Society of Civil Engineers conference Effects of Urban Runoff on Receiving Systems: An Interdisciplinary Analysis of Impact, Monitoring and Management, August 4-9, 1991, Mt. Crested Butte, Colorado, USA.

Brookes, A. 1994. Goals for geomorphology in UK river management, paper presented to British Geomorphological Research Group at IBG'94, January 4-7, 1994, Nottingham, UK.

Downs, P.W. 1992. *Spatial Variations in River Channel Adjustments: Implications for Channel Management in South-East England*, unpublished PhD thesis, University of Southampton, UK.

Downs, P.W. (in press) River Channel Adjustment Sensitivity to Drainage Basin Characteristics: Implications for Channel Management in South-East England. In: McGregor, D. and Thompson, D. (Eds) *Geomorphology and Land Management in a Changing Environment*, John Wiley & Sons, Chichester.

Downs, P.W. & Gregory, K.J. (in press) Evaluation of River Conservation Sites: The Context for a Drainage Basin Approach, in *Proc. Malvern Internat. Conf. on Geological and Landscape Conservation*.

Downs, P.W., Gregory, K.J. & Brookes, A. 1991. How integrated is river basin management?, *Environ. Manage.* **15**, 299-309.

Gregory, K.J. & Davis, R.J. 1993. The perception of riverscape aesthetics: an example from two Hampshire rivers, *J. Environ. Manage.* **39**, 171-185.

Hey, R.D. 1990. Environmental river engineering, *J. Instn Wat. and Environ. Manage.* **4**, 335-340.

Hey, R.D. & Winterbottom, A.N. 1990. River engineering in National Parks: the case of the River Wharfe, UK, *Regulated Rivers: Research and Mmanagement*, **5**, 35-44.

House, M.R. & Sangster, E.K. 1991. Public perception of river corridor management, *J. Instn Wat. and Environ. Manage.* **5**, 312-317.

Keller, E.A. 1975. Channelizaton: a search for a better way, *Geology*, **3**, 246-8.

Kellerhals, R., Church, M. & Bray, D.I. 1976. Classification and analysis of river processes, *J. Hydraul. Div. ASCE* **102**, 813-829.

Mosley, M. P. 1975. Channel changes on the River Bollin, Cheshire, 1872-1973, *East Midland Geographer*, **6**, 185-199.

NRA 1990a. River Stort Morphological Survey: Appraisal and Watercourse Summaries, compiled by Brookes, A. & Long, H. for National Rivers Authority Thames Region, September 1990.

NRA 1990b. Lower River Colne Catchment Morphological Surveys, prepared by Downs, P.W. for NRA Thames Region, November 1990.

NRA 1992a. River Blackwater Morphological Surveys, compiled by McEvoy, B. & Poole, K. for NRA Thames Region, December 1992.

NRA 1992b. River Corridor Surveys: Conservation Technical Handbook No.1, National Rivers Authority, Bristol.

NRA 1993a. Hogsmill Stream Geomorphological Evaluation, prepared for NRA Thames Region by Downsward, S.R., September 1993.

NRA 1993b. River Kennet Morphological Survey, compiled by Hills, K. for NRA Thames Region, November 1993.

NRA 1993c. River Landscape Assessment: Conservation Technical Handbook No.2, National Rivers
    Authority, Bristol.
RRP 1992. The River Restoration Project: working to restore and enhance our rivers, RRP
    Secretariat, Pond Action, Oxford, Oxford Brookes University.
Simon, A. 1989. A model of channel response in disturbed alluvial channels, *Earth Surf. Processes
    and Landforms*, **14**, 11-26.
Thorne, C.R. 1992. Field assessment techniques for bank erosion modelling, Final Report to US Army
    European Research Office, Contract no. R&D 6560-EN-09, Department of Geography, University
    of Nottingham.
Thorne, C.R. & Easton, K. 1994. Geomorphological reconnaissance of the River Sence, Leicestershire
    for river restoration, *East Midland Geographer*, **17**, 40-50.

PART VII

IMPACT OF URBAN AND INDUSTRIAL DEVELOPMENT

28 Integrated design approaches for urban river corridor management

J B ELLIS
Natural Environment Research Council, Swindon, UK
M A HOUSE
Urban Pollution Research Centre, Middlesex University, London, UK

INTRODUCTION

One of the principal causes of increased flood and pollution risk in river systems stems from increased surface runoff volumes and wash-off rates resulting from urbanisation, in addition to any superimposed problems due to sewer overflow discharges. The net effect of this enhanced risk has been a poor public perception of river water quality in urban areas. However, river water quality cannot be considered in isolation from river corridor management. Successful integrated and sustainable approaches to the management of river corridor environments in urban catchments requires a combination of organisational and institutional cooperation and planning as well as the implementation of sound and sustainable engineering and ecological principles.

Marginal river wetlands are ecosystems of particular value in urban areas as they are important wildlife corridors, often connecting isolated and therefore increasingly vulnerable habitats. Such 'umbilical' open spaces in urban environments offer both ecological sustainability and water quality benefits. In addition, their aesthetic appeal and central location in urban parklands and open spaces provide a strategic focus for the local community. Unfortunately they have suffered major losses in the UK as a result of flow regulation and speculative land-take, despite the growth of supporting control legislation such as the 1976 Land Drainage Act which required "due regard for flora and fauna". The 1981 Wildlife & Countryside Act helped to reinforce this conservation principle but only in so far as such protection was "consistent with other duties".

However, the last decade has witnessed a sea-change in both attitudes and practice, albeit a slow and progressive change. A number of reports have been released indicating how river works might be undertaken without causing excessive ecological damage (Newbold *et al.*, 1983; MAFF *et al.*, 1991). A few of these studies have specifically focused on the rehabilitation, restoration and integrated development of urban catchments (CIRIA, 1992; Hall *et al.*, 1993). There can be no doubt that there is a better awareness amongst river engineers for such ecological approaches and a genuine desire to integrate with the aspirations of many river corridor interest groups. In this context it is interesting to note that a recent "bible text" for river engineers (IWEM, 1989) devoted a considerable section to sensitive river management techniques which benefit wildlife. The launch of the River Restoration Project in December 1992 is further evidence that the water industry wishes to move towards identifying and implementing sensitive and effective river engineering approaches. The public have become much more environmentally-aware in recent years and are now more vociferous in their demands for natural-looking rivers and river corridors. Recent research has indicated that people have a clear idea of what they consider to be an 'attractive' or 'polluted' river and has suggested the type of river corridor features preferred by the local communities in their use of rivers for recreation and amenity. These include natural river banks and channels, trees and vegetational diversity (House & Sangster, 1991; House, 1991).

STRATEGIC ISSUES

There appears to be a general consensus that stormwater treatment and management should have a more vital role in future urban land use planning if we are to prevent the further degradation of our urban receiving waters. The inclusion and integration of ecological and environmental principles within engineering criteria provides a universally acceptable conceptual design framework. Such an approach is aimed at simulating natural water bodies possessing aquatic vegetation with both adjacent and fringing wetlands but also acknowledging their role in supporting local community interests and activities. The design goal should therefore be to create an environmental and recreational amenity for local communities that will at the same time provide a significant improvement in stormwater runoff quality as well as fulfilling a prime flood control function.

Such urban catchment management planning (CMP) describes a process or means of ensuring that all problems and opportunities resulting from competing functions and activities are reconciled within the context of a well-defined, structured and flexible planning framework. The developed procedure should be capable of enhancing, (if not optimising) the overall environmental well-being of the urban aquatic environment. Such CMPs must also be linked to (or be amenable to) the progressive or phased introduction of Statutory Water Quality Objectives (SWQOs).

The impending introduction of SWQOs (NRA, 1991) should not only serve to widen the debate on what level of water quality is desired to support potential river uses but also take into account what is achievable within technical and financial constraints. Arguably the greatest rewards and benefits of SWQOs will come from urban river reaches where improvement in water quality could lead to a more accessible and wider range of improvements. However, it can also be argued that the potential success of SWQOs is circumscribed in that they are achievable only through pollution control powers and apply specifically to point-source discharges. They may well be much less effective where pollution comes principally from non-point diffuse sources such as highway runoff, although the recent case successfully lodged against the Department of Transport (Doyle, 1994) in respect of receiving water impairments resulting from polluting discharges from the M25 motorway at Oxted in Surrey may argue otherwise.

The statutory framework provides an action-based coding from which the NRA can derive its pollution control powers. However, the impact of such empowerment is dependent on the degree to which water quality itself is the limiting factor. In terms of ecology, wildlife, aesthetic and leisure activities, water quality is only one of a range of factors which determines the health of these activities. Out-of-stream considerations do need to be addressed as part of an integrated management approach to urban river corridor enhancement. The only real procedure whereby the NRA can identify the applicable limiting factors and necessary follow-up activities is through integrated CMP. Such planning, combined with public and other interest group consultation, will allow inputs and contributions from those having specific or expert knowledge of the local urban catchment and river corridor. NRA statutory duties, if applied to the integrated catchment management process itself, would provide the legal basis to improve water quality through such considerations as:

● variations in water flows;
● channel morphology;
● floodplain activities.

The inclusion of such considerations could provide a more cost-effective solution than a mere tightening of discharge consents.

Integrated catchment management planning can, therefore, provide an entirely appropriate basis for wider, more coordinated environmental action planning. In this way it may be possible to extend the concept and binding powers arising from SWQOs to Environmental Quality Objectives (EQOs) that might be set in terms of targets relating to:

● wildlife and ecology;
● aesthetics and leisure uses;
● recreation and amenity, as *well* as
● flow quality and quantity limits.

The strategic concept of SWQOs envisaged by DOE takes no regard of this wider EQO concept as the proposed Fisheries Ecosystem (FE) classification scheme (DOE, 1993) is only concerned with limited *in-stream* ecology and does not even "come down" to invertebrate levels which are of more significance in terms of urban stream ecology and well-being. Most urban waters will be lumped into the proposed FE 5/6 categories and Annex B to the Consultative Paper in any case contains various escape clauses which allow affected determinands to be disregarded in compliance assessment where a solution can only be identified in the long term or where no practicable solution can yet be identified. In the latter instance, the determinands can be disregarded indefinitely.

Given the diffuse nature of urban runoff and the fact that the point of outfall can be "lost" within the network of sewers and culverted watercourses, the concept of integrated catchment management must endure if infrastructure problems are to be satisfactorily resolved. The absence (until recently at least), of national guidelines on urban stormwater management has inevitably led to inconsistencies in design parameters and the lack of coordinated catchment-level overviews. The recent availability of such documents as *Scope for Control of Urban Runoff* (CIRIA, 1992) and the *Design of Flood Storage Reservoirs* (Hall *et al.*, 1993) can now provide criteria and procedures to help achieve these long-awaited integrated, catchment scale objectives.

OBJECTIVES OF URBAN RIVER CORRIDOR MANAGEMENT

The objectives which integrated urban river corridor management is intended to achieve include:
● provision of an effective, efficient and safe drainage system.
● an assurance that satisfactory environmental protection is provided against environmental degradation due to increased water volumes, flow velocities and pollution discharges associated with urbanisation.
● integration of multi-purpose functions into the channel/waterway system, thus making the corridor a community asset and targeting aesthetic, recreational, wildlife and economic benefits including land enhancement.
● the implementation of operational systems capable of effective and efficient maintenance.

Urban river corridors — if properly developed and managed — can represent valuable local community resources and, particularly in a residential context, can:
● be landscaped into the urban context and incorporate "blue/green" features into an open space;
● support conservation principles through the establishment of a range of aquatic habitats and ecosystems;
● provide a range of recreational and leisure activities of both an active and passive nature;
● provide a potential source of water for re-use, e.g. park irrigation.

Conversely, if not properly managed, urban stormwater channels can seriously impair urban amenity, public safety, property values and environmental quality. The objective must be to attempt restoration of the complete river corridor system, not simply focusing on single-function improvements. The key working assumption here is that urban rivers have largely lost their corridor interest in addition to their channels being constrained and degraded by engineering works. Such in-channel "improvements" have prevented or suppressed normal self-cleansing capabilities which create and rejuvenate habitats and have also degraded adjacent floodplain interests.

INTEGRATED DESIGN APPROACHES

'Engineered' waterway corridors, prevalent in urban areas, require a far more integrated design approach to achieve an optimum multi-purpose system than many other situations. Engineering design must be sympathetic to planning and architectural requirements, landscape design and local cultural/historical requirements as well as to ecological principles.

Master drainage studies will in most cases have already established the function of the flow paths in the drainage hierarchy leading to the requirements for flow regulation, drainage resources and pollution control. However, reference should also be made to other parameters such as soil capability, aspect, ecology, cultural-historical factors as well as the need for service corridors and movement systems.

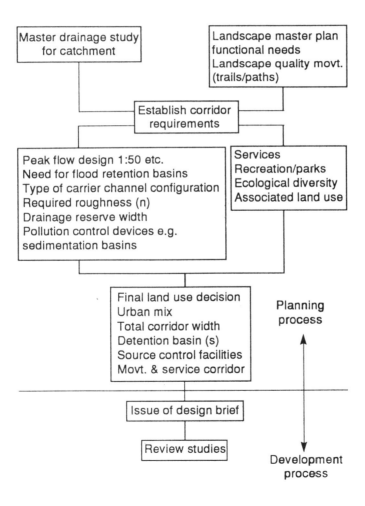

Figure 1   Planning and development process

Additionally, visual attributes and landscape quality should be identified and the *character* or nature of the local development area should also be an input to the final design decisions. Figure 1 provides a generalised overview of the planning context within which such an integrated design approach might be developed.

There may be a need to consider what type and range of inlet controls to receiving water bodies may be necessary as a means of reducing the corridor requirements and "reserve" area as well as serving as a means of flow attenuation. The corridor should be designed as a multi-purpose facility incorporating a channel "carrier" system having physical and biological components which may locally terminate or pass through an urban lake or pool section providing an on- or off-line quality as well as flow control function (Figure 2). These facilities would determine the forward flow and pollutant patterns discharging into the downstream river system. Shallow marginal shelves can be weaved between wet banks to stimulate both emergent and wetland plants in addition to the provision of linear ponds within the flood corridor.

The design approach would need to appraise site factors and functions, particularly as the corridor is likely to serve as a significant "heartland" to the surrounding urban community. From these functions the design should develop and generate site-specific design objectives and criteria which would need to be consistent with established guidelines in terms of agreed major parameters (see Sections 2.5 and 3.1 above). A range of corridor design options would be considered before selection and adoption of a preferred solution.

Figure 3 provides a more detailed design approach which would require multi-disciplinary collaboration between engineering, environmental, landscape architect

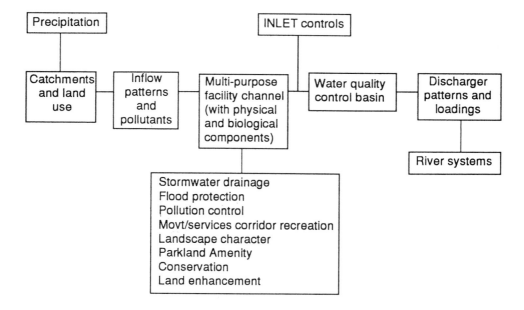

Figure 2   Corridor processes and functions

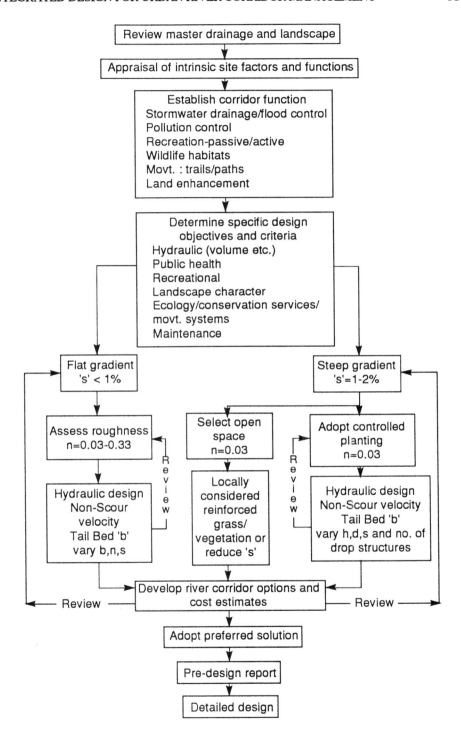

Figure 3 Design appraisal flow chart

and planning teams. The latter two groups would enable a full range of landscape amenity and recreational opportunities to be identified and additionally facilitate the incorporation of land enhancement criteria of advantage to the local community. In cases of flat gradients, landscape features would modify the Manning "n" friction values and in addition, there may be a need to consider sedimentation facilities which in turn would need to consider head-loss provision. There may also be a need to introduce shrubs/vegetation at a low level to interfere with flows and induce turbulence. Such shrubbery can introduce a high Manning "n" corresponding to assumed values of 0.04 to 0.06. Larger shrubs and thickets can achieve even higher "n" values of 0.06 -0.08. Overall, it is difficult to achieve "n" values greater than 0.05, since there will normally be some open grassed areas ("n" = 0.03) and footpaths (0.013) within the corridor. With steeper gradients, head losses and a controlled landscape need to be introduced to achieve a high Manning "n" value. A series of drops or artificial terraces can be used to flatten the gradient artificially to less than 1%. The hydraulic design process would involve "trialling" and varying the hydraulic factors and employing either a reiterative refinement process or the preparation of a range of waterway corridor configurations.

The prime function of the urban corridor is to act as a drain and floodway: this takes precedence over all other functions. However, a high standard of aesthetic and landscape design can be incorporated without prejudice to this prime function. In some locations a larger cross-section will need to be adopted to facilitate open-space active recreation and leisure amenity. Lower function values can be accommodated either by locally reducing the gradient to reduce flow velocity or by increasing resistance by use of appropriate vegetation. Drop structures can be introduced as shaped structures, cascades or terraces; small pools and ponds can be introduced provided sympathetic edge treatment is also introduced (Ellis, 1993). The engineered loss of deep water pools and slacks can be rectified through the creation of backwaters of varying sizes and depths as well as by provision of attenuation basins.

The most restrictive aspects of the urban corridor design will be those related to limited available width. Alignment curvature of the landscaped waterway should be incorporated wherever possible to improve the visual and environmental amenity. Lined inverts can meander across the waterway channel with appropriate changes in bed slopes. To maintain the meanders, dry (grass) relief channels can be created which will take much of the flow during high discharge periods. The landscape can be used to break up the waterway corridor visually and add interest, whilst still retaining the overall hydraulic design requirement.

Integrated planning and the inclusion of sports fields, open spaces, parklands, etc., adjoining the waterway will lead to a higher standard of visual amenity. These features can readily be made flood proof if required in the design, albeit at some cost. Trails and footpaths should be made into a "travelling experience" rather than being incorporated at a fixed offset with long uninteresting vistas. Such planned aesthetic, motivated changes add to the design effort, but provided they are introduced at the outset they can be accommodated without significantly adding to the construction cost.

Social surveys, undertaken on site at a number of river reaches in England and Wales, and aimed at ascertaining the public's choice of features which are important to their ideal river corridor setting and their preference for a range of river corridor features, have revealed that features which form part of the natural environment are considered to be of greater importance than those which form part of the built environment or relate to human activities (Table 1). The results in Table 1 show the mean scores given to each feature and their ranked order of importance. The latter is based on a significant difference being recorded between the mean scores obtained for each feature.

Table 1 "In thinking of your ideal river setting, how important would it be for each of the following to be present?"

Overall rank	Feature/Facility	Mean score	Standard deviation
1	* Peace and quiet	8.9	1.7
2.5	* Many kinds of small birds	8.4	1.8
	* Many kinds of waterfowl	8.4	1.9
5.5	* Trees	8.0	2.0
	* Otters & other water mammals	8.0	2.5
	* Many fish in water	7.9	2.3
	* Diversity of flowering plants and grasses	7.7	2.0
9	* Many dragonflies & water insects	7.1	2.8
	+ Easy access	7.0	2.8
	* Many butterflies	6.9	2.7
12	* Sedges, rushes and wetlands	6.4	3.1
	+ Car parking facilities nearby	5.9	3.4
	+ Seating	5.5	3.2
16.5	+ Picnic facilities	5.1	3.3
	+ Safe paddling or swimming areas	5.1	4.0
	+ Provision for fishing	4.6	3.6
	+ Provision for boating	4.5	3.4
18	+ Pubs & other social facilities	3.4	3.3

Scale: from 0 = unimportant to 10 = very important

*   Features of the natural environment
\+   Features of the built environment or relating to human activity
N = 376

Table 2   Strength of preference, expressed as a ratio, for river corridor features (1988 and 1989 Surveys)

1988 Survey	Ratio	1989 Survey	Ratio
dense forest	1:		
open forest	11	one bank tree-lined	1:
		open woods both sides	2
both banks tree-lined	1:		
one bank tree lined	1.2		
evergreen trees	1:		
deciduous trees	2.2	only short mown grass	1:
		mown grass & deciduous trees	9.6
uncut long grass	1:		
short mown grass	1.5		
plants and grasses	8.9:	flowering plants and grasses	1:
just grass	1	grass and trees	3.2
mix of grass and trees	1:	flowering plants and grasses	1:
either grass or trees	6.9	either grass or trees	1.2
grassy banks	1:		
tree-lined banks	1		
artificial paths	1:	dry gravel paths	1:
naturally worn paths	3.6	naturally worn paths	3.3
trees & veg'n overhanging	1:	trees & diverse veg'n overhanging	1:
trees & veg'n lining	1.2	trees & diverse veg'n lining	1
many plants in water	1:	many plants in river	1:
few plants in water	1.76	few plants in river	2.3
wide river (>4m)	2:5	wide river (>8m)	1.2:
narrow river (<4m)	1	narrow river (<8m)	1
fast flowing river	1:	fast flowing river	1:
slow flowing river	1.7	slow flowing river	1.6
rocks & stones in river	2.2:		
no rocks & stones	1	protruding rocks & stones	1:
		deep river (>1m)	1.5
shallow river	1:		
deep river (>1m)	2.1		
winding river	17:		
straight river	1		
natural river banks	15:		
artificial river banks	1		

Respondents were asked to indicate their preference for an initial list of 16 pairs of river and river corridor features (later reduced to 10 in the 1989 survey) on a scale of -3 ('much prefer first feature') to +3 ('much prefer second feature'). A score of 0 was given to the response of 'like both equally' and 8 to 'dislike both'. A weighted sum, indicating the strength of preference for each of the paired features, was obtained by multiplying the number of respondents to express a preference for one of the paired features by the ascribed preference rating. These weighted sums were then expressed as a ratio of the strength of preference for one of each pair of features in relation to the other (Table 2). The results reflect the public's ideal river landscape setting to be one with open deciduous forest, with a mixture of grass and plants or grass and trees, either overhanging or lining both banks of a deep, slow-flowing mature river more than 4 m wide, of natural form and without an overabundance of vegetation in the water, with naturally-worn paths following its course. This indicates a definite inclination towards a more natural environment and a move away from uniformity. Hence, far from desiring intensively managed and manicured landscapes, which is often assumed by engineers within the NRA and local authorities, the public showed a strong preference for natural river banks and channels, trees and vegetational diversity.

Figure 4 provides a generic cross-section for a corridor reserve design; vegetation selection should be based on local indigenous species which are tolerant to fluctuating flows and pollutant levels as well as requiring minimal maintenance and being able to withstand occasional inundation. The aesthetic vegetation qualities

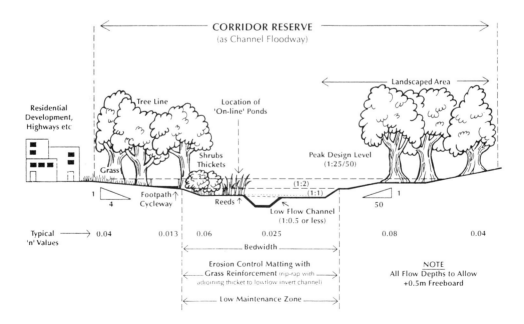

Figure 4 Typical corridor cross-section

should include the use of thick, impenetrable species to act as screens and buffers to prevent or divert access from designated sensitive areas. It would also be appropriate to include relevant source control devices and designs such as grassed swales (as relief channels), small retention ponds, infiltration trenches (lined if considered necessary), etc. Low flow inverts (inset into the channel), with allowance for frequent bed flooding with subsequent water table recharge, will lead to a good moisture regime for the bed and encourage adjacent plant growth stimulating a more diverse vegetation which in turn can provide opportunities for passive leisure uses. These design principles will help develop a linear park concept which fully integrates, enhances and services the adjoining urban land use.

The developer and landscape architect should actively seek designs which tie the waterway and included urban lakes into an open space network which complements the urban neighbourhood. Such integrated landscaping then not only offers potential for local amenity use but also helps to gain the acceptance of the local community. There should be a clear human involvement in the river corridor ecosystem. This can be encouraged by trails, paths, boardwalks, seats, jetties, attractive views, educational material and display interpretation boards. A sense of ownership can be increased through involvement of the surrounding community in the design process, planting days, educational trails and so on.

CONCLUSION

It must be recognised that in many urban situations, restoration design and related works must be conducted within a very constrained corridor environment. Such restrictions may well mean that a full range of wildlife habitats, restoration and landscaping is not possible. However, in most situations there will be at least some limited scope for amenity and land enhancement. Both in-stream and bank ecotones can be enhanced through sensitive improvements to and up-grading of the quality of those that currently exist. Additionally, creative biotechnical engineering can manipulate new improved conditions and aid bank stabilisation. Such in-channel bioengineering combined with out-of-channel landscaping can yield a variety of local benefits, including visual amenity, land enhancement, ecology, water quality as well as educational and local community benefits. Such integrated local-scale restoration and remediation is compatible with holistic conceptions of wider river catchment management as advocated by Gardiner (1991). The creation of self-sustaining, stable and diverse ecosystems and the restoration of "natural" features, processes and regimes to urban river corridors is a feasible and achievable objective. However, restoration goals must be clearly pre-defined and a sense of collective ownership must be generated if such integrated design approaches are to be successfully implemented and managed. The ultimate outcomes will not restore pre-urban idylls but the value of even relatively minor environmental enhancements to our urban river corridors must not be underestimated.

REFERENCES

CIRIA 1992. Scope for Control of Urban Runoff. Vols 1-4., Construction Industry Research and Information Association, London.

DOE 1993. River Quality. Draft regulations: The Surface Waters (Fisheries Ecosystem) (Classification) Regulations 1993. Consultation Paper. Department of the Environment, UK.

Doyle, Neil 1994. *New Civil Engineer*, 21 April, p. 3.

Ellis, J.B. 1993. Urban lakes; planning for amenity and leisure. In: Proc. Seminar, *Planning for Sustainable Development: The Water Dimension*. Smisson Foundation, Bristol.

Gardiner, J.L. 1991. *River projects and conservation: a manual for holistic appraisal*. John Wiley & Sons, Chichester.

Hall, M.J., Hockin, D.L. & Ellis, J.B. 1993. *Design of flood storage reservoirs*. Butterworth-Heinemann, Londonr

House, M. A., & Sangster, E. K., 1991. Public perception of river corridor management. *J. Instn Wat. & Env. Manag.*, **5**, 312-317.

House, M.A., 1991. Urban rivers: nature conservation and the use of rivers for recreation. In: Hall, M. and Smith, M.A. (eds.) Riverbank Conservation, University of Hertfordshire.

IWEM. 1989. Water practice manuals: River Engineering, Part II. Institution of Water and Environmental Management, London.

MAFF, DOE & WO 1991. *Conservation guidelines for rainage authorities*. HMSO, London.

Newbold, C., Purseglove, J. & Holmes, N. 1983. Nature conservation and river engineering. Nature Conserv. Council, Peterborough, UK.

NRA. 1991. Proposals for Statutory Water Quality Objectives. Report No. 5., Water Services, National Rivers Authority, Bristol, UK.

29 Integrated water quality monitoring: the case of the Ouseburn, Newcastle upon Tyne

D A TURNBULL and J R BEVAN
Department of Environment, University of Northumbria, UK

INTRODUCTION

The study area

The Ouseburn (Newcastle upon Tyne, UK) is a 12.5 km long lowland river that drains an area of 62.5 km². It has a predominantly rural upper section of relatively low relief and a lower somewhat steeper urban section (Figure 1). The solid geology comprises the Carboniferous Middle Coal Measures, including a mixture of resistant sandstones, softer shale marine bands and coal deposits. On top of this is a variable cover of glacial till, varying from 0 to 5 m deep on the watershed to a maximum of about 45 m in the

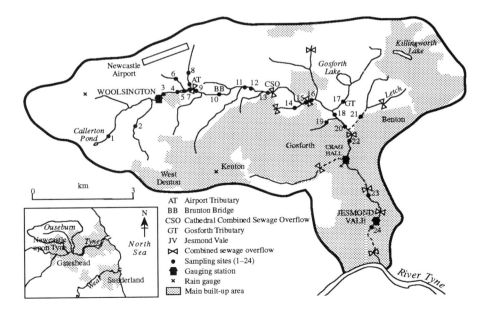

Figure 1 The Ouseburn Catchment: sampling stations and combined sewage overflows

centre of the catchment, where there is also a deposit of fluvio-glacial sands and gravels. Although deep coal mining has ceased, shallow seams are amenable to open-cast working. A small open cast site opened in 1992 drains into the Ouseburn upstream of Woolsington and periods of very high turbidity flow have been observed during this study. The land-use of the catchment as mapped in 1990/91 showed that the northern and western areas are principally rural, with 24 predominantly arable working farms covering approximately a third of the catchment. These may contribute to poorer river water quality via the effects of agricultural chemicals and other farm management practices, although the lack of extensive animal husbandry means that problems due to silage and slurry disposal are unlikely.

Newcastle International airport is a rapidly expanding regional airport located in 'Green Belt' land, 10 km north west of Newcastle city centre (Figure 1). It is within the watershed of a small tributary which drains a mainly rural area of about 2 km^2 and contributes between 3 and 5 per cent of the Ouseburn's average flow. The runway, taxiways and aprons are regularly de-iced during winter months. The lower catchment, comprising approximately half the total catchment area, is heavily urbanised, with housing being the main land-use. This affects the volume and quality of stream flow largely because of surface drainage and combined sewage overflow (CSO). Of the 39 consented discharges into the stream, a total of 22 are CSOs (National Rivers Authority, 1993). The poor aesthetic state of the river as a result of litter in its urban stretches reduces its conservation value. With the exception of the airport and open-cast site, industry, which comprises only 6 per cent of the catchment, discharges its effluent directly to sewer, therefore industrial contamination of surface waters within the watershed should only occur at times of CSO.

METHODS

Physico-chemical methods

Figure 1 shows the location of the three rain gauges and three runoff recorders. Minimum daily grass temperatures, provided by Newcastle University's weather station, indicated the likely times of urea application at the airport. Data loggers equipped with electronic sensors allow transient water quality events to be monitored. Two such instruments recorded temperature, pH, conductivity, dissolved oxygen, turbidity, ammonia and ammonium at 15-minute intervals. A suite of chemical parameters, including temperature, pH, dissolved oxygen, calcium, alkalinity, Biochemical Oxygen Demand (BOD), suspended solids (SS), nitrite-N, nitrate-N and total ammonia, were measured at approximately one-month intervals.

Biological methods

Densities of the faecal indicator groups faecal coliforms (FC) and faecal streptococci (FS) were measured according to the membrane filtration methodology recommended in the bacteriological examination of drinking water supplies (HMSO, 1982). Water samples were taken from ten sites located throughout the catchment during low intensity rainfall and three dilutions of each sample were filtered for FC and FS enumeration. Bacterial changes during storms were monitored by sampling at discrete time intervals throughout the storm using two automatic samplers (EPIC 1011T). Samples were kept cool using synthetic ice packs and were processed within 12 hr of the initial sample being taken. *In situ* bioassay experiments were used to demonstrate if there were any toxic effects upon introduced *Gammarus pulex* (L.) in the Ouseburn. The methodology was adapted from that of Seager and Abrahams (1990). 'Kick' samples were taken from 18 riffle sites spatially located throughout the catchment, during spring, summer and autumn 1991 and a further 7 (8 from Jesmond Vale) samples were taken between November 1992 and April 1993 at each of the bioassay sites. Samples were laboratory sorted, counted and identified to a taxonomic level at least suitable for Biological Monitoring Working Party (BMWP) scoring.

RESULTS

Spatial chemical monitoring
The NRA 1990 National River Quality Survey classified the upper reaches of the Ouseburn as 1B and 2, and the non-tidal urban section as 3. An analysis of the NRA's database of routine monthly chemical data for its four Ouseburn sites revealed winter peaks in total ammonia concentration in the mid and lower Ouseburn (illustrated by Brunton Bridge, site 10). Levels in the upper reaches (illustrated by Woolsington, site 3) did not follow this pattern (Figure 2b). The concentrations were far higher than those which could be expected from inputs such as sewage or silage, and the application of urea by Newcastle International Airport was thought to be a possible source.

Spatial macroinvertebrate sampling
The mean BMWP scores for the 1991 seasonal spatial macroinvertebrate survey were calculated. The River InVertebrate Prediction And Classification System (RIVPACS) (Wright *et al.*, 1989) was used to predict BMWP scores using a range of river characteristics, assuming unperturbed conditions. The sample scores were divided by the RIVPACS predictions and the means for these are shown in Table 1. Assuming that a ratio of one indicates an unstressed environment, all sites had lower than predicted scores, suggesting the influence of factors other than RIVPACS' variables. One-way analysis of variance showed significant between-site variations ($F=15.09$, df=52, $p<0.001$). The pairwise t-test significances of between-site differences are shown in Table 1. In making multiple comparisons, it is recommended (Wardlaw, 1989) that a more strict significance criterion be applied to reduce the possibility of a type 1 error. The Gosforth tributary (site 17, Figure 1) was different, with at least 95% confidence, from all but one of the other sites and with 99% confidence from all but three. It drained a SSSI listed area, and was the only site without obvious adverse anthropogenic influences upon its water quality. Because of similar land-use and

Table 1 Mean BMWP scores and t-tests for significance of difference between mean
Observed/RIVPACS Predicted BMWP ratios for the 1991 sampling period

Mean BMWP	Site	2	3	8	10	11	12	13	14	15	16	17	18	19	20	21	23	24	Mean Ratio
16.3	1	**0.039**	0.110	0.950	0.180	0.750	0.470	0.120	0.064	0.170	0.380	**0.003**	0.930	0.290	0.055	0.760	0.200	0.092	0.20
4.0	2		**0.034**	**0.002**	**0.043**	**0.034**	0.100	**0.013**	**0.008**	**0.001**	0.040	**0.004**	**0.002**	0.062	**0.005**	0.120	**0.003**	**0.014**	0.03
36.7	3			0.097	0.400	0.096	0.052	0.052	0.330	0.200	0.160	0.120	0.096	0.072	0.320	0.096	0.170	0.330	0.41
22.0	8				0.190	0.720	0.440	**0.016**	**0.023**	**0.025**	0.330	**0.007**	0.970	0.230	**0.015**	0.760	**0.025**	0.083	0.19
36.0	10					0.180	0.110	0.077	0.970	0.550	0.470	**0.023**	0.190	0.120	0.950	0.180	0.460	0.930	0.32
20.0	11						0.610	0.130	**0.036**	0.095	0.260	**0.002**	0.740	0.370	**0.029**	1.000	0.110	0.057	0.18
18.7	12							0.390	**0.044**	0.110	0.200	**0.003**	0.450	0.820	0.073	0.620	0.130	0.059	0.15
18.0	13								**0.016**	**0.003**	0.087	**0.005**	**0.018**	0.370	**0.003**	0.240	**0.002**	**0.028**	0.11
38.7	14									0.230	0.320	**0.005**	**0.023**	0.018	0.970	0.069	0.170	0.930	0.32
42.3	15										0.730	**0.011**	**0.025**	**0.050**	0.180	0.210	0.510	0.400	0.27
33.0	16											**0.008**	0.330	0.120	0.350	0.310	0.890	0.390	0.26
56.0	17												**0.007**	0.002	0.004	0.009	0.009	0.007	0.62
16.5	18													0.230	**0.015**	0.770	**0.026**	0.082	0.19
30.3	19														**0.015**	0.410	0.054	**0.029**	0.14
16.7	20															0.160	0.130	0.960	0.31
38.0	21																0.220	0.096	0.18
35.0	23																	0.280	0.26
38.0	24																		0.31

x $p < 0.05$

y̲ $p < 0.01$

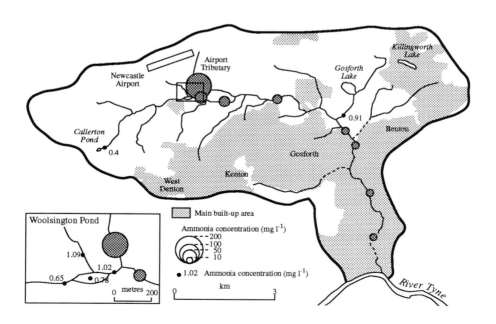

Figure 2 (a)   Total ammonia concentrations 26 February 1992

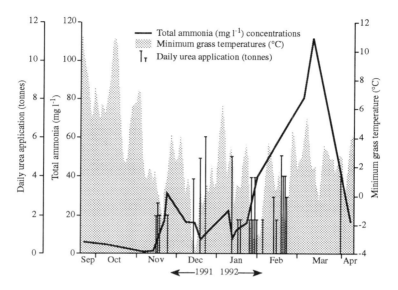

Figure 2 (b)   Minimum grass temperature, urea application and total ammonia concentrations in the airport tributary, winter 1991-2

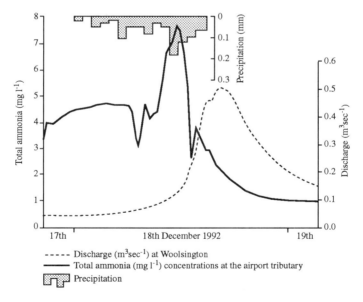

Figure 2 (c)   Total ammonia concentrations at the airport tributary during light rain, December 1992

hydrology it was expected that the faunas in the airport (site 8) and Gosforth (site 17) tributaries would be similar. The sites with the lowest ratios coincided with inputs of urban runoff, CSO and de-icing salt runoff.

Spatial faecal bacteria indicator survey

The geometric means of the FC and FS concentrations obtained at ten sites are shown in Table 2. The sites with the highest faecal coliform counts corresponded to known inputs of CSO. One exception to this was an input from a cross-connected sewerage system immediately upstream from site 1. Faecal coliform to faecal streptococci ratios (Geldreich *et al.*, 1968) were calculated for all sites. The ratio was above 4, indicative of human faecal contamination, for the CSO affected sites. The two sampled rural sites with no upstream inputs of CSO, site 8 - the airport tributary and site 17 - the Gosforth tributary, both had low ratios, 0.124 and 0.500 respectively, indicative of non-human faecal contamination. These results helped identify faecal inputs and verified sewage overflows postulated from the local authority drainage and sewerage maps (Figure 1).

Table 2 Geometric means (CFUs x 10^3 per 100 ml) of FC & FS and FC:FS ratios for 10 sites sampled on 14th June 1991

Site	1	3	8	10	16	17	20	21	23	24
FC	1030.000	19. 267	1.125	2.324	43.000	0.050	13.360	23.546	2.993	2.352
FS	5.400	0.653	9.099	0.470	3.069	0.100	1.720	9.050	0.406	1.045
FC:FS	190.741	29.505	0.124	4.945	14.011	0.500	7.767	2.602	7.372	2.251

In summary, the catchment survey suggested that episodic high concentrations of ammonia and sewage due to airport de-icer runoff and CSO were likely to be the primary factors causing the deterioration of water quality in the lower Ouseburn.

Results demonstrating the impact of the airport

To complement monthly water quality data, total ammonia was measured during seven freezing and eight non-freezing periods at 17 locations throughout the catchment. Ammonia concentrations were higher at the airport tributary (site 8), and all sites downstream Table 3(a), especially during times of freezing conditions (Table 3(b)). Peaks of ammonia were also observed during continuous chemical monitoring of cold weather runoff and snowmelt events (Figure 2(c)). The catchment's spatial variation following de-icing is demonstrated by the concentrations recorded on 26th February 1992 (Figure 2(a) and Table 3(c)).

Table 3 Total ammonia concentrations (mg l^{-1}) recorded at Ouseburn catchment sample sites

Site	1	3	4	5	6	7	8	9	10	13	16	17	20	21	22	23	24
(a)	0.85	0.50	0.36	0.42	0.13	0.60	2.52	1.60	0.23	0.78	0.73	0.29	0.35	0.27	*	0.26	0.44
(b)	0.40	0.18	0.61	0.48	0.75	1.02	105.2	3.6	15.1	8.67	10.4	0.67	7.84	0.49	5.45	7.41	6.62
(c)	0.40	*	0.65	0.78	1.09	1.02	173.3	6.7	29.5	22.3	21.0	0.91	15.4	*	15.1	14.4	13.8

(a) Mean total ammonia concentrations from samples taken during non-freezing conditions
(b) Mean total ammonia concentrations from samples taken during freezing conditions
(c) Total ammonia concentrations recorded on February 26th 1992
 * No readings available.

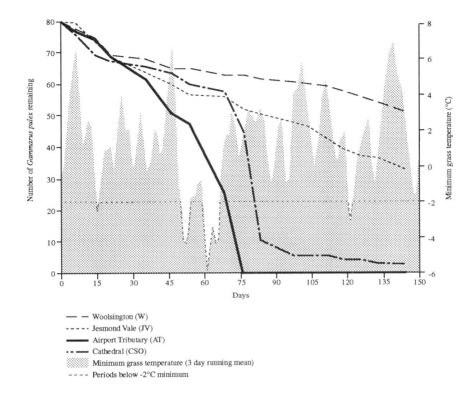

Figure 3 *In situ Gammarus pulex* bioassays in relation to minimum grass temperatures during winter 1992-93

Bioassays

The survivorship curves for an *in situ* bioassay experiment, together with minimum grass temperature, are shown in Figure 3. For all sites the regressions of cumulative mortality against time were significant ($p < 0.001$). An analysis of covariance (Zar, 1974) rejected the hypothesis that the slopes were equal ($F = 45.03$, $p < 0.001$). The minimum grass temperatures were used as a surrogate for urea application. *G. pulex* in the airport tributary experienced increased mortalities coincident with a prolonged period of hard frosts. The delayed mortalities at the cathedral CSO (site 13) appear to be due to a chronic response to the lower ammonia concentrations found there. Woolsington (site 3) had the fewest deaths and neither it nor Jesmond Vale (site 24) showed significant 'kills' during the study period.

Results demonstrating the impact of Combined Sewage Overflows

The large number and episodic operation made it difficult to investigate the impact of specific CSOs. The cumulative CSO responses during individual storm events were however studied at a downstream urban site (site 24) in comparison to a

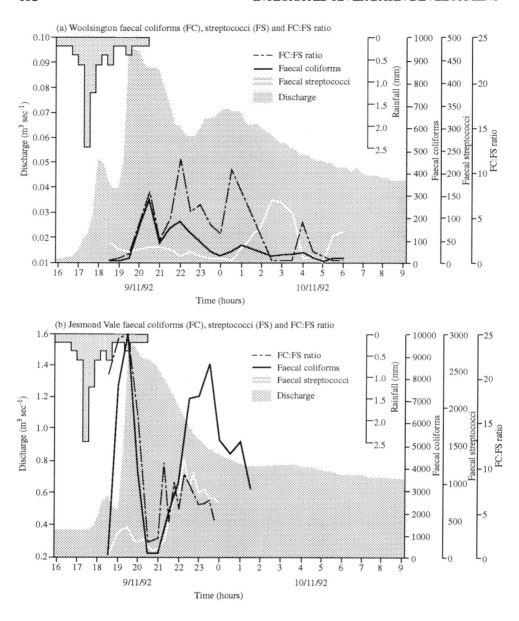

Figure 4 (a) & (b) Discharge and faecal bacteria concentrations for November 9th-10th 1992

relatively CSO unaffected rural upstream site (site 3) except for the sporadic inputs from the malfunctioning sewerage system near site 1 further upstream. As an example of CSO response in the catchment the results of chemical and bacterial sampling from one storm are presented below. The storm on 9/11/92 was the result of intense rainfall (peak intensity 4 mm h^{-1}) following eleven relatively dry days.

At Woolsington (site 3, Figure 4a) the rapid runoff was reflected by the pre-peak on the storm hydrograph. The first two samples had low FC counts and FC:FS ratios of less than one, suggesting that this initial runoff contained faecal matter of animal origin, most likely from the nearby cattle walkway. Immediately following the main discharge there was an increase in the FC counts and high FC:FS ratios, suggesting the input of human faecal matter, most likely from an upstream CSO. FC concentrations declined during the falling stage, however FS densities peaked following the third hydrograph peak. This delayed discharge is most likely the result of throughflow from the surrounding pasture areas and therefore may carry faecal matter of animal origin. The low FC:FS ratios support this hypothesis.

At Jesmond Vale (site 24, Figure 4b) there was also rapid runoff, generating the pre-peak on the storm hydrograph. The pre-peak discharge corresponded to a ten-fold increase in faecal coliform concentrations and a smaller increase in faecal streptococci numbers, indicating a rapid input of faecal matter into the river, and the high FC:FS ratios suggest that it was of human origin. This pre-peak discharge also coincided with a rise in water temperature, fall in pH and increase in turbidity (Figure 5). CSO effluent is extremely variable in its physico-chemical characteristics, however it frequently contains faecal matter and hence high concentrations of faecal indicator bacteria and is usually warmer than river water due to the input of water from municipal use and the warming/insulating effects of the sewerage system. The presence of organic acids in domestic sewage often results in a low pH and because of the high concentration of suspended solids it is highly turbid. These characteristics therefore strongly suggested CSO input. During the main discharge there was a significant drop in bacterial concentrations coincident with a fall in turbidity, suggesting a pulse of 'cleaner' water. After this diluting effect concentrations again peaked before declining at the latter stages of the recession curve.

To test if there were significant faecal bacterial differences between the two sites the data from seven sampled storm events was combined and subjected to an independent two sample t-test not assuming equal variances. The results (Table 4)

Table 4 T-tests for significance of difference between log transformed mean FC and FS concentrations and FC:FS ratios for 7 storms at Woolsington (W) and Jesmond Vale   (JV)

	Log_{10} Faecal coliforms	Log_{10} Faecal streptococci	FC:FS ratio
W Mean	1.735	1.151	8.0
W σ	0.576	0.469	11.2
W n	102	96	93
JV Mean	2.680	1.802	13.0
JV σ	1.060	0.906	15.6
JV n	130	137	127
t statistic	8.69	7.16	2.75
probability	0.0001	0.0001	0.0065
d.f.	206	214	217

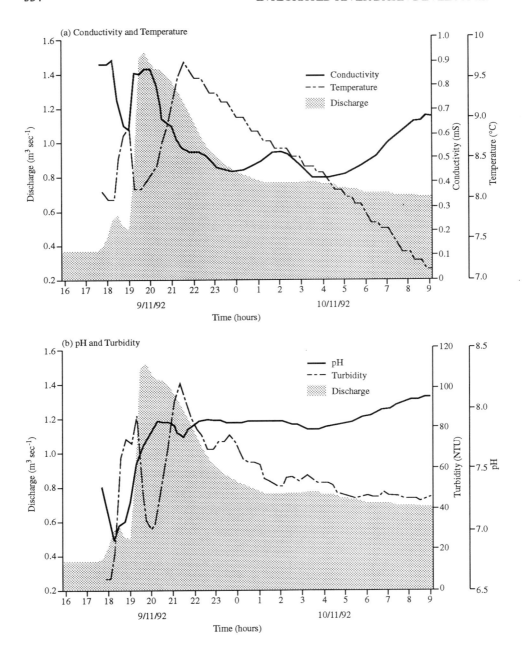

Figure 5 Discharge, temperature, pH, conductivity and turbidity at Jesmond Vale. November 9th-10th 1992

show that Jesmond Vale (site 24) had significantly greater concentrations of FC ($p<0.0001$), FS ($p<0.0001$) and FC:FS ratios ($p<0.0065$) than Woolsington (site 3).

MACROINVERTEBRATE IMPACTS

To study the effects of pollution, including urea de-icer and CSO on macroinvertebrates, four sites were sampled over a six-month period (19/11/92-26/4/93). The ratios of taxa found to RIVPACS predicted taxa for winter 92/93 indicate that Woolsington (site 3, mean ratio 0.678, std. dev. 0.105) had a more diverse fauna than the airport tributary (site 8, 0.367, 0.078), and other downstream sites (cathedral CSO, site 13, 0.369, 0.033 & Jesmond Vale, site 24, 0.369, 0.051).

The number of scoring taxa found in the airport tributary (site 8) fell progressively through the winter months. Using sample taxonomic composition, Two-Way INdicator SPecies ANalysis (TWINSPAN) (Hill, 1979a) classified the sites and highlighted

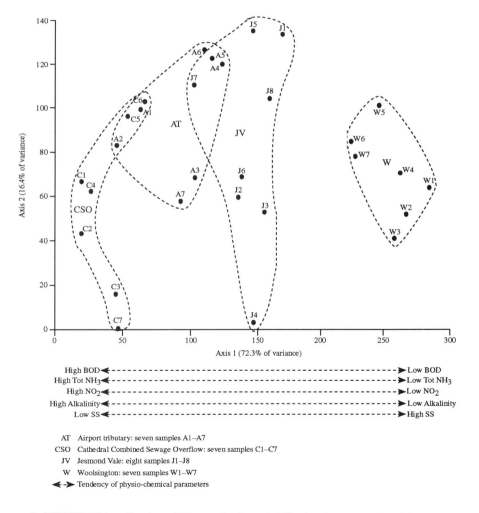

Figure 6 DECORANA ordination of 29 samples from 4 riffle sites between 19/11/92 and 26/4/93 based on TWINSPAN groupings

indicator species. TWINSPAN progressively divided the entire sample set into smaller sub-sets that formed the basis of the DECORANA groups (Figure 6) and it classified Woolsington (site 3) separately from the other sites. One way analysis of variance revealed significant (p<0.001) differences in Scoring Taxa and BMWP ratios between the sites. The Newman-Keuls multiple range test then isolated pairs of sites which were different (Table 5). Woolsington (site 3), upstream of the airport and CSO affected sites, was found to be significantly different (p<0.001) from the other sites, whilst the other episodically affected sites were not significantly different from each other.

Table 5   Results of Newman-Keuls test for the significance of differences between the means of observed / RIVPACS predicted ratios at four sites.

Taxa/RIVPACS Taxa					*q-Statistic*		
Mean	Std. dev.	n	Site	W	AT	CSO	JV
0.678	0.105	7	W	/	11.510***	11.425***	11.833***
0.367	0.078	7	AT	/	/	0.085	0.055
0.369	0.033	7	CSO	/	/	/	0.033
0.369	0.051	8	JV	/	/	/	/

One-way analysis of variance F = 33.24   Degrees of freedom = 28   P < 0.001

BMWP/RIVPACS BMWP					*q-Statistic*		
Mean	Std. dev.	n	Site	W	AT	CSO	JV
0.576	0.116	7	W	/	11.086***	13.622***	12.177***
0.278	0.063	7	AT	/	/	2.576	0.728
0.208	0.030	7	CSO	/	/	/	1.933
0.259	0.049	8	JV	/	/	/	/

One-way analysis of variance F = 38.62   Degrees of freedom = 28   P < 0.001.

*** P < 0.001. Other q-values are not significant

AT	Airport tributary	JV	Jesmond Vale
CSO	Cathedral Combined Sewage Overflow	W	Woolsington

DE-trended CORrespondence ANAlysis (DECORANA) (Hill, 1979b) ordinates samples according to their taxonomic composition. For the Ouseburn data it generated two primary axes that explained 88.7% of the sample variation. Correlations between these axial scores and the previous year's mean annual physico-chemical concentrations are shown in Table 6. The highest correlations were between axis 1 and BOD, nitrite, alkalinity and SS concentrations. High BOD and nitrite concentrations can indicate the breakdown of urea, which is present in both sewage and de-icing salt. Woolsington (site 3) was immediately downstream of an open-cast coal site and experienced sporadic discharges high in suspended solids. The correlation of total ammonia with both axes scores provided a distinguishing feature of the airport site (site 8) i.e.. low on axis 1 and high on axis 2 (Table 6 and Figure 6).

Table 6   Correlations (r) between DECORANA axis scores and mean annual water physico-chemical parameters (n=29).

Parameter	Axis 1	Axis 2
pH	-0.169	-0.080
Temperature	0.267	-0.134
Dissolved oxygen	0.128	-0.272
BOD	-0.450*	-0.393*
Total ammonia	-0.320	0.313
Nitrate-N	-0.308	0.298
Nitrite-N	-0.569**	0.219
SS	0.431*	-0.217
Alkalinity	-0.505**	0.183
Calcium	-0.005	-0.188*

$P < 0.05$ and ** $P < 0.01$. Other r-values are not significant.

CONCLUSIONS

The results of the study demonstrated the adverse effects of episodic discharges of CSO and urea de-icer on the biological quality of the Ouseburn. Due to the intermittent nature of the pollution events, routine chemical monitoring was not the most effective technique for detecting their occurrence. Continuous water quality monitoring has the advantage that it can record such events as they happen and proved useful in this study. However water quality loggers are costly, notoriously difficult to maintain and because of the limited range of available sensors, are unable to detect the majority of toxins. The two macroinvertebrate techniques responded to the intermittent pollution events. Multivariate classification techniques and use of the RIVPACS predictive model were valuable in identifying impacted sites. The bacterial analysis confirmed inputs of CSO. The integrated use of a range of techniques thus gave a better explanation of the causes, mechanisms and effects of episodic pollution compared with the use of any single method.

As a result of the water quality issues discussed and the Newcastle upon Tyne City Development Plans proposed in the upper catchment, the NRA Northumbria Region selected the Ouseburn as a pilot 'Catchment Management Plan'. Some improvements are already being implemented: (i) the enlargement of the trunk interceptor sewer that takes sewage from Ponteland (a small town neighbouring the north-western fringe of the catchment), which should reduce the frequency of CSO discharge at the cathedral site; (ii) a £2 m upgrading of the Letch sewerage system in the east of the catchment and (iii) Newcastle airport has recently changed from using urea to a 'less toxic' de-icer based on potassium acetate.

REFERENCES

Geldreich, E.E., Best, L.C., Kenner, B.A. & Van Donsel, D.J. 1968. The bacteriological aspects of stormwater pollution. *J. Wat. Pollut. Control Fed.*, **40**, 1861-1872.

Hill, M.O. 1979a. *TWINSPAN-A FORTRAN program for arranging multivariate data in an ordered two-way table by classification of individuals and attributes*, Ecology and Systematics, Cornell University, Ithaca, NY.

Hill, M.O. 1979b. *DECORANA-A FORTRAN program for detrended correspondence analysis and reciprocal averaging*, Ecology and Systematics, Cornell University, Ithaca, NY.

HMSO 1982. *The bacteriological examination of drinking water supplies*. Reports on Public Health and Medical Subjects. Number 71.

National Rivers Authority (Northumbria Region) 1993. *The Ouseburn : A catchment management plan - consultation report*. National Rivers' Authority, Newcastle upon Tyne.

Seager, J. & Abrahams, R.G. 1990. The impact of storm sewage discharges on the ecology of a small urban river. *Wat. Sci. Tech.*, 22, 163-171.

Wardlaw, A.C. 1989. *Practical statistics for experimental biologists*. John Wiley & Sons, Chichester.

Wright, J.F., Armitage, P.D., & Furse, M.T. 1989. Prediction of invertebrate communities using stream measurements. *Regulated Rivers : Research & Management*, 4, 147-155.

Zar, J.H. 1974. *Biostatistical analysis*. Prentice-Hall, Englewood Cliffs.

30 Preparing for change in the River Trent

C P CROCKETT and **D W MARTIN**
National Rivers Authority Severn-Trent Region, Solihull, UK

INTRODUCTION

The River Trent is one of the few major undeveloped surface water sources for potable water supply in England and Wales. A legacy of urbanisation over the years reduced the water quality of the River Trent to a low point in the 1950s and early 1960s. In the past, the relatively modest public supply demands, the ready availability of better quality resources and difficulties in treating the Trent river water to a wholesome quality, has meant that using the River Trent as a potable water supply source could not be justified. The dramatic improvements in river water quality over the last three decades and the increasing demands for water supply has meant that the licensing policy for future abstractions from the River Trent needs to be reassessed.

This potential new role for the River Trent could come into conflict with the existing use of the river. It is the role of the NRA to balance the needs of existing and potential future use. The first step in achieving this balance is to understand the existing requirements of the present use of the river, and to this end the NRA is undertaking a series of studies. These studies will serve to provide information on the minimum river flow requirements for the existing river use. This minimum river flow is that below which the ability of the river to support a defined use is affected. This flow may already occur naturally. A starting point of any new policy might reasonably be to ensure that the frequency of occurrence of river flow below this minimum requirement is not increased.

This paper reviews the quality improvements that have occurred in the River Trent over the last thirty years and the steps being taken by the NRA to accommodate the potential new role of water supply within the framework of existing river use.

THE RIVER TRENT CATCHMENT

The Trent is the major river of the East Midlands of England (see Figure 1). It drains an area of nearly 10,500 km². The River Trent rises 11 km north of Stoke-on-Trent and flows eastward for some 274 km to its confluence with the Humber Estuary at Trent Falls. The hydraulic tidal limit is at Cromwell weir, about 40 km downstream

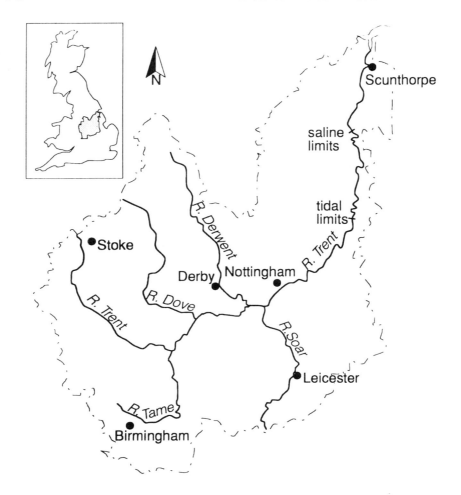

Figure 1 The River Trent basin

of Nottingham. The saline limit is normally at or near to Owston Ferry, approximately 26 km from Trent Falls, but this can extend further upstream during periods of low river flow and high spring tides.

The catchment contains a population of over 6.0 million people, almost half of whom live in the major urban conurbations of Birmingham, The Black Country and Stoke-on-Trent. It is this concentration of people and the associated industry in the headwaters, which have caused the major pollution problems in the Trent catchment. Further problems are caused by large population centres such as Derby, Nottingham and Leicester.

The Midlands conurbation straddles the Severn and Trent catchments and is served largely by resources from the River Severn but discharges much of its effluent to the Trent catchment. Natural flows in the River Trent are greatly modified by artificial inter-basin transfers from the River Severn and the Elan Valley. Much of this water is discharged as effluent and represents not only a significant proportion of the river flow during dry periods but also a large effluent load such that the upper reaches of the River Trent and the River Tame are heavily polluted. Both river reaches also suffer further pollution from storm overflows and urban runoff.

The River Trent is not used directly for public water supplies, although its cleaner tributaries the River Dove and River Derwent are important water supply rivers. These rivers are impounded or abstracted from directly or via pumped storage reservoir schemes. This includes the recently completed pumped storage reservoir at Carsington, which will be used to regulate the River Derwent.

The Trent catchment has a large amount of industrial activity, ranging from a number of major power stations to coal mining and metal finishing. Despite the pollution load and industrial activity, the River Trent also provides a major recreational activity area, most notably for coarse fishing.

ADMINISTRATIVE BACKGROUND

In 1974 the Severn Trent Water Authority (STWA) was created to provide an integrated management of all aspects of the water cycle on a catchment scale. The STWA assumed the responsibilities of the Severn and Trent River Authorities and also the duties of water supply from 26 undertakings and sewage disposal from 210 local authorities and 2 main drainage authorities (Martin & Woods, 1992).

The 1989 Water Act provided for the splitting up of the water authorities responsibilities between three new national regulatory bodies, the National Rivers Authority (NRA), the Office of Water Services and the Drinking Water Inspectorate and ten new regionally based Water Services Companies (WSCs). The NRA, as the environmental regulator, has responsibilities for pollution control, abstraction licensing, recreational fisheries, drainage and flood defence. The WSCs are responsible for the provision of water supplies and for sewerage and sewage disposal services. There are also a number of water companies which are involved only in the water supply. Within the Trent catchment the South Staffordshire Waterworks Company is an example of this type of organisation.

RIVER WATER QUALITY IN THE TRENT CATCHMENT

During the last century there was a rapid growth of industry and population in the Trent catchment. The river system was used extensively for the disposal of untreated or inadequately treated sewage and industrial effluent. Due to a lack of appropriate institutional structures and lack of investment, the water quality in the River Trent gradually declined. The Trent catchment was probably at its worst during the late 1950s.

Over the last 40 years, successive authorities in charge of pollution control have gradually improved river water quality within the catchment. This has been achieved

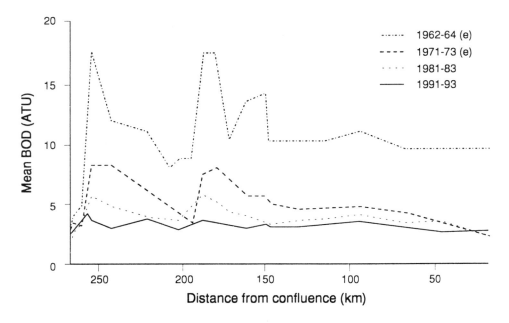

(e) BOD estimated

Figure 2   River Trent quality profile

through the introduction and enforcement of pollution prevention legislation, and the willingness of local authorities to invest in effluent treatment (Martin & Brewin, 1986).

The extent of the improvement in the water quality of the River Trent can be seen in Figures 2 and 3. These Figures show the longitudinal mean BOD (ATU) and

Table 1   Summary of River Classification Scheme

Class	Intended Use	DO (%Sat)	BOD(ATU)	Ammonia-N
1A	Water Supply/Game Fishery	>80	<3.0	<0.3
1B	Water Supply/Game Fishery	>60	<5.0	<0.7
2	Water Supply (a)/ Coarse Fishery	>40	<9.0	(b)
3		>10	<17.0	
4		<10	>17.0	

Note: Limits are 5 percentiles for dissolved oxygen (DO) percent saturation and 95 percentiles as mg l⁻¹ for BOD(ATU) and ammonia

(a) Class 2 suitable for water supply after advanced treatment

(b) Class 2 ammonia controlled by requirement to be non-toxic to fish

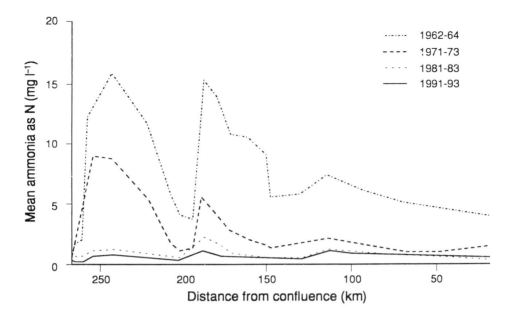

Figure 3 River Trent quality profile

ammonia for selected three-year periods from the early 1960s to the present day. There is a clear improvement over time, notably in the headwaters around Stoke and after the confluence with the River Tame (190 km upstream of Trent Falls).

Since 1979 the rivers in the UK have been classified into five main classes, as shown in Table 1. Figure 4 shows the percentage length of main river in the Trent catchment in each class over the last thirty years (Martin and Woods, 1992). The improvement in river water quality over time is again clearly illustrated. The final column in Figure 4 shows the River Quality Objective (RQO) for the Trent catchment which involves the elimination of all class 4 river reaches.

EXISTING USE OF THE RIVER TRENT

There are two categories of existing users of the River Trent: abstractors who take water from the River Trent to satisfy their requirements, and in-river users who rely on sufficient quantities of water being in the river. The requirements of these users are summarised in Table 2.

A recent review (Hall, 1993) undertaken on behalf of the NRA, has attempted to identify the critical river flows which are required to maintain the existing use of the River Trent. A profile of the estimated existing river flow requirements is presented in Figure 5. In this study the River Trent was spilt into fluvial and tidal reaches, above and below the tidal limit at Cromwell Weir.

Table 2   Existing Use of the River Trent (after Hall, 1993)

River User	*Requirements*
Abstractors	
Power stations	Net consumption is relatively small compared to the total abstracted volume. The NRA is not empowered to issue new licences that impinge upon the gross licensed abstractions which are held as a protected right.
Water Supply	At present the only major licensed abstraction direct from the River Trent is at Torksey for the Trent-Witham-Ancholme scheme. There are, however, major abstractions from some tributaries, such as the Dove and Derwent.
Agriculture	The volume of annual licensed abstractions is approximately 46 Ml day^{-1}, virtually all used for spray irrigation.
Industry	There are no large abstractions directly from the River Trent, most are from groundwater or from tributaries.
In-river	
Navigation	The River Trent is navigable to just upstream of the River Derwent confluence. Dredging is required to maintain a statutory minimum depth, particularly in the upper tidal reach.
Ecology	Given the large flow requirements of other existing users, the minimum ecologically acceptable flow is assumed to be the mean annual 7-day low flow.
Recreation	The river and associated water bodies are major attractions for water skiing, wind surfing, rowing, kayaking and pleasure cruising. Maintenance of a suitable water level is important for these activities.
Effluent Dilution	The River Trent and its tributaries receive more than 1000 Ml day^{-1} of effluent from Sewage Treatment Works. The flow in the river provides the dilution for these discharges to enable river water quality objectives to be met.
Fisheries	The River Trent is an important cyprinid fishery.
Flood Defence	Significant flood defence schemes exist along the full length of the river.

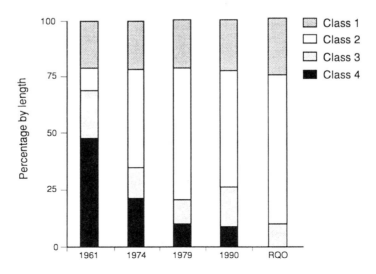

Figure 4   Classification of major rivers in the Trent catchment

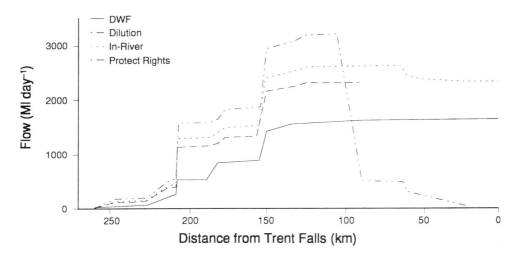

Figure 5 Estimated minimum flow requirements for existing uses of the River Trent

The protected right of one large power station abstractor approximately 100 km upstream of Trent Falls imposes the greatest minimum river flow requirement over much of the fluvial River Trent. The abstraction right is greater than the observed dry weather flow in the River Trent. This aside, the in-river navigational requirements have the greatest minimum river flow requirement over the fluvial and tidal reaches.

The minimum river flow requirements for the maintenance of navigation from the above study are only estimates. Further, more detailed, work will be required, particularly in the tidal River Trent, to gain a better understanding of the dynamic balance between sediment, river flow and tidal height. No estimate was made for the minimum river flow required to meet the water quality demands in the Humber estuary.

WATER RESOURCES DEVELOPMENTS

Towards the end of the 1960s there was a great deal of concern over future sources of water supply in the Trent catchment. The demand for water was increasing at such a rate that a major new source of supply was thought to be required within the decade. In order to look at the various options available, the Trent Research Programme (TRP) was initiated (Water Resources Board, 1973).

The results of the TRP showed that the optimal solution would involve the extensive abstraction of water from the River Trent and its treatment for supply. At that time it was felt that the public would not accept this as a reliable and 'wholesome' source. As a result the slightly more expensive strategic option of building Carsington Reservoir was adopted. Other smaller sources, mainly groundwater, were also extensively developed.

In 1990 a licence was issued for the abstraction of potable water from the River Trent. This involved a pumped transfer to the Fossdyke at Torksey and via the Rivers Witham and Ancholme to the Elsham treatment works. Previously, the licence had only been for industrial supply to South Humberside. This is not perceived to be a direct abstraction from the River Trent.

The River Trent is now being considered as an untapped source of water supply to meet the future demands in parts of the Anglian Region of the NRA, primarily Lincolnshire. Furthermore, the feasibility of abstractions of up to 600 Ml d^{-1} at Torksey have been investigated. Approximately 400 Ml d^{-1} of this could be fed south by a series of channels, pipes and rivers to meet a possible increase in public supply required by Essex and north-east London (Brown & Askew, 1993). A review of the demands has suggested that this need is unlikely before well into the next century. What is more likely is that the River Trent will be used to meet water supply demands within the Trent catchment itself. Proposals have been put forward by both South Staffordshire Waterworks Company and Severn Trent Water (Martin & Brewin, 1994). Such schemes could help ease the pressures on overused groundwater sources.

BALANCING RIVER USE

Under the present legislation the NRA is required to manage abstractions from surface and groundwaters through the abstraction licensing system, and to take action to conserve, redistribute and augment water resources and to secure their proper use (NRA, 1992). Proper use of water resources includes not only meeting the legitimate demands of abstractors but also the existing in-river demands, notably those of the aquatic life. Where these demands are in conflict, the NRA has the job of striking the right balance.

The improving water quality in the River Trent has enabled the possibility of using it for direct potable supply to be seriously investigated. In reviewing the potential of the River Trent to fulfil this new role, the requirements of the existing users identified in Table 2 need to be taken into account and to balance the potential conflict between existing and future use of the River Trent, the NRA must decide upon a licensing policy for future abstractions.

To formulate this licensing policy a number of related issues have to be addressed, including:

- What is the long-term flow regime of the River Trent ?
- What are the minimum prescribed flows required in the fluvial River Trent ?
- What is the minimum residual flow required by the tidal River Trent/ Humber Estuary ?
- What are the implications for future water quality standards in the Trent catchment ?
- What will be the implications for the potential return of salmon ?
- How will the licensing policy be implemented?

The Severn Trent Region of the NRA has begun a series of projects which aim to address these issues. Details of these studies are given below.

Long-term flow regime in the River Trent
The fundamental building block to predicting and controlling the changing role of the River Trent is a long-term sequence of naturalised flow data. This will enable the return period of flows under different licensing policies to be predicted. A study is under way to generate naturalised flows at various locations along the River Trent. These naturalised flows will be incorporated into an enhanced Regional Water Resources Allocation model, enabling the impact of any proposed scheme on the long-term flow regime of the river to be investigated

Minimum residual flow to the Humber Estuary
There is, as yet, no standard NRA methodology for calculating the minimum residual flow to an estuary related to the existing use and requirements of the estuary. The two primary requirements identified are the protection of the water quality standards in the estuary and the need to maintain navigation. The water quality standards are being investigated with the aid of a Humber Estuary water quality model that Water Research Centre have been contracted to develop. The impact of reduced freshwater flows, on sediment deposition and navigation, will need to be addressed separately through a more detailed study. This will involve a review of existing data and eventually the development of a mathematical model to look at the sedimentation process. The views of the navigational authorities along the River Trent are currently being sought.

Minimum prescribed flows in the fluvial River Trent
The initial investigation into the existing flow requirements in the fluvial reach of the Trent put forward some draft interim prescribed flows relating to the existing use of the river. Negotiations are proceeding to reduce the large licences of right which are currently restricting the licensing of any significant upstream abstractions. Further work is being carried out to confirm the estimated use related minimum flow requirements, particularly for navigation and dilution. This will involve the development of a water quality model of the fluvial Trent and investigating the flow and water level relationship along the navigable portion of the river.

Implication for water quality standards
Table 3 gives a summary of quality data from the River Trent at locations where future water supply abstractions are being considered. The determinands listed are those where there are potential failures of the standards in the EC Surface Water Abstraction Directive (75/440/EEC) or Drinking Standards (80/778/EEC). As Table 3 indicates, the use of the River Trent as a source of potable water will require increasing control of nitrate, sulphate, chloride and potassium at source (Martin & Brewin, 1994). The levels of pesticides and metals in the river may also limit future water supply developments.

There are no direct potable water abstractions from the River Trent. Consequently, there is no need to control nutrients from effluents prior to their discharge in order to meet the requirements of the EC Urban Waste Water Treatment Directive (91/271/EEC) through the designation of sensitive areas. However, as more use of the river is made, this may change and the potential implications need to be investigated further.

Table 3   River Trent: 95th Percentile Quality (mg l^{-1}) 1992 (after Martin & Brewin, 1994)

Location	NH_4 as N	NO_3 as N	SO_4	Cl	K	Na
Yoxall	0.71	12.1	238.0	199.0	19.5	150.0
Nottingham	0.63	11.8	196.0	159.0	13.2	143.0
Torksey	0.99	12.3	230.0	161.0	12.8	138.0
EC Standards	1.50	11.3	250.0	200.0	12.0	150.0

Potential return of salmon

With the improving water quality of the River Trent it may be possible for salmon to return. Recent observations indicate that fish are making their way upstream, despite physical and water quality barriers. The impact of physical, thermal and water quality barriers on the potential return of salmon or other migratory fish is being investigated by the NRA.

Implementation of licensing policy

A large number of the tributaries already have a formal licensing policy. In formulating the licensing policy for the River Trent, the impacts on these existing policies has to be understood. An internal review of the existing licensing policies in the Trent catchment is being undertaken by the NRA.

A new licensing policy, developed from a understanding of the flow requirements of existing users, could be applied, either by setting prescribed flows for individual licences or through the implementation of a Minimum Acceptable Flow (MAF) regime. No MAF regimes have been set in the UK even though there has been provision for such statutory control since 1963.

Given the multi-use nature of the River Trent and the need to develop a more formal licensing policy, the implementation of a MAF may provide the best way forward. The NRA is currently undertaking a research project to investigate carefully the implications of implementing a MAF.

THE WAY FORWARD

The dramatic improvement in water quality in the River Trent over the last 30 years has been a significant achievement. This improvement in river water quality means that the river is now more suitable for public supply (Waters & Woods, 1993). The estimated minimum flow requirements for the existing use of the River Trent illustrates the need to develop a more formal licensing policy for future abstractions.

In the last few years the NRA has adopted an integrated approach to catchment planning. This involves looking at all the current use of a catchment across all the functions (water quality, water resources, flood defence, fisheries, conservation and recreation) and possible future use, to develop a plan for future management. The results of the existing studies and the planned future studies of the River Trent will feed into this catchment planning process. With reference to the River Trent it is anticipated that the Catchment Management Plan for the Humber estuary will be completed this year.

Current knowledge of the flow requirements for certain river use is still incomplete. In particular, further investigations are required on the relationship between river flow to the tidal River Trent and sediment erosion, deposition and transport. It is only through undertaking these types of studies that the NRA will be able to strike the right balance between the demands placed on the River Trent by existing and future use.

ACKNOWLEDGEMENT

The authors are grateful to the Regional Manager of the National Rivers Authority for permission to present this paper. The views expressed in this paper are those of the authors and are not necessarily those of the National Rivers Authority.

REFERENCES

Brown, R.P.C. & Askew, T.E.A. 1993. Bulk water transfer. *British Hydrological Society Fourth National Hydrology Symposium*, Cardiff, 1.93-1.99.

Hall, J.K. River Trent control rules 1993. *British Hydrological Society Fourth National Hydrology Symposium*, Cardiff, 1.85-1.91.

Martin, J.R. & Brewin, D.J. 1986. River quality modelling in the management of the Trent catchment. In: *Water Quality Modelling in the Inland Natural Environment*, BHRA, The Fluid Engineering Centre, Cranfield, UK. 61-70.

Martin, J.R. & Brewin, D.J. 1994. Quality Improvements and Objectives in the Trent Catchment.

Martin, J.R. & Woods, D.R. 1992. Water Quality Management in the River Trent (UK): Thirty Years of Change. *Water Environment Federation, 65th Annual Conference,*New Orleans, USA

NRA 1992. *Water Resources Development Strategy — A Discussion Document*. NRA, Bristol.

Waters, B.D. & Woods, D.R. 1993. Water Quality Management of the River Trent Basin, *Proceedings of the 30th Japan Sewage Works Association Annual Technical Coference*. Tokyo, Japan.

Water Resources Board 1973. *The Trent Research Programme*, 12 Volumes. HMSO, London.

PART VIII

RIVER BASIN MANAGEMENT

31 The functions of wetlands within integrated river basin development: international perspectives

G E HOLLIS
University College London, UK
M C ACREMAN [1]
IUCN, Gland, Switzerland
[1] *On secondment from the Institute of Hydrology, Wallingford, UK*

INTRODUCTION

Now that the immense benefits of wetlands are being realised, policies and incentives to destroy wetlands have been removed, important wetlands are being protected and rehabilitated and in some cases artificial wetlands are being created. However, lack of understanding of wetland values is still the main threat to their conservation, and decision makers appear to believe that wetland destruction has little or no negative implications when designing engineering works such as dams, canals and drainage schemes. This paper provides examples of this range of vital functions, and describes how destruction of wetlands has led to environmental degradation and economic decline, how wise use of wetlands can provide sustainable benefits to society and wildlife and how wetlands should be part of the integrated development of river basins throughout the world.

WETLAND DEFINITION

Wetlands are transitional environments between terrestrial and fully aquatic ecosystems which normally include floodplains, fens, bogs, shallow lakes and salt marshes. The international definition within the Ramsar Convention (Convention on wetlands of International importance especially as waterfowl habitat - Ramsar, 1971), which is accepted by over 75 governments, gives wetlands as "areas of marsh, fen, peatland or water, whether natural or artificial, permanent or temporary, with water that is static or flowing, fresh, brackish or salt, including areas of marine waters, the depth of which at low tide does not exceed six metres" and they may include "riparian and coastal zones adjacent to the wetlands or islands or bodies of marine water deeper than six metres at low tide lying within".

The Canadian definition of wetland is more specific: "land that is saturated with water long enough to promote wetland or aquatic processes as indicated by poorly drained soils, hydrophytic vegetation and various kinds of biological activity which are adapted to a wet environment" (Environment Canada, 1991). The definitions in the USA emanating from different government agencies are reviewed by Mitsch and Gosselink (1993) who conclude that "applying a comprehensive definition in a uniform and fair way requires a generation of well-trained wetland scientists and managers armed with a fundamental understanding of (wetland processes)".

The issue of wetland definition can be even more important in those countries where the concept of wetlands does not appear in legislation. The traditional association of wetlands with wildlife and nature conservation in some countries means that water managers may ignore them in favour of a narrower definition of responsibilities related to rivers.

WETLAND FUNCTIONS AND VALUES

In the last 20 years, the vital functions performed by wetlands have become more fully appreciated (Adamus and Stockwell, 1983). However, it has been acknowledged that not all wetlands perform all of the functions, and that the degree of this performance varies (Table 1). The potential of wetlands to support and maintain economic development has been propounded by Davies and Claridge (1993). Wetlands provide natural storage areas for flood water, reducing the risk of inundation and damage downstream; their retention of water can also encourage recharge to underlying aquifers. Wetland vegetation, especially mangroves but also salt marsh to a lesser extent, can stabilise shorelines by reducing the energy of waves and currents and retaining sediment with their roots. Because toxicants (like pesticides) often adhere to suspended matter, sediment trapping frequently results in water quality improvements.

Water quality can also be improved by the ability of wetlands to strip nutrients (N & P) from water flowing through them. The facility of many wetland plants to pump oxygen to their roots also allows wetlands to treat water polluted with organic material. Mitsch and Gosselink (1993) refer to wetlands both as "the kidneys of the landscape", since they cleanse polluted water and play an important role in chemical cycling, and as "biological supermarkets" because of the extensive food chain and rich biodiversity they support. For centuries, wetlands have played a central role in the rural economy of many parts of the world, such as Sahelian Africa, where floodplains provide highly productive agricultural land, dry season grazing for migrant herds, fish, fuelwood, timber, reeds, medicines and other products in addition to important wildlife habitats. All these functions have direct economic benefits to human welfare (Dugan, 1990). The financial valuation of wetland functions and products is advancing rapidly (Barbier, 1989; Farber & Costanza, 1987; Pearce & Turner, 1990) and much of this paper is devoted to examples of wetland values within the context of integrated river basin management.

Table 1 Functions and values of wetland types (after Dugan, 1990)

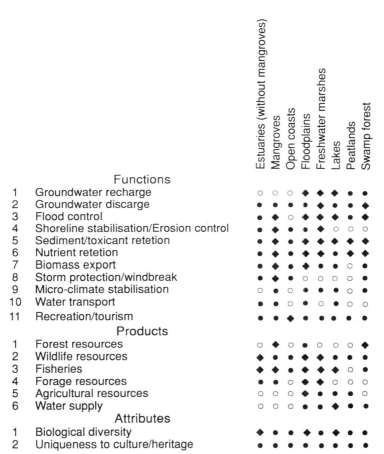

		Estuaries (without mangroves)	Mangroves	Open coasts	Floodplains	Freshwater marshes	Lakes	Peatlands	Swamp forest
Functions									
1	Groundwater recharge	○	○	○	◆	◆	◆	●	●
2	Groundwater discarge	●	●	●	●	◆	●	●	◆
3	Flood control	●	◆	○	◆	◆	◆	●	◆
4	Shoreline stabilisation/Erosion control	●	◆	●	●	◆	○	○	○
5	Sediment/toxicant retetion	●	◆	●	◆	◆	◆	◆	◆
6	Nutrient retetion	●	◆	●	◆	●	●	◆	◆
7	Biomass export	●	◆	●	◆	●	●	○	●
8	Storm protection/windbreak	●	◆	●	○	○	○	○	○
9	Micro-climate stabilisation	○	●	○	●	●	●	○	●
10	Water transport	●	●	○	●	○	●	○	○
11	Recreation/tourism	●	●	◆	●	●	●	●	●
Products									
1	Forest resources	○	◆	○	●	○	○	○	◆
2	Wildlife resources	◆	●	●	◆	◆	●	●	●
3	Fisheries	◆	◆	●	◆	◆	◆	○	●
4	Forage resources	●	●	○	◆	◆	○	○	○
5	Agricultural resources	○	○	○	◆	●	●	●	○
6	Water supply	○	○	○	●	●	◆	●	●
Attributes									
1	Biological diversity	◆	●	●	◆	●	◆	●	●
2	Uniqueness to culture/heritage	●	●	●	●	●	●	●	●

○ Absent or exceptional; ● Present ◆ Common and important value of that wetland type

HISTORICAL DEGRADATION OF WETLANDS

For much of history wetlands have been perceived as unhealthy waste-lands harbouring disease, insects and even monsters. Drainage and destruction of wetlands became accepted practices throughout the world as demand for agricultural and urban land increased. The battle against malaria often involved wetland drainage. However, as in Spain in the 1800s Montes and Bifani ,1991) and in many other instances throughout the developing world, the simultaneous promotion of irrigation can actually exacerbate the malaria problem.

Historically, river basin development has involved the eradication of wetlands for a variety of reasons either directly by drainage or bypassing, or indirectly by the construction of dams and river diversion works, or the over-exploitation of groundwater. This is because far from being integrated, the development of river basins has often been uni-sectoral, concentrating on hydro-power generation, irrigation or flood control at the expense of other sectors. The catalogue of wetland destruction is remarkable. In England the area of freshwater marsh fell by 52% between 1947 and 1982 and in the early 1980s the rate of drainage continued at between 4,000 and 8,000 ha per year. France's largest wetland type, the freshwater meadows, bogs and woods, once covered 1.3 million ha but it has been declining at a rate of 10,000 ha per year (Baldock, 1990). In Roman times, 10% of Italy (3 million ha) was wetlands. Only 764,000 ha remained by 1865, and by 1972 this had diminished to only 190,000 ha. In the Castille-La Mancha region of Spain, 60% of the wetlands have been lost with three quarters of this loss (20,000 ha) taking place in the last 25 years (Montes and Bifani, 1991). In Greece a 60% loss of wetlands, mainly lakes and marshland, took place by land drainage for agriculture since 1925. In the same way, 28% of Tunisian wetlands have disappeared in the last 100 years (Maamouri and Hughes, 1992). In Asia, about 85% of the 947 sites listed in the Directory of Asian Wetlands are under threat, 50% of them seriously (Scott and Poole, 1989).

Hadejia-Nguru Wetlands, Nigeria

Africa in particular has borne the brunt of uni-sectoral river basin development (Adams, 1992). The Hadejia-Nguru wetlands in northern Nigeria provide an outstanding example of the wide ranging social, economic and environmental effects on wetlands upstream of hydraulic projects promoted without regard to downstream effects (Hollis *et al.*, 1993). In common with other wetlands, they are an area of high agricultural productivity which supports a large human population as a result of periodic inundation from river flooding. The flood waters bring large numbers of young rapidly-growing fish into the wetlands and essential moisture and nutrients for the soil to grow crops. Water which soaks through the soil is not lost, but recharges the underground reservoir which supplies water to wells in the valley and the surrounding Sahelian drylands. Agriculture is at its peak as the flood water recedes, but some soil moisture persists through the dry season to allow a flood recession crop and to stimulate the growth of grasses which provide grazing for cattle. The wetlands also provide an essential all year supply of firewood, timber and other tree products, such as fruit and medicines. The establishment of the Chad Basin National Park shows the importance of the area for wildlife, particularly resident and migratory bird populations. The traditional agricultural uses of the wetland worked in harmony with nature and produced a sustainable system for both the people and wildlife of the region.

However, in recent times, the Hadejia-Nguru wetlands have come under increasing pressure from droughts in the 1970s and 1980s and from rising populations of people and livestock and over exploitation of resources, especially fuelwood. The solution to this was seen as the development of major river engineering schemes including dams to divert the water for intensive cereal cultivation. The retention of flood water behind the dams, has had disastrous effects on the traditional rural economy reducing the inundated area from an average of 2,500 km² in the 1960s to only 470 km² in the wet year of 1975 (Hollis *et al.*, 1993). This small area of flooding still persists because the irrigation scheme which will use water held behind one dam (Challawa Gorge) has not yet been completed and some flood waters are still brought down by the Jama'are River. A computer model of a river basin developed by Adams and Hollis (1988) and Thompson (1993) has been used to demonstrate the catastrophic effect that completion of a new dam on the last remaining tributary will have on the wetland system. In order to evaluate economic implications of these developments to the region, Barbier *et al.* (1991) undertook an economic analysis of the floodplain agriculture, fishing and fuelwood. They then compared this with the returns from a major irrigation scheme, the Kano River Project, which is in direct competition for water with the wetlands. They found that economic returns from the wetland were substantially higher both in terms of crop production and water use.

Ichkeul National Park, Tunisia

Many wetlands have high use-values. The establishment of the Ichkeul National Park in northern Tunisia, its designation as a World Heritage Site, a Biosphere Reserve and Wetland of International Importance under the Ramsar Convention, was based primarily on non-use, existence values of its wetlands deriving largely from their large tracts of *Potamogeton pectinatus* and *Scirpus maritimus*, and the hundreds of thousands of waterfowl wintering there. However, the lake and marshes have important values and functions which include livestock grazing on the marshes, fisheries in the lake, tourism, the role of rivers in groundwater recharge, and the treatment of sewage by the marshes.

The Ichkeul Wetland is threatened by a scheme which will dam all the influent rivers, irrigate parts of the upper catchment and divert large volumes of water out of the basin. As part of an effort to bring more multi-purpose perspectives to water management in the region, Thomas *et al.* (1991) have shown that the economic value of the fisheries and grazing that could be sustained by releasing the minimum 12 10⁶ m³ of water from dams on the various rivers feeding the wetlands, would outweigh the benefits of using that water in agricultural irrigation. In fact, the maintenance of fisheries and grazing would produce a return of 0.049 Tunisian Dinars (TD) per m³ whilst the Ghezala Irrigation scheme will lose 0.063 TD/m³. The sewage treatment function alone was valued at over US\$ 170,000 per year, since this would be the cost of an industrial sewage treatment works, whilst the lake fisheries were valued at US\$650,000 per year.

State Hydraulic Works, Turkey

In Turkey wetland destruction was institutionalised in 1953 by Law 6200 which established the General Directorate of State Hydraulic Works (DSI). Article 2 c requires DSI "to drain swamps" and other sections cover work on flood control, irrigation, hydro-power and navigation: each of which has its direct and indirect effect upon wetlands (DSI, 1953). The Environment Foundation of Turkey (1993) reports that, outside the former Soviet Union, Turkey has the largest expanse of wetlands of

European and Middle Eastern countries. However, 190,000 ha were drained up to 1986, initially for health reasons but more recently the rate of loss has accelerated in an effort to gain new farmland. However, the Environment Foundation of Turkey (1993) reports that "only 35% of the farmland gained as a result of these drainage projects has actually become useful for farming ... and large parts of [the lands unsuitable for farming because of] salinisation, burning of peat and wind erosion ... have rapidly turned barren".

Indus Delta, Pakistan

The mangroves of the Indus River delta wetland in Pakistan have many vital functions. By breaking the force of wind and waves they protect the coast from erosion. Wave height can reach six metres in the open sea beyond the mangroves, but in the sheltered creeks the maximum recorded has been 0.5 metres. Mangroves also stabilise the creek banks which maintains channel width. The creeks are thus self-cleaning and able to maintain their geometry naturally. Without mangroves Port Qasim would need expensive engineering works such as sea walls and constant dredging and thus would not be economical. The mangroves also support extensive fisheries. In 1988, 29,000 tonnes of shrimps were landed, making up 68% of Pakistan's US$100 million fish export. In addition, the delta has a very diverse wildlife ranging from crabs to dolphins and herons and has a high tourist potential. Inhabitants of the delta harvest leaves for cattle fodder which are very nutritious and branches for fuel wood exceeding 18,000 tonnes per year; the 16,000 camels who graze on the mangroves for six months of the year are estimated to eat around 86,000 tonnes of leaves in a season.

The 1990 SPOT satellite image showed around 160,000 ha of mangrove forest now growing on the mudflats. This compares with 263,000 in 1977 estimated from a LANDSAT image. These differences may be partly due to differences in resolution of the images and in interpretation, but there seems little doubt that the mangrove forests are reducing significantly in extent. Although mangroves grow in a basically salt water environment, they are most healthy where there are repeated inflows of fresh water. Average flows from the Indus to the delta used to be around $5,800 \, m^3 \, sec^{-1}$ but the construction of dams and barrages has reduced this to a present flow of only $780 \, m^3 \, sec^{-1}$. Now freshwater from the Indus reaches the seas through only a single creek rather than spreading out through the delta's many distributaries and in dry years the delta may receive almost no freshwater. Recently, a further dam on the Indus, at Kalabagh, has been proposed which may reduce flows further.

THE RENAISSANCE OF WETLANDS

Groundwater recharge

There seems to have been little if any work on the economic valuation of groundwater recharge function of wetlands. The example of Garaet El Haouaria in Tunisia illustrates the adverse social costs of ignoring this function. It covers 3600 ha on the tip of the Cap Bon peninsula and used to have an area of open water every winter which gave it great importance on one of the major bird migration routes between Europe and Africa (Hughes *et al*. 1994). It was used for hunting, grazing and fishing. Drainage freed the Garaet of surface water in the winter by the early 1960s thanks to two drainage canals, each about 4 m deep and flowing directly to the sea. Landless farmers were moved in by the State and well digging and drilling began. SCET (1962) described the groundwater resources of the aquifer and the deeper and thicker Mio-Pliocene aquifer which was recharged largely by percolation from the higher aquifer. Projects for citrus orchards and market gardening with a demand of $7.6 \ 10^6 \ m^3$ per year of irrigation water were launched and small farmers irrigated peppers and tomatoes from shallow wells into the surface aquifer. The water supply for the tomato processing plants and the town of El Haouaria (population 10,000) comes from wells in the deeper aquifer. The volume is adequate but the taste is unpalatable.

No attention seems to have been paid to the function of the original wetland as a recharge area for freshwater moving into the surface aquifer and thence to the deeper aquifer. The rapid drainage of the fresh water from the area by the drainage canals has radically altered the water balance through a reduction in recharge and an increase in abstraction for the irrigation of the dry bed of the Garaet. Groundwater levels began to fall in the area in the early 1970s and accelerated during the 1980s. A fall of 9 m in level during the last 15 years is not uncommon but more importantly a large number of shallow wells have been abandoned because of salinity problems. It would appear that marine intrusions into the surface aquifer are occurring. When wells become saline, the younger farmers or family members tend to leave for work elsewhere, leaving the old people to scratch a living from rain-fed cultivation. Such is the social cost of destroying the groundwater recharge function of wetlands.

Flood control

The most famous and often quoted example of the economic value of wetlands for flood control is the 3,800 ha of mainstream wetlands along the Charles River in Massachusetts. The US Corps of Engineers (1972) estimated that had they been filled, as so many other wetlands in the USA have been until a change of government policy, the increase in annual flood damages downstream would have been US$17 million. The study valued the retained natural wetland at $1,203,000 per year, this being the difference between present annual flood losses based on current land use and the projected flood losses in 1990 if 30% of the wetlands had been lost.

The River Quaggy in south-east London drains a heavily urbanised basin (18.4 km²) (National Rivers Authority, 1988). Severe flooding of property led to a scheme to channelise the river with extensive use of concrete. In spite of substantial advice to avoid such an environmentally damaging scheme, the National Rivers Authority launched the project to a barrage of objections. In addition, funding for the scheme was restricted because its high cost gave it a cost-benefit ratio below the Government's norm. A further tightening of spending guidelines, and the discovery that flood storage in a partially excavated public park with a couple of small restored wetlands would be an effective first step in solving the flood problems, has led to the promotion of the storage option (Table 2). The cost-benefit ratio is now more favourable and it is appreciated that there are additional benefits to fisheries, wildlife, water quality, low flows and amenity. Design work and negotiation with the landowners is now under way.

Table 2 River Quaggy flood alleviation options (National Rivers Authority, 1993)

Scheme	Downstream flow (m³ s⁻¹)	Cost (£ million)	Cost/Benefit Ratio
Original channelisation scheme	20.2	6.23	1.08
All feasible storage sites	15.8	4.92	1.36
Largest storage site only	17.5	6.00	1.12

This more environmentally-friendly scheme for the Quaggy including elements of briefly-flooded football pitches and small restored wetlands is in line with the London Planning Advisory Committee's (1993) view that London's rivers are one of "its greatest but most undervalued resources ... and all proposals for development must ... safeguard the ecological and conservation value of the water, banks and associated open spaces".

The summer 1993 floods in the Mississippi (Parrett, Melcher & James, 1993) have initiated a debate about the appropriateness of a flood control strategy based on ever higher levees. Allen (1993) argued that "increased numbers and heights of levees and massive loss of natural safety valves along the Mississippi, Missouri and Illinois and other rivers were to blame, at least as much as the record rain, for the severity of the flood". Even Time Magazine (July 26th 1993) stated that human activity has "robbed the Mississippi basin of its most precious resource: the wetlands and riparian forests that once absorbed excess rainwater like so many giant sponges". Time (July 26th 1993) quotes Larry Larson, head of Wisconsin's floodplain programme: "we need to start giving land back to the river. If we don't, sooner or later, the river will take it back".

Shoreline stabilisation and erosion control

Costanza *et al.* (1989), in a paper on economic valuation methodology, applied a range of techniques to value the coastal wetlands of Louisiana (Table 3). They found that storm protection was the dominant value, comprising around 80% of the overall value of the coastal wetlands using the "willingness to pay" methodology. Importantly, their alternative "energy analysis-based gross primary productivity conversion" methodology values the wetlands at about four times that derived from the "willingness to pay" approach.

An example of the value of coastal mangrove systems for storm protection is the scheme to protect the new Brisbane airport cited by Hamilton & Snedaker (1984).

Table 3 Summary of wetland values estimates (1983 $) (after Costanza *et al.*, 1989)

Method	Per acre present value ($) at specified discount rate	
	8%	3%
WTP method		
commercial fishery	317	846
trapping	151	401
recreation	46	181
storm protection	1915	7549
TOTAL	2429	8977
Option and existence values	?	?
EA-based GPP conversion	6400 - 10600	17000 - 28200
BEST ESTIMATE	2429 - 6400	8977 - 17000

Some 51,000 mangrove seedlings were planted on a 10 metre wide bench, 12.4 km long, to provide protection from tropical storms at the modest cost of A$228,271. In the less stormy domain of England, English Nature (1992) have argued that "natural systems like sand dunes, mudflats, saltmarshes and shingle bars absorb and dissipate the energy of wind, waves and tide and work as effective sea defences ... man-made defences destroy yet more wildlife habitat and by their own effects fuel an escalating, never-ending struggle to protect the land".

Sediment and toxicant retention
In the heavily urbanised centre of London, an artificial wetland (0.7 ha) has been dug in a school playing field on the floodplain of the Pyl Brook to provide enhanced flood storage because of the rapid runoff from a nearby shopping complex. The planting of the site with reeds and the creation of a deeper water pond near the outfall has been shown to strip many of the pollutants from the inflowing stormwater. Cutbill (1993) found that, on the basis of weekly sampling of the inflow and outflow during the whole of 1991, dissolved oxygen increased by 4.6% and sodium and potassium levels increased marginally. The important pollutants all declined; phosphate by 35.9%, nitrite by 20%, nitrate by 68.9%, turbidity by 55.8%, faecal coliforms by 77.5%, copper by 71.4%, zinc by 36.8%, lead by 6.8% and cadmium by 5.9%.

Nutrient retention
In Sweden, traditional water meadows are being re-established in order to strip nutrients from river water and to reduce the algal blooms which afflict the coastal zone of the Kattegat. It has been shown that the marginal economic cost of removing 1 kg of nitrate nitrogen by the water meadows was 67 cents compared to $33 by a conventional sewage works (Eriksson, 1990). Fleischer, Stibe and Leonardson (1991), also working in south-west Sweden, have shown a linear relationship on log-log paper for nitrogen loading and nitrogen retention on shallow wetlands. They found

a ten fold increase in maximum nitrogen retention when there was harvesting of plant material. They measured maximum rates of denitrification in summer with values of 16 mg N m^2 ha^{-1} for a farm pond and between 4 and 7 mg N m^2 ha^{-1} for farmland wetlands, farm lakes and forested wetlands.

Waste water treatment

The water cleansing function of wetlands has recently attracted great attention as a low cost alternative to industrial sewage treatment in developing countries. An excellent example is the wetland complex east of Calcutta which has been in operation for over 100 years. It receives solid waste from the city and absorbs all the urban sewage and recycles the nutrients to support a highly productive fishery and agricultural land growing vegetables for Calcutta (Ghosh & Sen, 1987). Papyrus swamps are particularly good at absorbing sewage and purifying water, consequently the National Sewage and Water Corporation of Uganda is supporting the conservation of wetlands around Kampala to perform this role (Mafabi and Taylor, 1992).

Table 4   Economic values placed on mangrove systems and mangrove ecosystem products (after Hamilton and Snedaker, 1984)

Resource or product	Location	Date	Value ($/ha/year)
Complete system	Trinidad	1974	500
	Fiji	1976	950 - 1,250
	Puerto Rico	1973	1,550
Forestry	Trinidad	1974	70
	Indonesia	1978	10-20 (charcoal and wood chips)
	Malaysia	1980	25
Fishery	Trinidad	1974	125
	Indonesia	1978	50
	Fiji	1976	640
	Queensland	1976	1,975
	Thailand	1982	30-100 (fish)
			200-2000 (shrimp)
Tourism	Trinidad	1974	200

Fisheries

Maltby (1986) cites two telling examples of the fisheries productivity of coastal wetlands. First, the marshy Nantic River in Connecticut in the USA yields an annual scallop harvest greater than that of prime beef on an equivalent area of grazing land. Second, of the commercially important North Sea populations, 60% of the brown shrimp, 80% of plaice, 50% of sole and nearly all of the herring are dependent on the shallow waters of the Dutch Wadden Sea at some part of their life history. Table 4 gives some examples of the direct economic values placed on mangroves following a literature review by Hamilton and Snedaker (1984).

Agricultural resources

The Organisation pour le Mise en Valeur du Fleuve Senegal (OMVS) manages the high dam at Manantali in Mali. In order to permit a phased agricultural transformation from traditional flood recession sorghum to double cropped irrigated rice it was decided to implement an autumnal artificial flood from Manantali to inundate 50,000 ha of floodplain each year for ten years after the dam's completion. A major planning issue was the "cost" of this artificial flood in hydro-electric power foregone. Gibb *et al.* (1987) claimed that a 7.5 $10^6 m^3$ artificial flood, beginning on September 5th and rising to 2,500 m^3/sec, would "cost" 103 GWh of power which was worth 2.68 10^9 FCFA or 53,600 FCFA/ha of flooded land. Since the yield of sorghum is 34,000 FCFA/ha for a yield of 600 kg/ha (unit price of sorghum of 95 FCFA/kg less the cost of inputs at 22,600 FCFA/ha) it was claimed that the generation of hydro-power was more economic than the sorghum derived from an artificial flood. This analysis was deeply flawed because it failed to recognize the multiple products emerging from the floodplain wetlands.

Salim-Murdoch and Horowitz (1991) showed that the restoration of floodplain inundation through the artificial flood was more valuable than the electricity foregone because of the multiple products of the floodplain ecosystem. Fish production was worth 35,000 FCFA/ha and livestock activities contributed 17,500 FCFA/ha. Even without monetary values for the fuelwood, building timber and underground water supplies, the artificial flood was worth 1.6 times the hydro-power not produced.

Biological diversity

The maintenance of natural and wild wetland areas can have enormous significance for the preservation of genetic diversity. Maltby (1986) reported that in the 1960s, scientists from the International Rice Research Institute in The Philippines screened almost 10,000 varieties looking for a gene giving resistance against grassy stunt virus, a disease then devastating rice crops. Eventually just two seeds of *Oriza nirana*, an Indian wild rice, were found to have the resistant genes. Researchers returned to the wild habitat to collect more but found none with the correct gene. Today, all the world's rice crop has genes from those two wild rice seeds.

ROLE OF WETLANDS IN THE INTEGRATED RIVER BASIN DEVELOPMENT

The concept of integrated river basin planning throughout the world really has its root in the 1930s when the Tennessee Valley Authority (TVA) was formed. However, although widely used as a model, the TVA became a uni-sectoral power generation organisation, with much of the energy coming in recent years from nuclear power, rather than hydro-power stations. In Africa, integrated river basin planning is made more difficult by the fact that all large rivers cross international boundaries. In 1972 the Organisation pour la Mise en Valeur du Fleuve Senegal (OMVS) was established to coordinate the interests of Senegal, Mali and Mauritania in the Senegal River. However, little progress on true integrated management has been achieved. The Manantali dam in the headwaters was constructed firstly to generate hydro-electricity and secondly to control water levels in the river downstream for irrigation. To date little of the planned 250,000 ha of irrigation plots have been constructed, but the dam prevents flooding and thus traditional recession agriculture of 100,000 ha of floodplain cannot be practised. Salem-Murdock and Horowitz (1991) have calculated that the best economic returns from the river basin would result from utilising the wetlands. They showed that 95 per cent of the time (based on data from 1904 to 1984) there would be enough water to generate 74 megawatts of electricity and to release an artificial flood sufficient to inundate 50,000 ha of the floodplain. Unfortunately there are many political barriers to this since electricity benefits the urban élite whilst flooding benefits the rural poor. However, Vincke and Thiaw (*in press*) have shown that rehabilitation of wetlands, including the Djoudj National Park in Senegal, is now an important part of the thinking of OMVS.

Various other international river basin development authorities exist, such as the Chad Basin Authority with representatives from Chad, Cameroon, Nigeria and Niger, but this has rarely met in its 20-year existence and has achieved little as the wetlands in these countries have been destroyed or degraded. Eight of the ten members of the Southern Africa Development Community (SADC) are riparian states of the Zambezi river basin and thus have legitimate claims on its water resources. In a recent workshop (Matiza and Dale, *in press*) it was recognised that wetland functions of flow regulating, groundwater recharge and provision of wildlife habitats were crucial to the well-being of the Zambezi and should be integrated into the implementation of the ZACPLAN as a regional action plan for environmentally sound water resources management. The Mekong Secretariat has existed for many years but the wars and unrest in the countries comprising the catchment area have limited its effectiveness. Dugan (1993) has pointed to twin threats to the river system. Schemes to construct further hydro-electric dams will reduce beneficial flooding, reduce freshwater outflow to the delta and interrupt fish migration. Clearance of swamp forest for firewood, agriculture and fish ponds around the invaluable Ton le Sap Lake (2,500 to 13,000 km²) in Cambodia will harm regional fisheries productivity. Whilst the destruction of the *Melaleuca* forest in the Mekong Delta during the Vietnam War is gradually being reversed, there is an urgent need to begin work to plan the rational and integrated exploitation of the Mekong by the six states that occupy the basin of this sixth longest river in the world (Dugan, 1993).

One of the main problems with advocating integrated development of river basins is that in many countries the responsible authorities are uni-sectoral, with autonomous ministries of water, environment, agriculture, forestry, who rarely meet. The effective integration of wetlands into catchment management requires:

- an inventory of wetlands in a basin, whether extant, degraded and lost;
- an assessment of their potential functions and values;
- recognition via formal classification within land use assessment and planning;
- enhancement of public and political awareness — especially involving young people..

Integrated catchment management requires both a series of technical steps and the creation of an appropriate inter-disciplinary basin-wide authority. Perhaps more importantly, a process of collaboration must be initiated between the "technocrats" of the basin management authority and the people of the catchment.

The technical measures underpinning integrated catchment management are:

- identification and quantification of all water resources and water related resources (**including wetlands**) in the entire basin,
- identification and quantification of all existing demands on water with attention given to **the ecological needs for water in various parts of the basin's hydrological cycle,**
- estimation of future demands under different scenarios including policies to limit or reduce demand,
- assessment of water-related problems, their root causes, and **options for sustainable solutions**,
- working with, and for, the entire community of water users in the basin, including **wetland users and species of flora and fauna**,
- identification of management strategies that maximise benefits, minimise conflicts and profit from the multiple use of water-related resources, including **wetlands,**
- a focus on sustainability, i.e. the maintenance of the natural capital of the river basin, its water stores and transfers, its ecological structures and processes, and long-term options for management including the maintenance and **enhancement of wetlands ,**
- linkage of integrated river basin planning to territorial planning in relationship to urban development, industrial location, forestry, agriculture, recreation.

Such steps for integrated catchment management clearly require appropriate organisational structures with adequate finance and a suitable legislative framework. However, in addition to these bureaucratic procedures, there is a need for **the process of integrated management** to be conducted:

- in an open and flexible manner,
- with a free flow of information between all parties involved.

CONCLUSIONS

Wetlands are buffers, regulators, stores, filters and producers. They are an integral component of the hydrological system which operates within the geomorphological unit of the drainage basin. Existing wetlands need to be conserved because losses worldwide have been so great. In addition, degraded wetlands may merit rehabilitation, "lost" wetlands can be restored and artificial wetlands can be created. They are not merely wastelands, as they have sometimes been portrayed, but a valuable natural resource requiring integrated planning on a catchment basis with appropriate investment of time and money.

REFERENCES

Adams, W.M. 1992. *Wasting the rain: Rivers, people and planning in Africa*. Earthscan, London. 256pp.

Adams, W.M. & Hollis, G.E. 1988. *Hadejia-Nguru Wetlands: Hydrology and sustainable development of a Sahelian floodplain wetland*. Report to IUCN and RSPB, Departments of Geography, University of Cambridge & University College London.

Adamus, P. and Stockwell, L. 1983. *A method for wetland functional assessment*. Vols I and II. Reports FHWA-IP-82-23 and 24 US Department of Transportation, Federal Highway Administration, Washington. 181 and 134pp.

Allen, W.H. 1993. The Great Flood of 1993. *Bioscience*, 43, 732-737.

Baldock, D. 1990. *Agriculture and Habitat Loss in Europe*. World Wide Fund for Nature and European Environmental Bureau, London. 60pp.

Barbier, E.B. 1989. *The economic value of ecosystems: 1 Tropical Wetlands*. Gatekeeper Series No LEEC 89-02, IIED-London Environmental Economics Centre, 15pp.

Barbier, E.B., Adams, W.M. & Kimmage, K. 1991. *Economic valuation of wetland benefits: the Hadejia-Jama' are floodplain, Nigeria*. London Environmental Economics Centre Paper DP 91-02. International Institute for Environment and Development, London. 26 pp.

Costanza, R., Farber, S.C. & Maxwell, J. 1989. Valuation and management of wetland ecosystems. *Ecological Economics*, 1, 335-361.

Cutbill, L.B. 1993 Urban stormwater treatment by artificial wetlands: A case study. In: *Proceedings of the 6th International conference on Urban Storm Drainage*, Niagara Falls, Canada. Volume II, p1068-1073.

Davies, J. and Claridge, G. 1993. *Wetland Benefits: The Potential for Wetlands to Support and Maintain Development*. Asian Wetland Bureau Publication No. 87, IWRB Special Publication No. 27, Wetlands for the Americas Publication No. 11. AWB, Kuala Lumpur; IWRB, Slimbridge, UK; WA, Manomet, Mass., USA. 45pp.

Directorate of State Hydraulic Works (DSI) 1953. *Law concerning the organization and duties of the General Directorate of State Hydraulic Works - Law 6200*: Date Passed 18th December 1953, Date Published 25th December 1953. 3pp.

Dugan, P.J. (Ed.) 1990. *Wetland Conservation: Review of Current Issues and Required Action*. IUCN-World Conservation Union Publication, Gland, Switzerland, 96pp.

Dugan, P. (Ed.) 1993. *Wetlands in Danger*. Mitchell Beazley, London. 192pp.

English Nature 1992. *Coastal Zone Conservation: English Nature's rationale, objectives and practical recommendations*. English Nature, Peterborough. 20pp.

Environment Canada 1991. *The Federal Policy on Wetland Conservation*. Environment Canada, Ottawa, 14pp.

Environment Foundation of Turkey 1993. *Wetlands of Turkey*. Environment Foundation of Turkey, Ankara. 398pp in bi-lingual edition.

Eriksson, S. 1990. Re-creation of wetlands for nitrogen retention. In: Ramsar Bureau, *Proc. fourth Conference of the Contracting Parties to the Ramsar Convention*, Montreux, Switzerland, Volume II Conference Workshops. Ramsar Bureau, Gland, Switzerland, p 178-180.

Farber, S. & Costanza, R. 1987 The Economic value of wetland systems. *J. Env. Manag.* 24, 41-51.

Fleischer, S., Stibe, L. & Leonardson, L. 1991. Restoration of wetlands as a means of reducing nitrogen transport to coastal waters. *Ambio* **20**, 271-272.

Ghosh, D. & Sen, S. 1987. Ecological history of Calcutta's wetland conservation. *Environmental Conservation* **14**, 219-226.

Gibb and Partners, Electricité de France and Euroconsult 1987. *Etude de la Gestion des Ouvrages Communs de l' OMVS*. 4 volumes.

Hamilton, L.S. & Snedaker, S. 1984. *Handbook for Mangrove Area Management*. IUCN/MAB/UNESCO, 123pp.

Hollis, G.E., Adams, W.M. & Aminu-Kano, M. (eds) 1993. *The Hadejia-Nguru Wetlands: environment economy and sustainable development of a Sahelian floodplain wetland*, IUCN Gland Switzerland and Cambridge, UK. 244 pp.

Hughes, J., Maamouri, F, Hollis, G.E. & Avis, C. 1994. *A preliminary inventory of Tunisian wetlands*. Department of Geography, University College London. 640pp.

London Planning Advisory Committee 1993. *Draft 1993 Advice on Strategic Planning Guidance for London: For Consultation*, June 1993. LPAC, London.

Maamouri, F. & Hughes, J. 1992. Prospects for wetlands and waterfowl in Tunisia. In: Finalyson, M., Hollis, G.E. and Davis, T. (eds.) *Managing Mediterranean Wetlands and their Birds*, IWRB Special Publication 20, Slimbridge, England, p47-52.

Mafabi, P. & Taylor, A.R.D. 1993. The National Wetlands Programme, Uganda. In: Davis, T.J. (Ed.) *Towards the wise use of wetlands*. Wise Use Project, Ramsar Convention Bureau, Gland, Switzerland. p52-63.

Maltby, E. 1986. *Waterlogged Wealth*. Earthscan, London, 200pp.

Matiza, M. & Dale, P. In press. *Zambezi basin water projects: Proceedings of the SADC/IUCN workshop held at Kasane, Botswana. 28 April-2 May 1993*. IUCN, Gland Switzerland.

Mitsch, W.J. & Gosselink, J.G 1993. *Wetlands*. 2nd Edition. Reinhold, New York. 722pp.

Montes, C. & Bifani, P. 1991. Spain. In: Turner, K. and Jones, T. (Eds.) *Wetlands: Market and Intervention Failures - Four Case Studies*. Earthscan, London. p144-195.

National Rivers Authority 1988. *Ravensbourne Catchment study: Quaggy River - Clarendon Rise to Kidbrooke Park Road - Implementation Report*. NRA London Appraisal Team, London. 43pp.

National Rivers Authority 1993. *Quaggy River Project Liaison Group: 17th June 1993*. National Rivers Authority, London. Mimeo.

Parrett, C., Melcher, N.B. & James, R.W. 1993. Flood discharges in the Upper Mississippi River Basin, 1993. *US Geological Survey Circular 1120-A*, 8pp.

Pearce, D.W. & Turner, R.K. 1990. A case study of wetlands. Chapter 21 in *Economics of natural resources and environment*, Harvester-Wheatsheaf, London, p320-341.

Salem-Murdock, M. & Horowitz, M. 1991. Monitoring development in the Senegal river basin. *Development Anthropology Network* 9, 1, 8-15.

SCET 1962. *Ressources en eaux souterraines de la Region d' El Haouaria: Estimation du Bilan*. Direction des Ressources en Eau Library, Tunis, Mimeo. 9pp + maps.

Scott, D. & Poole, C. 1989. *A status overview of Asian Wetlands*. Asian Wetland Bureau Publication 53, Kuala Lumpur, Malaysia. 140pp.

Thomas, D.H.L., Ayache, F. & Hollis, G.E. 1991. Use and non-use values in the conservation of Ichkeul National Park, Tunisia. *Environmental Conservation* 18, 2, 119-130.

Thompson, J.R. 1993. *Hydrological model of the Hadejia-Nguru wetlands, Northern Nigeria*. Wetlands Research Unit, Department of Geography, University College, London. 27pp.

US Corps of Engineers 1972. Cited in Sather, J.M. and Smith, S.D. *An overview of major wetland functions and values*. Report to US Fish and Wildlife Service, FWS/OBS-84/18, 1984.

Vincke, P.P. & Thiaw, I. In press. Water resources management and protected areas the case of the Senegal River. In: Acreman, M.C., Pirot, J.-Y. & Dugan, P.J. (Eds) *Proceedings of the Workshop on Protected Areas and the Hydrological Cycle - IV World Congress on National Parks and Protected Areas, Caracas, Venezuela. 10-21 February 1992*. IUCN, Gland Switzerland.

32 The Green River in Bangladesh: the Lower Atrai Basin

T R FRANKS

Development and Project Planning Centre, University of Bradford, UK

THE NORTH WEST REGION OF BANGLADESH

The North West region of Bangladesh lies to the west of the Brahmaputra and the north of the Ganges (Figure 1). The geomorphology of the region is set by its deltaic nature: it is extremely flat, varying in elevation from 10 m above sea level in the south, to about 30 m in the north where it meets the Himalayan piedmont fan. Within the region there are a number of medium and small rivers, most of which drain south and then east into the Brahmaputra, as shown in Figure 1.

The hydrological regime over most of the southern part of the region is determined by the river Atrai. This rises in India, enters Bangladesh in the far north west of the country, then flows back into India before re-entering Bangladesh in the Lower Atrai basin. Here it collects flows from several tributaries, notably the Little Jamuna and the Bangali, before discharging to the Brahmaputra through the Hurasagar. River slopes vary from 1:3000 upstream to 1:17 000 downstream.

The Hurasagar forms an important drainage constraint, due partly to lack of capacity but more particularly due to high levels in the Brahmaputra. During the monsoon, when flows in the Atrai and its tributaries are large and drainage requirements greatest, levels in the Brahmaputra are also at their peak, resulting in backing up in the Atrai, and, in turn, extensive flooding in the Lower Atrai basin.

In earlier times the regular deep and persistent flooding led to the development of large perennial wetlands, called the Chalan Beel. Changing natural and anthropogenic conditions have reduced the size of the Chalan Beel over the years but there are still significant parts which remain flooded throughout the year and a very large area which is subject to seasonal flooding.

INFRASTRUCTURE DEVELOPMENT IN THE REGION

There is evidence of efforts towards hydraulic control in the region over the past few centuries. However, the main boost to development came in the 1960s with the preparation of a master plan for water resources development for the whole country. Out of this came the Chalan Beel projects (EPWAPDA, 1970), which were a series

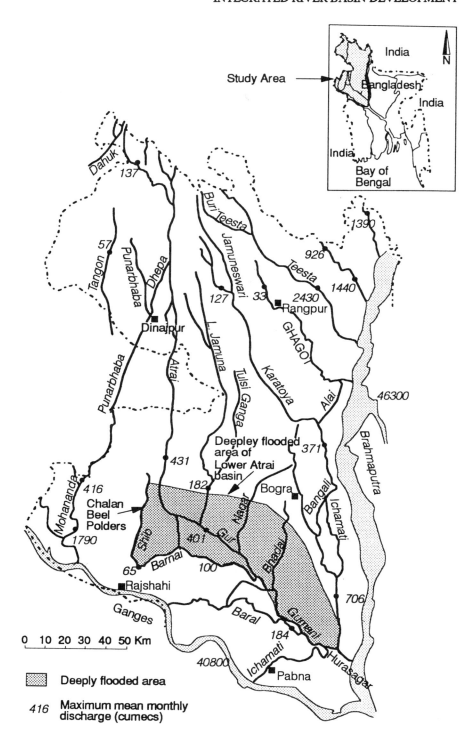

Figure 1   The North West region of Bangladesh

of flood protection polders along the right bank of the Lower Atrai. The main objectives of these projects were the provision of flood control and irrigation to increase agricultural production. The basin was to be divided into a number of large areas which would be completely surrounded by a protective embankment. This would exclude unwanted flood waters, and allow the close control of water levels which is required for the growing of transplanted rice: regulators in the embankment would allow drainage and inflows for irrigation as required.

The Chalan Beel polders were conceived as a number of independent developments, and there was little attempt to assess the impacts of the developments on adjacent areas upstream and downstream. However, it was noted that "leaving (the area of flood-plain on the opposite bank) open to flood flows...would substantially reduce the height of embankment required for the Chalan Beel project" (EPWAPDA, 1970, p. VII-4). Planning at this time was thus prepared to see benefits gained in one area at the expense of deeper flooding in another area.

Failure to take account of the impacts of flood protection measures on adjacent areas has resulted in disappointing performance from most of the developments that have been undertaken in the Lower Atrai. Practically all the embankments are now breached or, more commonly, cut by local people. The frequency of cutting has increased dramatically in the past few years: a recent study of the four Chalan Beel polders recorded that there have been over 100 cuts in their embankments since 1988 (BWDB, 1992). This indicates that local people, and particularly farmers, now have a very low tolerance to the damaging effect of increased depths of flooding if they perceive it to be caused by an embankment which is protecting other people. It is said that a head difference of as little as 0.3 m is sufficient to lead to public cuts by those on the flooded side. Underlying these problems is the fact that one person's flood control is another person's flood. In rural Bangladesh, prosperity is directly linked to agricultural production through control of water levels. Developments intended principally or solely for flood control are therefore bound to face social and political problems.

THE FLOOD ACTION PLAN FOR BANGLADESH

Water resources planning in Bangladesh has recently received considerable prominence under the Flood Action Plan (FAP). The origins of FAP lie in the severe floods of 1987 and 1988. In consecutive years large areas were deeply flooded, considerable damage was done and, particularly in 1988, many lives were lost. This resulted in a number of separate efforts by the international donor community, in conjunction with the Government of Bangladesh, to study the causes of floods and possible solutions for dealing with them. Finally, in 1989, UNDP prepared the Flood Policy Study (UNDP, 1989) which integrated many elements of the previous studies into the Flood Action Plan. This took the form of regional studies covering most of the country, specific local studies such as protection for Dhaka and other major towns, and supporting studies on related aspects. At its start, FAP emphasised the identification of the causes and solutions to flooding because of its origins in the floods of 1987 and 1988. It was primarily concerned with water resource planning during the monsoon season, and tended to look towards structural solutions to the problems. Thus, embankments were considered along many of the larger rivers, and drainage and large hydraulic structures also figured prominently in many of the suggested solutions. Although the difficulties of providing full protection from floods in a country like Bangladesh was recognised, nevertheless the driving force continued to be, at that time, the perceived need to protect agricultural (rice) production to the greatest extent possible.

In spite of its emphasis at the start on structural solutions, FAP always took a broad view of the impacts and scope of flooding. Supporting studies undertaken included an evaluation of past flood protection schemes, studies of the social impacts of floods, studies of the environment and fisheries, mapping, river surveys, pilot river training schemes, and hydraulic modelling. The supporting studies were not well synchronised with the regional studies (many of which came to an end before results from the supporting studies were available), but nevertheless had an important influence on a change in approach to FAP as it progressed. The evaluation of past schemes showed that many of them had not performed as well as had been hoped in boosting agricultural production, for reasons which were as much social as technical. In addition, considerable prominence was given to environmental aspects, and in particular the growing importance of the fisheries sector, so that it became clear that a single over-riding objective of increased rice production had to be modified to take account of other needs and priorities. This change in direction was much assisted by non-governmental organisations, both in Bangladesh and overseas: these highlighted FAP as one in the succession of major water projects (of which the Narmada project in India is the most well-known) which had apparently serious environmental disadvantages and they strongly criticised it for proposing major structural interventions to control floods. "Living with floods" was suggested as the viable alternative, and more appropriate for a poor and densely-populated country like Bangladesh. Influences both within and outside FAP therefore combined to shift its approach to one which relied less on major engineering, (though this was not entirely discounted) and more on an integrated approach which took into account the needs of other sectors and the environment. Subsequent FAP studies therefore became concerned more with integrated natural resource (land and water) planning over the whole year, rather than concentrating primarily on control of flood waters during the monsoon.

THE GREEN RIVER PROPOSALS FOR THE LOWER ATRAI

The North West region was included within the original FAP plans as the FAP2 study (Mott MacDonald, 1993). The Lower Atrai basin was known to be a key component of the hydrology of the region, for the reasons discussed above. Therefore the original terms of reference for FAP2 required the study of certain very large engineering interventions. Two of these were based on the idea of massive drains, with capacities of the order of 1000 cumecs, which would divert a significant portion of the flow away from the basin and channel it direct to either the Brahmaputra or the Ganges. A third alternative suggested was a large regulator, at the confluence of the Atrai/Hurasagar and the Brahmaputra. All of these alternatives were designed to reduce the depth of flooding in the basin and to allow much greater control of flood water. As for other similar developments earlier, the main benefits would be seen in increased rice production and reduced crop and non-crop damage. The first part of the FAP2 study showed all these proposals to be non-viable. They would be very costly (around $500 millions at 1991 prices) and only partially effective because the flooding over the downstream part of the basin is due to backwater effects from the Brahmaputra, rather than flows entering the basin from upstream.

The changes in the overall direction of FAP were also reflected in changes of direction within FAP2, as the non-viability of major structural options for the Lower Atrai became clear. Instead, planning options began to crystallise round a concept which balanced the needs of the different sectors in the river basin in a manner which would be feasible and practicable, and which would not result in the extensive public cutting that had been experienced hitherto. This concept became known as the "green river" and is shown in Figure 2. The basic principle of the green river is

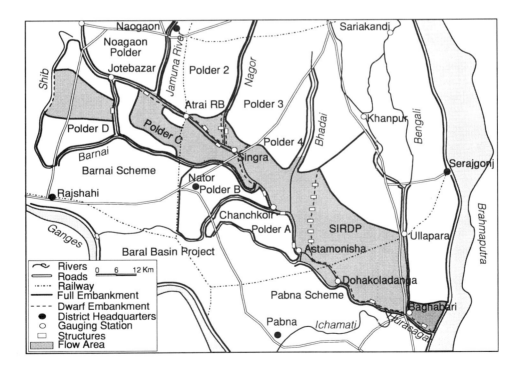

Figure 2   The Lower Atrai basin: the "green river"

not to attempt to confine all flood waters within the low flow course of the river by means of embankments (thus excluding flood waters from the surrounding flood plain), but to allow some overbank spillage and flow during the peak monsoon. In a densely-populated rural area like the Lower Atrai, it is not possible to exclude people or agricultural production from the flow area. The principle for these areas is therefore to provide sufficient protection from the early season flood to enable farmers to harvest the dry-season irrigated rice crop, and then to allow overland flow and flooding to take place. In certain parts of the flow areas, depending on the conditions and the type of protection provided, it may be possible to cultivate a lower-value broadcast monsoon rice crop. In others, no agricultural production would be possible during the monsoon (though flood-plain fisheries provide an important alternative production option). Outside the flow areas of the green river, the possibility of providing full flood protection exists, though here as elsewhere the benefits of doing this have to be balanced with the needs and priorities of other sectors such as fisheries, groundwater recharge for irrigation, and the environment.

Although much planning under FAP had been primarily concerned with the productive sectors of agriculture and, to a lesser extent, fisheries, it was always recognised that the safety and convenience of the human population was also an important factor. This had been addressed through the concept of "flood proofing", which comprised both structural and institutional measures to improve people's abilities to "live with floods". It was recognised that the principle of the green river would require that some part of the local population would live in areas where flooding was expected during the monsoon (though it was also thought that this was preferable to the uncertainty of living within an area which enjoyed theoretical protection but which was susceptible to sudden rapid rises of water level when breaches or cuts took place). For the people living within the flow areas, flood proofing would be an important measure to improve their quality of life.

THE RATIONALE BEHIND THE GREEN RIVER CONCEPT

A number of changed priorities have helped to influence alternatives to a policy of full flood protection in the lower Atrai basin. One such change is in the political and developmental situation. In the year following the formulation of FAP, the Government of Bangladesh changed from a military autocracy to a parliamentary democracy. This transformation was reflected in the increasing emphasis being put on popular participation in the process of development, and the need to modify previous "top-down" planning approaches. "Top-down" had naturally tended to support traditional developments aimed at full flood control, since it strengthened the alliance between the articulate and respected engineering profession, and influential landowners who stood to gain most from interventions which increased agricultural productivity. The widening of the process to include poorer groups, such as fishing communities, landless and women weakened this approach: fishing communities in particular were vulnerable to loss of income and often livelihood as a result of flood control.

Another factor which began to be increasingly considered was risk and hazard assessment. Observation and calculation had indicated the widespread damage and losses attendant on a cut in the upper polders in the Lower Atrai. It was pointed out that embankments share many of the characteristics of small dams and should therefore be subject to the same rigorous scrutiny for safety and the impacts of failure. Both types of development may result in increasing development for as long as the protection is secure: losses may then be all the greater if failure does occur. An additional factor to be taken into account is that Bangladesh lies in an area of significant seismological risk. Historical accounts describe dramatic changes in river courses as a result of earthquakes, with consequent devastation downstream. Under present and foreseeable future conditions it is necessary to try to ensure that failure of flood protection infrastructure under seismic shock would not result in unnecessarily high damage: in this respect the green river concept is considered to be a lower risk strategy than a policy of full confinement.

The current situation with regard to food supply and security in Bangladesh has also resulted in changes in planning policy. Bangladesh is a rice country: like others in south and south-east Asia, it has been very successful in boosting productions mainly through the irrigated dry-season crop of high-yielding varieties, to the extent that it is now more or less self-sufficient in its staple food. The last few seasons have not witnessed any unduly harsh climatic conditions and harvests have been good, with the result that farm-gate prices have been falling. One side-effect of this is the increasing emphasis being put on diversification out of rice, though natural conditions and markets make the identification of alternative crops difficult. As rice prices drop, returns to monsoon flood protection aimed at boosting rice production also naturally fall. At the same time fish production has been increasing in importance, partly because its nutritional importance has been recognised, and partly because flood plain fisheries have been an important source of food for the rural poor who have no land.

Other benefits of a less restrictive policy of flood control that have been noted include increased groundwater recharge. Since it is irrigation from groundwater that has boosted production of the dry season crop to the extent that it is the main dependable crop for most farmers, efforts should be made to maintain or increase this resource to the extent possible. There is already evidence that groundwater supplies are beginning to run short towards the end of the dry season in some areas. This is mainly due to increased abstractions as groundwater irrigation extends but any exacerbation of this effect by increased flood control is clearly undesirable, and must be accounted an important disadvantage.

Likewise, access to areas on the flood plains by means of small and medium-scale country boats can also be considered a significant potential impact of the green river strategy. Inland navigation in Bangladesh suffered a decline over the past few years as sail-powered boats were unable to compete with the speed and convenience of trucks. This trend may, however, be reversed as more and more boats are equipped with cheap diesel engines (which are also used to power the irrigation pumps) which can be readily acquired and maintained by local rural people.

Many of these impacts can be subsumed within the general heading of environmental impacts, if the environment is very widely defined to include both the bio-physical and socio-cultural systems. There are other effects of the flooding regime within the basin which lie specifically within the bio-physical environment, and also need to be taken into account in its planning. In view of the complexity of this environment, there are potentially a very large number of these impacts. Key ones relate to the role and importance of perennially flooded wetlands, both for the local eco-system of the river basin and internationally, the loss of biodiversity as agricultural production is increasingly concentrated on a few high-yielding varieties, and the impact on soil and fertility of reduced silt deposits brought down by floods. It should be noted that all these issues, and many others in the biophysical environment, remain very contentious, both in relation to the scientific data that supports them and to the moral and ethical judgements that are involved. Nevertheless their importance in influencing the overall planning process can not be underestimated.

The consideration of the variety of factors such as these leads inescapably to the conclusion that planning for a river basin like the Lower Atrai must take an integrated view, balancing the needs of people for convenience and safety, agricultural production, fisheries and aquatic resources, the needs for navigation, and other environmental concerns. This is done partly through the use of a multi-criteria decision matrix which ranks projects on the basis of a number of criteria which includes, but is not limited to, economic returns. In the case of the planning for the Lower Atrai, these criteria included socio-economic indicators such as impacts on rice and fish production, agricultural and construction employment, impacts on the biophysical environment, and other factors such as institutional complexity and susceptibility to risk and hazard.

COMPARATIVE EXPERIENCE AND CURRENT DEVELOPMENT IN EUROPE

In conclusion, it is instructive to compare European experience with that in Bangladesh and to look at some interesting developments in current river basin policy in Europe. Both along the Rhine in Germany, and in Holland, plans to change river basin management policies are under discussion. As with Bangladesh, these reflect changing priorities as development takes place.

The Green River was a convenient short title given to a new concept in planning for the Lower Atrai in Bangladesh. The concept has been used over many years in Europe. Here, however, it refers to an unimpeded floodway, which allows over-bank flood flows to take place without constraint. While the basic principle of over-bank flood flows is common to the green rivers of Europe and Bangladesh, in the case of the latter there are important differences. These are due to the densely-populated and intensively-cultivated nature of the landscape, which means that it is not possible to set aside significant areas to act as floodway, without allowing for alternative uses at the same time. Most importantly this means allowing for hydraulic structures for the protection of property (and perhaps agriculture), and particularly physical infrastructure. In the Lower Atrai basin there are two major rail crossings, one major road crossing, and innumerable smaller sealed and dirt village roads, nearly all of which run on embankments to provide some degree of protection from floods. All these embankments have to be allowed for in the green river, a fact which very much complicates both its planning and its management.

Turning now to Europe, the Upper Rhine has undergone far-reaching changes over the last 170 years (Dister *et al.*, 1990), starting with the "rectification of the Upper Rhine" by Tulla in the nineteenth century. The main objective of the rectification programme was to protect human lives and settlement from flood waters and to gain land for agriculture (a direct comparison with the objectives of the recent pattern of flood control works in Bangladesh). This programme was successful in achieving its objectives but was accompanied by major morphological changes, leading to drainage of large areas of former flood plains and a drastic drop in groundwater levels. A further phase of development was initiated in the mid twentieth century with a programme of hydropower development; this again was successful in its own terms but was also accompanied by impacts on hydrology, morphology, ecology and navigation. In particular increased flood peaks and speed of propagation have increased the risk of flooding downstream, leading to the requirement that peak flows be reduced to the values obtaining in the 1950s (Larsen, 1993) through the requirement of the re-establishment of a retention capacity. A plan to provide this retention capacity has been put forward through the integrated Rhine programme, under the concept of ecological flooding. This concept tries to find a compromise between the hydrological and ecological requirements of flooding, and provides a good parallel with the Bangladesh situation in developing an integrated approach to the sometimes conflicting requirements of different sectors.

The situation in Holland is slightly different. Over the centuries, the Dutch have been extremely successful in reclaiming land from below sea level, through a process of empoldering, drainage and pumping. This has resulted in the development of a very intensive agriculture, but the cost has been high and there have been other environmental effects, such as lowering of ground levels as the soil compresses, and the leaching of agrochemicals into the drainage water. Because of concerns about the cost and the environmental damage attributed to drainage, the Dutch Government has developed a plan to return about 250 000 ha acres to lakes, wetlands and forests over time, through a programme of controlled flooding, and the linking of several small lakes into larger ones through the breaking of embankments (Simon M, 1993).

The driving force in these different situations is the change in the balance between economic and environmental considerations as development takes place, and the consequent need to take an integrated view of river basin planning and management. It seems likely that similar changes will take place elsewhere in the world as development priorities change.

REFERENCES

Bangladesh Water Development Board (BWDB), 1992. *The Assessment and Hydrological Studies of Chalan Beel Polders A,B,C and D*.

EPWAPDA, 1970. *Chalan Beel Project Feasibility Study*. East Pakistan Water and Power Development Authority.

Dister, E. *et al.*, 1990. Water management and ecological perspectives of the Upper Rhine's floodplain, *Regulated Rivers: Research and Management* **5**, 1-15.

Larsen, P. 1993. The "Integrated Rhine Programme": a contribution to an improved ecology at the Upper Rhine. *Rehabilitation of the River Rhine Conf. Proc.* Institute for Inland Water Management and Waster Water Treatment, The Netherlands.

Mott Macdonald, 1993. The north west regional study (FAP2): final report.

Simon, M. 1993. A Dutch reversal: letting the seas back in. *New York Times*, 7th March, 1993.

UNDP, 1989. *Bangladesh Flood Policy Study*. United Nations Development Programme.

33 The Medway River Project: an example of community participation in integrated river management

BRIAN J SMITH

Medway River Project, Maidstone, Kent, UK

Perceptions of rivers range from the totally utilitarian to the purely aesthetic. But whatever view we take there can be no doubt that rivers and their associated landscapes play a vital role in the quality of everyone's life. Throughout the 1970s and 1980s concern over the loss of landscape and amenity resources increased dramatically. In respect of the water environment this has not been solely in reaction to the capital works schemes that have canalised rivers and denuded many miles of river bank. Of equal importance has been the gradual loss of flood meadows, pollard willows and riverside footpaths, and the damage caused by increasing recreation pressures.

The need to develop an integrated approach to river basin management is widely accepted: "Uncoordinated use....is at best inefficient and at worst extremely damaging"; "We [must] formulate acceptable strategies for the integrated development of river basins."; "We need to manage rivers, to restore degraded ecosystems ... [so] that they can sustain themselves naturally" — all phrases used in the *Call for Papers* for this Conference.

"We must..." and "We need to.." Perhaps the most fundamental issue that needs to be addressed is who are "We"? Who is to participate in the management process?

Countryside Management Projects (CMP) are responsible for promoting community awareness of, and participation in, the management of the local environment. The river environment is a public resource and the active participation of the general public in its management must be seen as a primary objective. The purpose of this paper is to show by example that a positive and proactive partnership with the local community is a key element in the integrated river basin management process.

BARRIERS TO EFFECTIVE COMMUNITY PARTICIPATION

One factor that may severely limit the effectiveness of public involvement in the management process is the misconception, on both sides, that their concerns and aspirations are not recognised or properly understood.

The term "General Public" is often used by those in authority to define everybody else. It implies, indeed states, an "Us and Them" attitude, in which "They" become the General Public and "We" are the professionals and managing authorities.

Criticism from the "General Public" is often emotively based and may be rejected as showing a lack of understanding of the "real" issues. However, to the General Public, the loss of their favourite pollard or shingle beach *is* a very real issue.

Inevitably such reactions create a barrier to effective working relationships and lead to a sense of frustration amongst communities, who feel unable to influence the management decisions which determine the quality of their river environment.

If "General Public" is replaced with Local Community, "They" become landowners, river users and individuals with a direct and personal association with the river. Most importantly, professionals will wish to see themselves as part of that community and mutual understanding may then become easier to achieve.

THE MEDWAY RIVER PROJECT: ORIGINS AND BACKGROUND

Established in March 1988, the Medway River Project (MRP) resulted from community criticism and has become the catalyst for community action.

The Medway valley is predominantly rural, with a resident population of around 400 000, concentrated in Tonbridge, Maidstone, Rochester and Chatham (see Figure 1). About 4.97 million people live within a one-hour drive of the Medway (Grant, 1986) and many communities in south and east London have strong cultural links with the river. The recreational pressure from these communities is immense and is a strong base for opposition to changes which may be to the detriment of their river amenity.

The most accessible section of the Medway is the Navigation, which extends for 28 km (17.5 miles) from Tonbridge to Maidstone (Figure 1). The navigation was created in the 1740s to enhance commercial trade between the Weald and coastal ports of Kent, today it serves the combined purposes of recreation, flood defence and water resources. It is used by approximately 2000 boats every year, over 75,000 anglers and countless thousands of canoeists, rowers and walkers. Situated at the heart of the Medway catchment the Navigation is also a vital link in the chain of aquatic and river edge habitats that enable flora and fauna to migrate freely throughout the catchment. Inevitably such intense recreation pressures lead to conflicts of interest.

In 1979 the Southern Water Authority (SWA) were faced with the need for substantial capital investment to prevent the imminent collapse of the Oakweir lock gates. With no realistic prospect of gaining an economic return on their capital, and ever-increasing maintenance costs, SWA attempted to close the upper 5.2 km (4.5 miles) of the Navigation. Public opposition was overwhelming, not only from boaters and anglers, but also from parish councils and individuals concerned at the loss of the river amenity. SWA withdrew their proposal and invested in a ten-year programme of lock gate renewal and navigation improvements. But they failed to

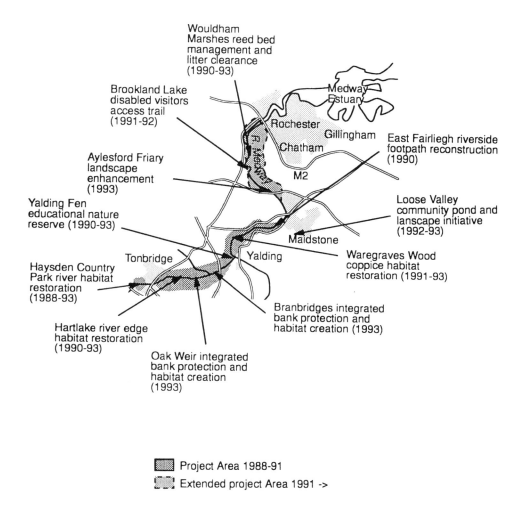

Figure 1 The Medway catchment and project area

understand the breadth of community concern: by focusing solely on the Navigation, they provoked criticism over their neglect of other amenities, particularly access and landscape maintenance.

In 1986, in partnership with Kent County, Tonbridge and Malling Borough and Maidstone Borough Councils, SWA appointed consultants to evaluate the overall needs and opportunities for recreation on the Medway. A report "The Medway Navigation: an assessment of leisure and tourism potential" (Grant, 1986) recommended that "the Navigation would benefit from a consistent river management policy being implemented". It also advocated that "a second major improvement would be to adopt the same approach with regard to implementing a consistent environmental and landscaping programme".

Grant's final recommendation was that the emphasis should be placed on low key enhancements which reflect the interests and concerns of the local community and that they should be coordinated by a Project Officer. The Medway River Project was established in March 1988 when a full time Project Officer was appointed. It is sponsored by the National Rivers Authority (NRA) Southern Region (originally SWA), the Countryside Commission, Kent County, Tonbridge & Malling Borough and Maidstone Borough Councils.

PROJECT AIM, OBJECTIVES AND INITIAL ACHIEVEMENTS

The aim of the Medway River Project, set out in the Memorandum of Agreement (Countryside Commission, 1987), is to undertake a programme of improvements and initiatives to enhance the environment and develop the potential of the Medway corridor between Allington [Maidstone] and Tonbridge for informal recreation (Figure 2).

Specific objectives include activities which:

● Manage and enhance the landscape and wildlife of the Medway
● Maintain and enhance the access and recreational use of the Medway
● Promote local community awareness of, and involvement in the enhancement of the Medway's environment
● Encourage landowners to take a positive role in enhancing the Medway and the surrounding countryside.

The first two of these objectives are the basis for management actions, the second two the means by which they can most effectively be achieved.

The first year of the Project was spent establishing links within the local community, organising small scale initiatives and demonstrating the scope for community participation. During the year volunteers from ten community groups contributed 483 days of work to Project-led schemes, encompassing both landscape and amenity enhancements. Landscape improvements included tree planting, restoration of hedgerows through traditional laying, and pond clearance. Access enhancements included replacing stiles with kissing gates to improve access for less able walkers, waymarking and bridge replacement. Some minor bank restoration schemes brought community volunteers into working contact with SWA river staff. This proved to be a novel and enlightening experience for all concerned!

The achievements of the first year demonstrated to the Project sponsors that close links could very quickly be built with the local community. It became clear that people wanted to play a positive role in the management of their river environment, and that communities have a very strong sense of association with their local river. To the local community the Project's first year showed that there were ways in which they could make a positive contribution. Local people quickly came to see the Medway River Project as "their" project and a personal link into the otherwise

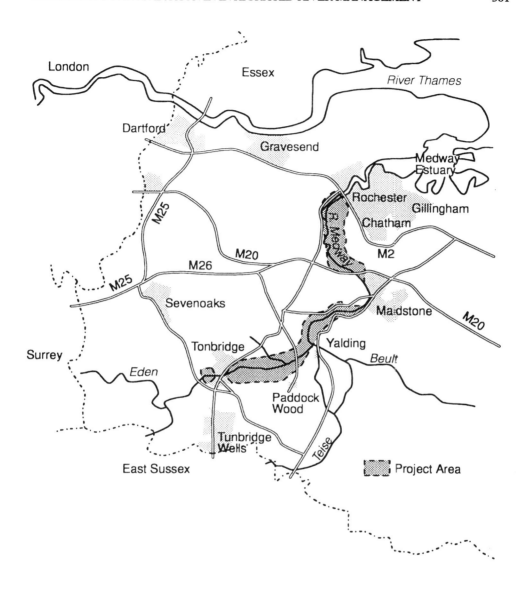

Figure 2 Key management sites in the Medway River Project area

inaccessible authorities.

In 1989 the Project, in partnership with the Kent Trust for Nature Conservation (KTNC), carried out a comprehensive habitat survey of the Medway Navigation. The survey report; "The Medway Navigation, a landscape, habitat and amenity survey", MRP/KTNC (1989) was based on the NCC phase one format, enabling the Project to identify the priority sites for conservation or enhancement management.

In addition to providing a concise evaluation of the Medway's environmental resources, the survey generated considerable interest and support from local industry and led to one of the Project's most exciting and rewarding partnerships.

INTEGRATING HABITAT MANAGEMENT AND ENVIRONMENTAL EDUCATION

Yalding Fen (see Figure 2) was highlighted by the survey as a complex of fen/ grassland, willow carr and scrub habitats that was unique on the Medway. Adjacent to the five-hectare (12-acre) fen is a four-hectare (ten-acre) traditional apple orchard which by virtue of its age and lack of chemical treatment, was also of high wildlife value. Both sites were owned by ICI (Zeneca) Agrochemicals.

To the Project these sites had immense potential for both habitat/landscape conservation and environmental education. The Fen contains a wide range of flora but is dominated by Reed Sweet Grass *(Glyceria maxima)*, resulting in a gradual loss of diversity. Situated less than 16 km (10 miles) from both Maidstone and Tonbridge, the site was easily accessible to 200 primary schools and 120 secondary schools, giving almost unrivalled potential as an educational resource for local schools.

To ICI the site was valued as a buffer zone between the chemical plant and the general public. ICI made no attempt to manage the sites and were unaware of its ecological importance. The MRP approached ICI with proposals to prepare and implement a management plan for the site which integrated conservation and educational objectives. These proposals were enthusiastically received by ICI, a sponsor of the original 1989 survey, and the development of the Yalding Fen Educational Nature Reserve began in 1990.

Conservation management has centred on restoring water levels throughout the fen, reintroducing grazing and bringing the overgrown fruit trees back into a healthy state. Access has been created though a network of boardwalks, waymarked trails, dipping platforms and bridges.

Educational support is provided through the Study Pack (Figure 3) prepared by KTNC and a team of enthusiastic voluntary fen wardens. The study pack and field equipment enable pupils to explore ecological themes such as camouflage, food chains, predator/prey relationships and seasonal cycles. Since it opened to schools in summer 1992 over 600 pupils have visited the reserve.

The success and value of the reserve is manifold, including:

- Enabling the Project to undertake, and demonstrate the viability of, integrated habitat and access management;
- Creating a community resource which enables the Project to promote greater awareness and appreciation of the Medway's natural environment;
- Encouraging schools to look for opportunities to explore other aspects of the river environment, and ways in which their pupils can make a positive contribution;
- Providing a source of immense pride amongst ICI (Zeneca) staff: in 1992 the Fen won the ICI international environmental project award.

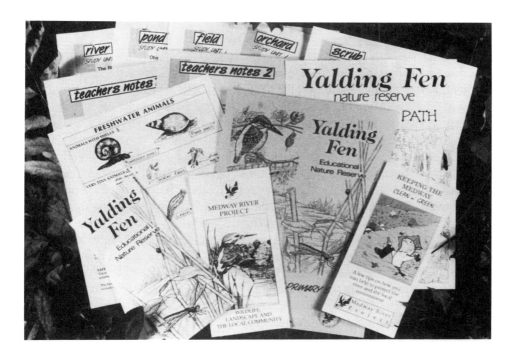

Figure 3 Yalding Fen study pack

ENVIRONMENTAL EDUCATION PARTNERSHIPS BEYOND THE FEN

In response to requests from schools, the Project (in partnership with the KTNC) has prepared a series of study packs based on the river environment. Two packs, sponsored by Kimberly-Clark, based at Allington Lock and the Teston Picnic Site, explore issues related to the impact of recreation on the wildlife and general environment. A third pack, to be published in the spring of 1994, investigates landscapes, what attracts people to an area and ways to enhance a degraded landscape.

It is an old maxim that "to do is to understand". The Project has established a series of work experience and extra-curricular activities to exploit this direct approach to education. Every year the Project involves students from senior schools, principally from the industrial Medway Towns, in working alongside Project staff. Tasks range from litter sweeps to building stiles and footbridges, but all the students gain experience of a range of countryside management skills. Comments from students, many of whom may rarely visit the countryside, such as "I will never drop a crisp bag again" or "Next time I go for a walk I will remember how much work went into building a simple stile!", clearly prove the value of this approach.

LITTER: THE UNENDING CHALLENGE AND EVERYONE ELSE'S PROBLEM

Litter is an issue that is guaranteed to raise strong public reaction, yet no-one is ever responsible. It is always someone else's fault and someone else's responsibility.

In 1990 the Project launched the first of two initiatives to combat litter. The Clean 'n Green campaign, in partnership with the Clean Kent Campaign (CKC), was an attempt to promote greater awareness and cooperation amongst river users. 20,000 copies of the leaflet have been distributed via navigation and fishing licences, angling shops, boat yards, libraries and other public centres. The effectiveness of this approach is hard to gauge and is always open to the criticism that it is "preaching to the converted". But, if it has changed the attitude of just a few people then it must be regarded as a success in a long and difficult campaign.

The success of the Project's second initiative has been far more obvious. The Medway Litter Wardens were launched in May 1991, with support from the CKC and the NRA. The dedicated team of volunteers have walked the riverside path collecting litter and talking to river users. Most of the wardens report that they have influenced the attitudes of river users: anglers stop fishing and collect litter around their swims and walkers collect handfuls of rubbish. Small changes perhaps, but important ones and the commitment of the wardens has now been rewarded by the Tidy Britain Group, with two national awards.

PROJECT DEVELOPMENT, 1988-93

The Medway River Project aim, set out in the five-year development plan, has developed from the original memorandum of agreement to reflect community

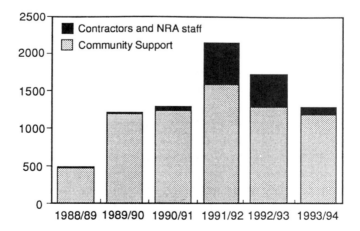

Figure 4   Community support, 1988-94

interests more accurately, and to incorporate the extended Project area. The Medway River Project now aims to "maintain and enhance the Medway valley, from Tonbridge to Rochester, as a green corridor for the benefit of wildlife and the local community, through the promotion of local community awareness and action" (Smith, 1992).

Community participation and support has grown dramatically (Figure 4) to around 1500 volunteer work days per annum. By the end of 1993/4, over 8000 volunteers from 70 community groups will have participated in MRP-led enhancement schemes.

The Project's development plan and priorities for action have evolved, since 1988, in response to local community interest and concerns. Initially access was the key issue and during the first three years dominated the work programme (Figure 5). As broken stiles were replaced and paths cleared, the Project promoted discussion on the landscape and river environment. Community concern quickly turned to issues such as wildlife, landscape degradation and litter. Where erosion threatened the continuity of the riverside footpath, the community pressed for solutions which integrated bank protection and habitat creation.

The emphasis of the Project's work has taken a subtle but important shift. Encouraged by their experience with the Project, and supported by a range of grant schemes, community groups are now developing their own initiatives — with support and advice from the Project. MRP is no longer taking the lead in every initiative but is enabling communities to identify and resolve their own priorities.

Since 1991 a range of new government grants, most notably Countryside Stewardship, Hedgerow Incentive and Rural Action have prompted significant changes in the

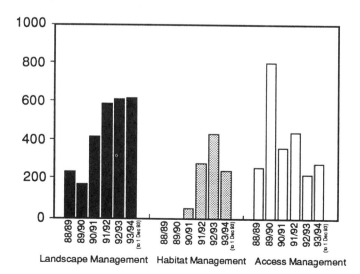

Figure 5 Project activity analysed by management tasks, 1988-93

Project's role. Rural Action, introduced in 1992, has greatly acceler-ated the development of community-led schemes described above. Countryside Stewardship and Hedgerow Incentive grants for landscape conservation have given the Project an advisory and management planning role, bringing us into closer working relationships with landowners. To date these two grants have enabled the project to secure the conservation of 30 ha of traditional orchard, 1.8 km of historic hedgerow and 10 ha of ancient coppice woodland, in addition to the creation of approximately 5 km of new permissive access.

The integration of bank protection and habitat enhancement has been a key element of the Project's riverside schemes and reflects the balance needed to achieve the broad, and potentially conflicting, aims of the sponsoring authorities. To achieve this integration the Project has promoted a return to traditional techniques and natural materials. Small schemes, carried out by volunteers, proved the potential and the cost effectiveness of Sweet Chestnut *(Castanea sativa)* faggots and Osier *(Salix viminalis)*, in bank restoration schemes. In 1993, the sponsoring authorities funded, through the Project, two major bank restoration schemes based on the traditional techniques promoted by the Project. These schemes are important not only because of the techniques employed, but also because they are achieving objectives which are beyond the resources and statutory obligations of any individual sponsoring authority.

CONCLUSIONS:

THE BENEFITS AND FUTURE FOR COMMUNITY PARTICIPATION IN INTEGRATED RIVER MANAGEMENT

The river environment is a resource of immense value to all sectors of society. If future policies are to achieve the goal of fully integrated management, then community participation must be seen as an essential element of the integrated river basin development process.

It could be said that 8000 days of free labour is a clear measure of the Medway River Project's success, and the value of community participation. But community participation is not about bald statistics or free labour, it is about people, promoting understanding and participation.

Participation in the management of their local river environment enhances people's understanding of the wider management issues and enables them to contribute in a more effective and constructive manner. Countryside Management Projects (CMP), such as the MRP, offer a mechanism by which such participation can be developed and sustained. In short a means by which "they" can become a valued part of "us".

For the Project sponsors the benefits lie in closer cooperation between authorities, and with the local community. The Project provides:

- a high profile and sustained point of contact with the local community;
- a means of identifying and coordinating the implementation of mutually agreed

targets;
- a cost-effective means of achieving important, yet peripheral, management objectives;
- a basis for financing and achieving initiatives that are beyond the resources of any one authority.

For the local community the Project provides:
- an accessible source of advice and practical support;
- a means by which the concerns and aspirations of the whole community, not just the active and vocal lobby groups, can be fed into the planning and management process;
- a catalyst promoting community awareness and, by example, the opportunities for active participation.

Community participation is not an easy option; neither is it the whole story, but perhaps the clearest measure of its potential is the way NRA regions have become increasingly involved in schemes similar to the Medway River Project. When the MRP was established in 1988 it was unique — no other water authority was directly supporting a CMP. Now there are three in Kent, and most NRA regions are supporting similar initiatives.

REFERENCES

Countryside Commission 1987. Medway River Project, Memorandum of Agreement.
Grant, A.Y. 1986. The Medway Navigation, an assessment of the leisure and tourism potential.
MRP/KTNC 1989. The Medway Navigation, a landscape, habitat and amenity survey.
Smith, B.J. 1992. Medway River Project, five-year management plan 1992/96.

34 Salmon in the river basin

COLIN G CARNIE
Crouch Hogg Waterman, Glasgow, UK

INTRODUCTION

The fundamental fact that the river is the lowest point in any cross section of a valley results in both every natural condition — and every action of man in that valley — impacting on the water in the river. Water carries with it all the characteristics of the ground over which, or through, it flows, picking up chemical components, flow conditions, organic material, particulate matter and even flotsam and jetsam.

It is the effect of all these characteristics taken from the flow of water in the river and impacting on the fish it contains, particularly on salmon, which gives the greatest importance to Integrated River Basin Development. The potential for damage is enormous, and our history shows how easily we have done this to rivers like the Tees and the Clyde; but today we face the challenge for beneficence which we know is laborious and expensive. Although it requires an enormous co-ordinated effort to create improvements, such improvements are readily apparent; look again at the Tees and the Clyde, or the Thames and the rivers of South Wales, all of which have modest runs of salmon and sea trout after many years when no fish could survive in their lower reaches.

There is still a great deal of work to do in improving our "industrial rivers" but, because their condition has been such an insult to us all, appropriate legislation is in place to tackle the problem. However the more difficult challenge is not so much in the urban as in the rural areas. It is in these parts of the valley where things are happening which appear to be benign, or at least to be appropriate for the rural environment, but are impacting to the dis-benefit of salmon and other salmonids.

Fisheries — the environment in which fish live — are at the very heart of Integrated River Basin Development and it is from the standpoint of conservation related particularly to the freshwater life cycle of the salmon that the parameters for this paper are set.

SALMON

The anadromous life cycle of the salmon creates an enormous range of requirements on a river system. The spawning area must have clean gravel and well oxygenated water which is largely free from sediment and does not become too acidic. The nursery areas for the juvenile fish must also have good quality water with an adequate flow over gravel and cobbles where plants and invertebrates can grow, where fish can establish their territory and find shelter and protection with bankside cover. During the process of smolting and migration to the sea, the young salmon are delicate little animals and require easy downstream access. The returning adults must have protection and shelter so that they can ascend many kilometres of river to reach spawning areas perhaps 500m above sea level, without feeding and while going through the physiological changes of maturation.

MAJOR IMPACTS

The Salmon Advisory Committee has written two reports (Anon, 1993a and 1993b) outlining the factors which affect natural smolt production and migration within rivers. These factors are reviewed below:

Afforestation
Afforestation and associated ditching can adversely affect the hydrological characteristics of the smaller streams in a catchment, increasing the flows at times of spate causing erosion and sediment transport, and reducing the flow at times of drought resulting in dry stream beds with no fish life. There will be reduced flow yields from the ground once the trees start to grow (the Forestry Commission's guidelines suggest that water losses will amount to 2% of the total yield for every 10% of the catchment which is affected).

Acidification
Salmonid eggs do not hatch if the pH falls below 4.5 and the percentage which successfully hatch at pH 5.0 is low. In addition, acidic water leaches chemicals which are toxic, particularly to newly-hatched and very young fish.

Fertilizers
Of the major constituents of fertilisers used for agriculture, nitrogen is likely to be washed into the river systems in significant quantities and in some cases phosphorus may be a problem. It is suggested that there is a link between agricultural fertilizer and the growth of algae slime on the bed of the Tweed and its tributaries.

Pesticides
Many of the pesticides used in agriculture, horticulture and silviculture have been screened to reduce the likelihood of damage to non-target organisms. The change from the organo-chloride pesticides has been helpful in limiting the damage and there are EC Directives which limit the concentrations of pesticides in order to protect fish and aquatic life.

Agriculture
Agriculture produces 200 million tonnes of livestock slurry and over 40 million tonnes of silage each year. Slurry has a BOD of 30,000 mg l^{-1} and silage effluent a BOD of 80,000 mg l^{-1}; thus any spillage is likely to be disastrous. Improved handling and farm yard care is reducing the number of pollution incidents but they continue to amount to about 3,000 per year in Great Britain.

River regulation and dams
There is a range of problems for salmon which are associated with river regulation, dams and barrages. These include impediments to migration, flooding of spawning and nursery areas, alteration to the natural flows, and modifications to water quality.

Water abstraction
There are few catchments which are immune to water abstraction and very few which do not suffer from this effect. Where this occurs it results in a reduction in flow which is likely to be particularly noticeable at times of low flow when abstraction also has the potential to decrease dilution of effluent. In addition, the flow of water into intakes may draw fish against screens.

River and riparian management
There are a number of factors related to riparian management which adversely affect salmon. Such factors include upland drainage which may benefit grazings but deplete low flows, enlargement of the river channel for flood prevention, overgrazing and physical damage to river banks by sheep and cattle.

Gravel extraction
The commercial extraction of river gravel can impact on salmon by the removal of spawning gravel, by channel degradation and by the release of fine sediments when the gravel is worked. Furthermore the removal of gravel from one point in a river is likely to accelerate the process of gravel transport from the river bed upstream, to the detriment of plant, insect and fish life.

Waste disposal
The control which is placed on the quality and quantity of effluents from industry of all types, and from domestic sewage works, is fairly strict, but there are still problems with storm water overflows, with inert suspended solids, and from some chemical processes. In many cases even very low concentrations of effluent may be toxic to fish.

Fish farming
The intensive production of trout and salmon smolts in fresh water has expanded substantially in recent years and fish farming is now a major producer of food. But with intensive farming goes the problems of disease, of solid waste and waste feed, and of escape, all of which impact on wild stocks. The control of disease uses chemicals which may be harmful to small fish, solid and liquid waste lead to

eutrophication and escapes may be adversely affecting the genetic characteristics of the wild fish.

Water quality
The quality of the water in which salmon migrate is the most important single factor which demonstrates man's impact on our rivers. While a number of the factors referred to above have a detrimental effect, in most cases the effect is through deteriorating water quality. Waste from industry, domestic sources, and fish farms, hot water from power stations, plant nutrients from agriculture, silts, sludges, and sediments, oils and toxic substances — all are likely to create conditions which are unacceptable to fish. In England and Wales (NRA, 1991) there was a decline between 1980 and 1990 of some 1100 km of river and canal which were classified on good, and a comparable decline of 65 km of estuary. In Scotland (Anon, 1990) there has been over the same period an improvement of approximately 1000km of rivers of good quality.

REMEDIATION

There are solutions to all of the problems associated with man's activities in the catchment listed above. Some are quite simple, some involve significant works and some are expensive. None, however, should be looked upon on its own but rather should be seen as part of a widely encompassing approach to catchment management.

This process has to start at the watershed on the highest points of the catchment and extend down to the sea. But to achieve the ideal, which will ensure enhanced stocks of salmon of all ages in the river system, the fisheries manager has to be part of a wide group of managers who can work together so that as many as possible of all their individual objectives can be achieved.

Fisheries cannot be looked at on their own, nor can land use or flood prevention; they must each be integrated so that all users of water are contributing to effective management.

Taking the factors of man's impact in turn:
- New techniques of forestry ploughing can prevent severe alterations to the hydrology, less intensive planting with open areas and deciduous trees near river banks provide favourable buffer zones.
- A reduction in sulphate and nitrous emissions will help to reduce the acidity in rain, wide buffer zones on the edges of all streams will reduce the acidity in the water and in extreme cases dumping limestone in the water course will help to neutralise some of the acidity.
- The reduction of the pressure on farmers for higher production will reduce the amount of fertilizer being put into land, and the introduction of set-aside allows a positive approach to be adopted to the protection of banksides and adjacent land.
- The progression of controls and the increased knowledge of the effect of pesticides is reducing their impact particularly in fish life, but there are unknowns

about the effect of pesticides on invertebrates other than those for which they are intended.

- Better control and better management of slurry and silage storage is leading to fewer incidents of spillage.
- A better understanding now exists about appropriate compensation flows and the times when the majority of fish move upstream so that improved operating rules can be developed. Also, knowledge of the behaviour of salmon at fish ladders has improved so that they can be more effective.
- An approach to abstraction which takes account of both environmental and water user needs, and the establishment of a limit of abstraction at flows where a prescribed minimum flow is set, both help to create more favourable conditions for salmon.
- While upland drainage is no longer being extended and some drained areas are reverting to a more natural condition, there is still extensive hill grazing which prevents regeneration of moorland and river banks. But these are problems which can readily be overcome.
- The effect of gravel extraction can be minimised if it is confined to areas away from the bed of a river and worked with proper control of stilling ponds.
- The improvement in industrial rivers shows what can be done to overcome the impact of waste disposal on salmonids and continuing steps towards improved treatment processes and better waste management will carry this improvement further.
- The fish farming industry recognises that there are problems associated with their production methods which impact on wild stocks, and they, more than most, understand the steps needed to overcome these.
- The challenge of upgrading the water quality in our rivers is one of the overriding reasons for the need to integrate all aspects of the management of activities which impact on the catchment. The fact that only 63% of rivers in England and Wales are classified as good emphasises this point and while in Scotland the percentage is much higher (97%) there are still many aspects of water quality there which need to be improved.

INTEGRATION

If a typical valley in the Highlands of Scotland is taken as an example, with peat overlying acidic rock, the land uses and activities which impact on the salmon begin on the high ground where deer are likely to be a more important sporting asset than salmon, where afforestation is an important long term investment for the upland owner and where sheep graze on the grass and heather of the stream banks. A little further down the glen a dam on a side stream impounds water and has intakes on other streams in the catchment but it is deemed to be so far up the catchment that fish passes were not incorporated at the time of construction. Still further downstream there may be more afforestation, some arable land, an intensive dairy unit, a small town of

10 000 people, an abattoir and a leather works.

From this brief description a range of interests can be identified:

- The Scottish Office Department of Agriculture and Fisheries
- Scottish Natural Heritage
- The Forestry Authority
- River Purification Board
- District Fishery Board
- Regional and District Councils
- Farmers and landowners
- Anglers

All of these people are likely to have different approaches to the use of land in the valley but all can contribute to better use of the water in the river.

An alternative example might be an English river in a more populated area which retains an industrial heart. Afforestation may be less of a problem, and underlying calcareous rock may prevent acidic river flows. However impoundment and abstraction both from surface water and deep wells are likely to be a feature as will the discharge of municipal and industrial wastes. Farming will extend far up the valley and the high ground will be grazed by sheep. Extraction of gravel, fish farming and intensive agriculture are likely in the valley above a town with some industry, discharging treated sewage and industrial waste into the river before it reaches the sea.

Significantly, in England and Wales there are fewer organisations which are concerned with the river basin than there are in Scotland. The National Rivers Authority (NRA) plays a major role in the flood control, water abstraction, effluent reception, fisheries and recreation and thus solutions to problems in catchment management should be found more readily.

There will still however be a number of organisations concerned with land use which will include:

- Department of the Environment
- Ministry of Agriculture, Food and Fisheries
- The Forestry Authority
- Local Authorities
- National Farmers Union
- National Trust and English Heritage
- National Rivers Authority (NRA)
- Water companies
- Farmers and landowners
- Anglers

Other users of rivers and lakes

Indeed, the NRA has advanced quite far in its programme of catchment management studies, and defines catchment management planning as "An agreed strategy for realising the environmental potential of a catchment within prevailing economic and political constraints". (NRA, 1993). While these guidelines refer to optimising the overall future well-being of the water environment and seeking to develop and implement the principles of sustainable development, perhaps this approach does not go far enough in the small streams of the catchment.

The process of integration is an iterative process which requires an enormous amount of goodwill and commitment to achieve its objectives. It is one in which probably no-one is entirely right, but also no-one is entirely wrong, where no-one can expect to take at the expense of others but no-one should be expected to give everything. In the past "take" has been too easy an approach. Present salmon stocks suggest that this has been to the detriment of this particular species. The adoption of an integrated approach will ensure that the interests of the salmon receive due consideration so that this environmental resource is sustained.

REFERENCES

Salmon Advisory Committee. 1993. *Factors affecting natural smolt production*. MAFF, London.

Salmon Advisory Committee. 1993. *Factors affecting emigration smolts and returning adults*. MAFF, London

National Rivers Authority. 1993. Catchment management planning guidelines.

National Rivers Authority. 1991. The Quality of Rivers, Canals and Estuaries in Western England. *NRA Water Quality Series* No. 4, December 1991.

Water Quality Survey of Scotland. 1990.

35 A hydrological model application system — a tool for integrated river basin management

D A HUGHES, K A MURDOCH and K SAMI
Institute for Water Research, Rhodes University, South Africa

INTRODUCTION

The Department of Water Affairs and Forestry of the Republic of South Africa is committed to ensuring the sustained provision of adequate quantities and qualities of water to competing users at acceptable costs and with assurances of supply (Forster, 1992). The high variability in the hydrology and development status of South African catchments means that catchment specific management initiatives are often necessary. There is an equally high variability in the amount and type of information that is available to develop an understanding of the natural and modified dynamics of catchment hydrological regimes. This understanding is essential to the development of planning and management strategies and where existing information is inadequate, it is frequently necessary to resort to simulation tools. Even where information on existing conditions is available, models can be useful to explore different development scenarios.

There are a wide variety of hydrological models available and the choice of which to employ can depend on several factors. Hughes and Beater (1989) identified some of these and refer to the type and resolution of output required, the catchment response and climate characteristics and the amount of information available for defining the input data and quantifying model parameter values. For integrated river basin management applications there are often a wide variety of questions related to natural hydrology, water supply potential, environmental impacts and water quality for which answers are required. No single model is usually capable of supplying all the answers, yet an array of models will inevitably share information requirements.

This paper discusses some of the elements of a PC-based, integrated *HY*drological *M*odel *A*pplication *S*ystem (HYMAS). A number of different types of models have been incorporated into a single package with shared procedures for the interpretation of catchment physiographic data, estimation of model parameters, compilation of time series data, model running and analysis of the model results. Seven models have currently been incorporated into the system and cover many of the hydrological estimation requirements that may be needed for integrated river basin management.

All models have been established in semi-distributed formats and some have several optional methods for simulating part of the hydrological system.

HYMAS — A BRIEF DESCRIPTION

There are four major tasks that have to be carried out during any model application and these form the basic components of HYMAS. They are accessed via a shell program which displays a menu of the available processes. 'Project files' are used to store the name of the application and reference to the locations and names of the files which need to be accessed to run any of the HYMAS procedures. The project file also contains flags to indicate the status of the files referred to and is used to determine automatically whether a particular procedure can be run (i.e. if the files exist or have been specified with legitimate names).

Parameter estimation
The first group of parameter estimation procedures is associated with establishing values for a set of standard physiographic variables which describe the catchment, while a second deals with actual parameter value estimation and editing.

The standard physiographic variables include topographic, channel, soil, vegetation, geological and climate indices. Some are primary variables, estimated from maps, field work, air photos, etc., while others are secondary indices estimated from the primary variables. The estimation equations are contained within a text definition file in Reverse Polish notation form and can be readily edited if necessary. There are several programs within HYMAS that allow digitised topographic data to be used to estimate some of the primary variables. The developers are currently expanding these aspects of the system to allow more direct interfaces with GIS and DTM tools and to permit improved automated estimation of the physiographic variables.

Having established a physiographic data file for the sub-areas of a distribution system, initial values for some parameters of some models can be determined. Some models do not have direct relationships between their parameters and physical catchment data and it is therefore not possible to estimate default values. However, wherever this is feasible, it is carried out using a file of Reverse Polish notation equivalents of the relationships. All the model parameter files can be fully edited using a spreadsheet-type utility.

HYMAS contains additional facilities to allow parameter values to change within a period of modelling, referred to as 'time slicing.' Time periods within the total period where one or more parameter value will change are defined and the change can be specified as linear over the slice, or instantaneous at the start of the slice.

Establishing time series data inputs to models
There are several types of time series data that hydrological models commonly require as input and establishing these can form a major component of any modelling task. They include rainfall, observed discharge, upstream inflow, evaporation, reservoir

draft and observed storage volume. As the original data may be available in several different formats, HYMAS contains a number of routines to convert differently formatted data to a standard format used by all the models. An emphasis has been placed on facilitating conversion from several common formats available within South Africa.

The rainfall data compilation routines are used to create an input time series for each of the sub-areas of the distribution system. An inverse distance squared interpolation procedure is used to determine average sub-area rainfall depth from individual gauge data based on a matrix of distances from gauge locations to sub-area centres. A further facility is available to weight the sub-area rainfalls by the ratio of the average of the median monthly rainfalls for all elements of a grid within the sub-area to the median monthly rainfall for the gauge. This has been added to make use of the gridded median monthly database established for South Africa by the University of Natal, Pietermaritzburg, and to improve the estimation of sub-area rainfalls in catchments with steep rainfall gradients where some areas are poorly represented by raingauges.

The input data files for other time series data are usually only built after the rainfall files as the latter determine the period over which modelling will take place and the other data are compiled to be coincident in time. In the case of observed discharge data, it is also possible to create a HYMAS data file of a complete observed record. This allows some results analysis routines (flow duration curves, low flow analysis, etc.) to be used on observed discharge data without modelling.

Facilities are also provided for combining two or more data files to create, for example, a sum of discharges from several catchments.

Running the models
Having established the time series input data and the parameter data file, running models is normally a simple operation. However, in some cases, the models contain optional approaches to some of their components and the selected choices need to be linked together with the main model code to form the final executable program.

It is often convenient to run several applications of the same or different models as a 'batch process'. This is particularly relevant to integrated catchment management applications, where individual models of part of the system may be set up separately at first, but later a change may be made to one part and the effect on the whole system analysed. HYMAS allows for this by including a facility to link several projects (model runs, data file combinations, etc.) and run them together as a batch.

Assessing the model results
This is clearly one of the most important elements of any system designed to provide information about available options in a catchment management scheme. While the results analysis procedures are being continually updated and added to, HYMAS already includes a fairly comprehensive array of model output display, listing and analysis procedures. These include simple time series data listing and graphing, comparative plots, statistical summaries, probability analyses as well as a range of low flow analysis procedures. Most procedures allow the user to select for analysis

the output from different sub-areas, all or part of the time series or specific months or seasons of the time series.

The data output from the models has been designed to include information on as much of the internal state variables (soil moisture, groundwater levels, etc.) for each sub-area of the system as is practical without creating massive output files. This allows the results analysis procedures to be used with more than just the final catchment runoff or water quality time series output. The internal operation of the system being simulated, either at a component level, or at different points within the spatial distribution system, can therefore be analysed. These facilities can assist with model calibration, especially when several gauging stations are nested, or when observed information on other aspects than just total streamflow output is available.

If some analysis requirements are not met by the existing HYMAS facilities then the output data can be written to text files in several formats. These may then be imported into other software (statistical packages, spreadsheets, etc.) for further analysis or graphical display. It is straightforward to modify the text file that controls the menu options in the main HYMAS shell program to add direct calls to load other software packages and still remain within the HYMAS environment.

THE MODELS CURRENTLY INCLUDED IN HYMAS

There is insufficient space to include detailed descriptions of the seven models that currently form part of HYMAS but brief outlines and reference to relevant literature is given below:

Variable Time Interval (VTI) model
This model has been developed as a compromise between strongly physics-based, fully distributed models and more empirical approaches to simulating catchment hydrology (Hughes and Sami, 1994). It utilises a semi-distributed, sub-catchment approach and the model time interval structure is variable (five minutes to one day) and determined by the rainfall intensity and a set of user defined thresholds (Hughes, 1993). It contains the majority of the usual components of a catchment hydrology model, including groundwater recharge and interactions between surface and groundwater. Although this is a semi-distributed model, attempts have been made to allow for smaller scale ('sub-grid') effects by including algorithms based on probability distributions of variation within the sub-areas. While some of the parameters can be considered to be physics-based, others are more empirical. Initial, generally acceptable, values for many parameters can be derived from the standard HYMAS data file of physiographic indices using relationships developed by the authors. However, where some form of observed data are available, further calibration is always recommended.

RAFLES model
This is a daily time step version (Hughes, 1994) of a kinematic modular rainfall-flow

model originally developed at the University of the Witwatersrand (Stephenson and Paling, 1992) which includes algorithms for simulating soil erosion, sedimentation and reservoir storage. It has been included for its sediment routines but has many empirical parameters and remains largely untested at present.

PEXP model

This is a daily time step model designed primarily as a management tool to provide estimates of stormwater volumes and nutrient (specifically phosphorus) loads from developing urban areas within southern Africa (Hughes and van Ginkel, 1993). It makes use of a modified SCS (US Soil Conservation Service) approach to simulate runoff and empirical relationships to define the proportion of stored nutrients that are exported during runoff events of different 'power'. The nutrient inputs to storage are defined on the basis of socio-economic surveys of the urban area (Grobler, *et al.*, 1987).

Pitman model

This is a monthly time step, rainfall-runoff model that has been widely used within South Africa (Pitman, 1973; Pitman and Kakebeeke, 1991) for water resource assessments. Although a highly empirical conceptual model, a number of guidelines have been developed for determining the parameter values for ungauged catchments.

Multiple reservoir simulation model

This is a monthly time step reservoir water balance model that is frequently used in conjunction with the Pitman model for evaluating reservoir water supply dynamics. This version of the model allows several linked reservoirs to be simulated together (Hughes, 1992).

Simple Muskingham routing design flood model

This is an hourly time step design flood model based on several options to define the proportion of the design storm hyetograph that contributes to the design flood and then uses Muskingham routing to define the flood hydrograph shape. As with the other models it is semi-distributed and can be used to account for spatial variations in design rainfall input.

Daily raintank resource model

This is a simple roof runoff and raintank water balance time series model that was developed to evaluate the potential of raintanks to supplement the clean water supply of developing communities (urban and rural) in southern Africa.

APPLICATIONS OF HYMAS

Buffalo River catchment

The Buffalo River catchment is situated in the eastern Cape Province of South Africa

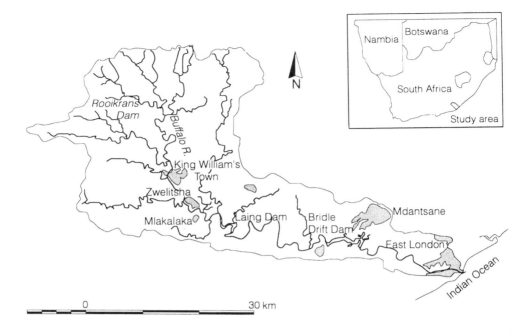

Figure 1 Buffalo River catchment

and has a total area of over 1600 km² (Figure 1). There are two small water supply reservoirs in the upper mountainous regions but the main modifications to the natural regime occur at Laing Dam and Bridle Drift Dam, in the middle and lower reaches respectively. The major urban centres are King Williams Town (developed with some industry) and Zwelitsha (mixed developed and developing), above Laing Dam, and Mdantsane (largely developing) lying adjacent to Bridle Drift Dam. The difference between developed and developing mainly refers to the extent of services (water supply and sewerage) available. The upper reaches of the river system consist of perennial streams flowing from a higher rainfall (> 1000 mm) mountain area, while the middle and lower reaches experience a semi-arid climate (rainfall < 700 mm), as well as being heavily utilised for water supply. The main problem was to develop a water quality management strategy based on the receiving water quality approach and from a hydrological point of view this meant that spatially distributed information was required on the low flow regime of the river and its current water quality status.

The water quantity regime was simulated using a combination of the Pitman and multiple reservoir simulation models. Data available for establishing the model consisted of observed records from over 20 daily raingauges, median monthly rainfalls for a 1' * 1' grid covering the area, monthly runoff volumes for eight streamflow recorders (many in the upper to upper-middle reaches), monthly reservoir storage and abstraction volume data, some pan evaporation data and some data on effluent volumes released to the river. The total catchment was divided up into some 40 sub-areas, whose locations were chosen on the basis of a combination of the position of gauging stations, reservoirs and points of concern within the river system. The Pitman model was calibrated on the upstream gauged sites initially and a combination of regional parameter values and the calibration experience used to determine parameter values for the remaining sub-areas. As the calibration exercise moved downstream, simulations of the major reservoirs were incorporated and 'time slicing' of the storage volume and abstraction patterns was necessary due to the changes that had taken place during the calibration period (1967-1976).

A 40-year rainfall record was then used to generate a representative flow record for the system as a whole, given the present day status of the reservoir storage capacities, abstraction and effluent return flow patterns. A combination of this flow record and empirical relationships between discharge volumes and some water quality variables, developed from limited observations, was then used to estimate the patterns of the water quality regime, concentrating on nutrients and total dissolved solids.

Because of the perceived importance of nutrient washoff from the larger developing urban areas, where formal waste disposal services are limited, and the lack of water quality observations to quantify such washoff, the PEXP model was applied to these areas. Unfortunately, the empirical nature of this model, limited experience with its application, and the lack of observed information, meant that the patterns of nutrient washoff could only be quantified approximately. The combined, spatially distributed, simulated information on water quantity and quality is now available to form the basis of the development of a management strategy.

Southern Cape coastal lakes region

This region is located in the southern Cape Province between George and Knysna (Figure 2) and physiographically consists of an inland mountain area, a coastal plateau region and a series of natural lakes along the coastal strip. It is a developing tourist area where the ecology of the lakes has been identified as being important and

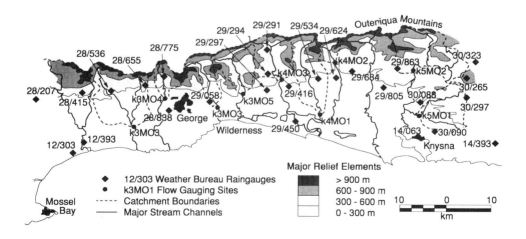

Figure 2 Southern Cape coastal lakes region

sensitive. The catchments have experienced land use changes in the past with the development of commercial pine plantations in the mountain and foothill areas and irrigated agriculture in the plateau areas. The perceived problem is related to the effects of the reduction in natural flow caused by the land use changes, the construction of many small farm dams, as well as some direct river abstractions for domestic supply purposes. There are several daily recording rainfall stations and three flow gauging sites within the area.

The initial modelling exercise consisted of applying the Pitman monthly model in such a way as to ensure that the results were sensitive to the major land use types prevailing in the catchments, as well as the impacts of the small farm dam developments and irrigation abstractions from them. The land use parameters of the

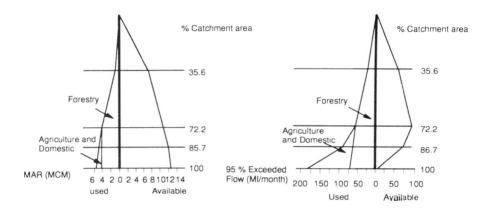

Figure 3 Current use of water in four sub-areas of the Diep River, based on MAR and 95% exceeded monthly flow

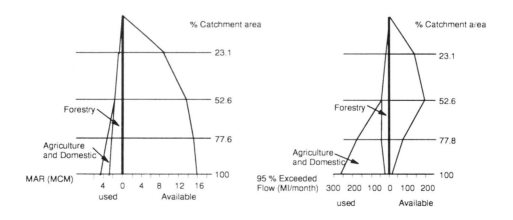

Figure 4 Current use of water in four sub-areas of the Karatara River, based on MAR and 95% exceeded monthly flow

model were then changed to estimate virgin conditions, water abstractions removed and the impacts assessed in terms of both mean annual runoff and the 95% exceeded monthly flow. Illustrations of the results are provided in Figure 3 for the Diep River (gauged at K4M03), which is heavily forested in the mid to upper catchment areas, and Figure 4 for the Karatara River (gauged at K4M02) which is still largely covered by natural vegetation but is influenced by domestic abstractions in the lower reaches. The VTI model has also been initially established for this area as there is some question about whether monthly streamflow volumes are adequate to assess the impacts of flow regime change on the ecology of the lakes. The two models share the same basic input time series data and use the same physiographic index data file to assist with parameter estimation.

CONCLUSIONS

Although the development of HYMAS is expected to continue well into the future it has already proved to be a useful tool for applying a range of different hydrological and water resource estimation models within a single integrated environment. The range of models, as well as the comprehensive results analysis procedures, allow many of the requirements of a number of disciplines (water quantity and quality managers, water supply engineers, ecologists, etc.) to be satisfied using the same basic system. It has also proved useful as a model development tool because all the pre- and post-modelling procedures are well established. Incorporating new models into HYMAS is therefore relatively straightforward and the developers are continually on the alert for new modelling approaches that can extend the range of applications for which HYMAS can be of value.

Future development of HYMAS is expected to concentrate on improved automated parameter estimation procedures through interfaces with external DTM and GIS packages, as well as research into better approaches to the estimation of model parameters from catchment physiographic data. Further work is also being carried out on expanding the output analysis procedures.

ACKNOWLEDGEMENTS

The development of the HYMAS software was funded through a research grant from the Water Research Commission of South Africa. Dr Vladimir Smakhtin of the IWR at Rhodes is responsible for some of the low flow analysis routines forming part of HYMAS.

REFERENCES

Forster, S.F. 1992. Integrated catchment management: justification, feasibility and limitations.

Water Sewage and Effluent, **12**, 30-34.

Grobler, D.C., Ashton, P.J., Mogane, B. & Roosebomm, A. 1987. Assessment of the impact of low-cost, high-density urban development at Boshabelo on water quality in the Modder River catchment. Report to the Dept. of Water Affairs, Pretoria, South Africa.

Hughes, D.A. 1992. A monthly time step, multiple reservoir water balance simulation model. Water SA, 18(4), 279-286.

Hughes, D.A. 1993. Variable time intervals in deterministic models. **J. Hydrol.**, **143**, 217-232.

Hughes, D.A. 1994. Soil moisture and runoff simulations using four catchment rainfall-runoff models. *J. Hydrol.*, In press.

Hughes, D.A. & Beater, A.B. 1989. The applicability of two single event models to catchments with different physical characteristics. *Hydrol. Sci. J.*, **34**, 63-78.

Hughes, D.A. and Sami, K. 1994. A semi-distributed, variable-time interval model of catchment hydrology - structure and parameter estimation procedures. *J. Hydrol.*, In Press.

Hughes, D.A. and Van Ginkel, C. 1993. Nutrient loads in runoff from developing urban areas, a modelling approach. Proc. Sixth South African National Hydrological Symposium, Pietermaritzburg, South Africa, Sept. 1993, 483-490.

Pitman, W.V. 1973. A mathematical model for generating monthly river flows from meteorological data in South Africa. Report No. 2/73, Hydrological Research Unit, Univ. of the Witwatersrand, Johannesburg, South Africa.

Pitman, W.V. & Kakebeeke, J.P. 1991. The Pitman model - into the 1990's. Proc. 5th South African National Hydrological Symposium. Stellenbosch, South Africa, Nov. 1991.

Stephenson, D. & Paling, W.A.J. 1992. An hydraulic based model for simulating monthly runoff and erosion. *Water SA*, **18**, 43-52.

36 Approaches to integrated water resource development and management of the Kafue basin, Zambia

J J BURKE
Water Resources Branch, United Nations, New York, USA

M J JONES
Coode Blizard Ltd, Consulting Engineers, UK

V KASIMONA
Department of Water Affairs, Lusaka, Zambia

INTRODUCTION

The Kafue catchment has an area of 155 000 km² and occupies 20% of Zambia's total land area. It forms one of the principal sub-catchments of the Zambezi basin and has a mean annual flow of some 300 m³s⁻¹ near its confluence with the Zambezi. The mean annual flow represents only 6.2% of the mean annual rainfall of 1057 mm falling over the catchment. This low-yielding hydrological regime of the Kafue is greatly influenced by the geological and geomorphological setting of the basin. The topography is very subdued: the landforms are largely developed by chemical weathering and erosion, as opposed to mechanical processes which dominate land-form development in most other large catchments. This suggests a very active and widespread role for groundwater circulation in the hydrology of the Kafue basin. Minor down-warping and tilting occurred in the late Pleistocene, forming swamps and drainage reversals. The net effect has been high open water losses from areas of impeded drainage and high evapotranspiration losses from groundwater seepage zones associated with saprolite and karstic aquifers.

An outline of the physical setting of the catchment is given in Figure 1, together with the principal Department of Water Affairs (DWA) hydrometric stations that are referred to throughout the text. The Kafue basin plays a central role in Zambia's economy: most of the mining, industrial and agricultural activities and approximately 50% of Zambia's total population are concentrated within the catchment area. However, continued economic development of the basin is not sustainable if inter-sectoral competition continues over limited surface water resources by consumptive

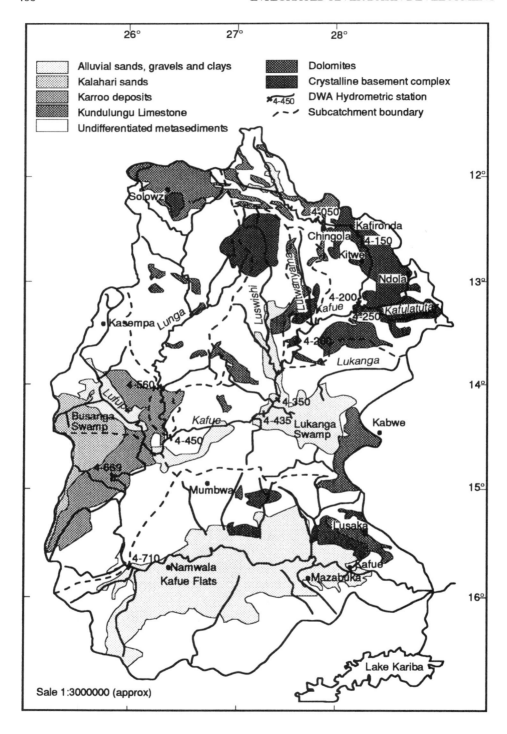

Figure 1 Kafue basin geology, hydrology and selected discharge monitoring stations

and non-consumptive users. Such competition has recently involved interagency disputes over flow allocations during times of low flow.

Despite the detailed work that has gone into the basin, particularly during the late 1960s and early 1970s, immediately following independence, plans for the basin's development are currently based on very limited understanding of its resources. Given this legacy, and the need to indicate a direction for the basin's development, management and conservation, a rapid diagnostic study was carried out in April 1993 by the United Nations in co-operation with the Department of Water Affairs, Lusaka (UNDESD, 1992). Trust funds were made available by the Swedish International Development Agency (SIDA). This paper summarises the study's findings and outlines approaches for integrated development of the catchment.

From this diagnostic study it became apparent that past approaches to water resource development of the basin have generally failed to view the basin as an integrated system with distributed inputs, outputs and storage elements.

Water supply schemes in the post-independence years have generally concentrated on stored surface water or run-of-river solutions and ignored the potential of some very prolific and ideally located aquifers. More recent attempts to look at the system in environmental terms have been very limited in their technical scope.

Clearly, the challenge for the basin is to find some way of resolving the economic, social and environmental demands placed upon it. The first requirement is to establish a physical framework leading to an understanding of the catchment hydrology. Second, the main water resource issues facing the catchment need to be examined. Third, the approaches for integrated resource management must be considered.

ESTABLISHING A FRAMEWORK

An examination of the physical characteristics of the Kafue catchment has revealed a geological/geomorphological framework in which to discuss water resource development and management (UNDESD, 1992). Figure 1 illustrates a set of principal sub-catchments with well-defined geological, geomorphological and hydraulic characteristics which, contributing to the catchment hydrological response, provide a template on which to discuss water resource planning. Without such a framework, it is difficult to explain adequately the hydrological response of the catchment and account for the distribution of the available surface and groundwater resources. The 1:1 500 000 hydrogeological map published recently as part of the World Bank/UNDP Sub-Saharan Hydrological Assessment for Zambia (World Bank/UNDP, 1990a) provides a clear and up-to-date basis for such a framework. Figure 1 is an attempt to summarise this mapping and to bring out the salient hydrological and hydrogeological features of the Kafue basin.

Previous water resource studies
Initial surveys in the catchment emphasised the role of surface water in satisfying the irrigation needs of the lower catchment (Hawes, 1951) and the water supply needs

Table 1 Principal subcatchments and incremental catchments in the Kafue Basin

Sub-catchment or basin increment	Area	Hydromet. station	Mean flow ($m^3 s^{-1}$)	Rainfall (mm)	Runoff (mm)	% Runoff	Mean flow (MCM/y)	Geology and Geomorphology (BC=Basement Complex)
Upper Kafue-Raglan Farm	5000	4-050	37	1300	231	18	1154	Upper catchment on Kundulungu limestones, shales, etc.
Upper Kafue-Smiths Bridge	8599	4-130	80	1300	291	22	2507	Onto BC. Granitic gneiss, north of Kitwe
Upper Kafue-Wusakile Bridge	9195	4-150	95	1306	257	20	2996	Still on BC (gneiss and metasediments south of Kitwe)
Upper Kafue- Mpatamato	11655	4-200						Just coming off BC onto lower Palaeozoic Kundulungu shales surrounded by Katanga subgroup quartzites and dolomites
Kafulafuta	4817	4-250	27	1275	178	14	857	BC gneiss and metasediments
Lufwanyama	3000	4-272						
Middle Kafue - Machiya	22922	4-280						Set in Upper Roan dolomites & Lower Road quartzites of Katanga supergroup. Hydrology strongly influenced by Mpongwe block storage and losses
Luswishi	9000	4-340						
Lukanga	16000							
Middle Kafue - Chilenga	34178	4-350	189	1257	175	14	5967	On Kundulungu shales & alluvium associated with Lukanga depression
								— as above —
Middle Kafue - Meswebi	50479	4-435	203	1189	127	11	6402	Just upstream of Lunga confluence on the controlling edge of the BC culmination (Hook granites)
Middle Kafue - Lubungu	54442	4-450	199	1170	115	10	6282	
Lunga confluence	24268	4-595	117	1250	152	12	3700	Principally on Kundulungu shales with isolated limestone outcrops in central and lower catchment and more complete Kundulungu limestone cover in upper catchment
Lufupa								Principally Karoo and Kalahari deposits
Middle Kafue - Kafue Hook	95053	4-669	336	1160	112	10	10602	Entry of Kafue onto Hook culmination. Some inflow from Karoo (Mesozoic) and Kalahari/Karoo (Cenozoic) catchments to the west.
Middle Kafue at Itezhi-Tezhi	105620	4-710	317	1151	95	8	9997	Entry on to the Kafue flats with upstream lateral inflow from granitic subcatchments probably accounting for some flashiness
Lower Kafue at Kasaka	150971	4-977	314	1057	66	6	9902	Kafue flats on Karoo and alluvium surrounded by dolomite to the north and folded BC to south. Lateral inflow probably insignificant
Lower Kafue at Zambezi	154829							Kafue Gorge set in BC and Mine Series. Falls onto Karoo

of the Copper Belt (Duff, 1959). At this stage a comprehensive review of the groundwater potential of the catchment's dolomite and limestone aquifers was not considered necessary, despite the exploitation of the Lusaka dolomite which commenced in the early 1930s to serve the newly-established capital.

Development plans in the mid-1960s still concentrated on the role of surface water (FAO, 1968) and stressed the role of deforestation in altering the hydrological regime. Experimental catchments were set up in 1964 in the upper catchment at the Luano Forest Reserve, near Chingola, to research this issue (NCSR, 1971) and similar concerns are perpetuated in the more recent literature (IUCN, 1985). However, the experimental catchment work was not completed and, despite the collection of many data, the results remain poorly documented and inconclusive.

The potential use of groundwater within Zambia as a whole was outlined in the mid-1960s (Lambert, 1965) and further elaborated at the end of the 1970s (Chenov. 1978). In addition, the significance of geological controls on sub-catchment water balances and flow durations had been appreciated by the early 1970s (Mander, 1968; Warren, 1972). The Department of Water Affairs initiated groundwater studies for town supplies in Lusaka and the Copper Belt (Tague, 1965; Jones, 1970; Jones, 1971; Jones & Topfer, 1972; Hadwen, 1972) However despite confirmation of the prolific groundwater resources of the dolomitic aquifers in both regions, surface water schemes predominated in the Copper Belt (Gibb, 1971; BCP, 1975) and Lusaka was provided with a surface water intake to serve the rapidly growing metropolitan area as part of the Kafue Gorge Scheme (SWECO, 1968). It was not until a set of expensive surface water schemes had been constructed at Lusaka, Kabwe and Ndola that cheaper groundwater alternatives were proposed and developed (FIGNR, 1978; BCP, 1980). It is therefore difficult to understand why the further work in the 1980s and 1990s has concentrated purely on river flow interpretations and analysis (DHV, 1980; Sharma, 1984; Shawinigan, 1990).

Whilst elements of each individual study referred to above are useful, none provides a comprehensive and sufficiently detailed account of the sub-catchments throughout the whole basin, on which to base, first, an understanding of the hydrological and hydrogeological sub-systems operating within the basin and, second, a framework for water resource planning. To this end a more comprehensive breakdown of the basin's sub-catchments is suggested in Table 1.

Catchment yield

The yield of the basin is limited. Mean annual rainfall over the catchment amounts to 1060 mm but is subject to distinctive temporal and spatial distribution (Torrence & Lineham, 1963). However, at the entrance to the Kafue gorge at Kasaka (catchment area 151 000 km^2) mean annual flows amount to approximately 66 mm or 10 000 MCM, which represents only 6.2% of the catchment rainfall. This contrasts with the adjacent Luangwa catchment which over an area of 144 000 km^2 yields 13% of its mean annual rainfall in flow.

Three principal physical explanations may be invoked to explain the low yield of the Kafue in relation to the regional hydrology:

- The low topographic relief within the catchment, coupled with a thick mantle of permeable soils, which contributes to the attenuation and diffusion of hydrograph response;
- The spatial and temporal variability of rainfall over the catchment, which causes related variations in the river flows and recharge and consequent changes in groundwater storage and discharge;
- The geological and geomorphological evolution of the catchment, which has led to high open water and evapotranspiration losses from areas of impeded drainage and seepage zones associated with saprolite and karstic aquifers. The consequence is a low contribution from the aquifers to river base flow.

With these physical controls, the annual regime of the Kafue at Kafue Hook follows a remarkably consistent pattern, with a single principal peak at the end of February or early in March, followed by a long uniform recession until the onset of the early rains in late November. Figure 2 shows a plot of the monthly flows at Kafue Hook and Pontoon stations covering 1905 to 1993, to illustrate the flow regime over the longest period available.

However, the Kafue Hook hydrograph masks significant variation in upstream routing and lateral inflow. To illustrate this feature, a composite plot of daily discharge data for 1962/3 is presented in Figure 3 for a set of representative mainstream and tributary stations. The 1962/3 data were chosen for their depiction of a water year with near mean annual discharges and the reliability and availability of the data.

The implications for water resource development and management in such a low-yielding complex catchment are manifold, but the choice of approaches to water resource development and management must be based on a sound understanding of these hydrological limitations if the approach is to be considered at all "integrated".

Developing a framework

Approximately 65% of the catchment area is underlain by fissured crystalline and metasedimentary Basement Complex. Dambo drainage systems are common over the Basement Complex outcrop. Chemical weathering *in situ* of the rocks underlying the African and Post-African surfaces has produced a continuous cover of thick saprolite soils. These weathering horizons range from 25 m to over 60 m in thickness and play an important role in the hydrological cycle of the catchment.

The small areas of limestone and dolomite outcrop associated with the complex folding on the eastern margin of the catchment form distinctive elevated limestone pavements. The morpho-structural setting of these outcrops is significant. In the upper catchment they strongly influence the character of baseflow contributions to the Kafue system and the configuration of the groundwater occurrences and flow regimes, as recognised by Warren (1972).

The Kalahari Beds and associated alluvium in the west of the catchment form low undulating relief on highly permeable soils with shallow water tables and locally impeded drainage where fine sediments have accumulated.

The broad alluvial plains associated with the major areas of impeded drainage such as the Lukanga Swamp and the Kafue Flats are of neo-tectonic origin. With the exception of the Kafue Flats, these features would have probably been infilled had natural rates of erosion been higher during the Pleistocene.

Most of tropical Africa has been exposed as continental land above sea level for over 200 million years. In the Kafue basin the resulting modern landforms are extensive old land surfaces interrupted by isolated highlands, scarps and broad linear valleys of tectonic origin. This landscape is the result of two main erosion cycles, the African and the Post-African.

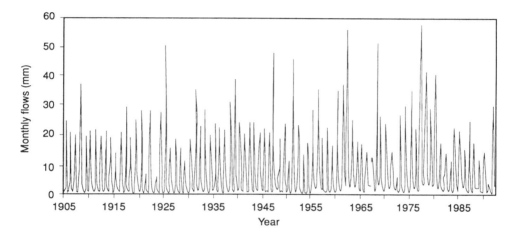

Figure 2 Kafue Hook monthly flows October, 1905 – September 1993

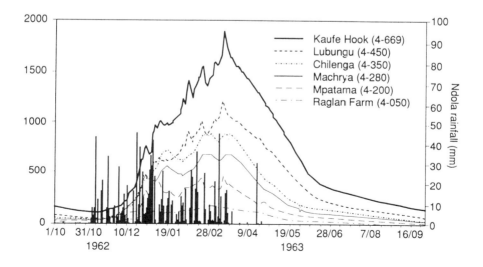

Figure 3 Selected hydrographs for the water year 1962/3

The upper basin largely lies on the African erosion surface. The African surface in the Kafue basin is characterised by an extremely smooth plateau surface with subdued slopes, shallow and wide drainage patterns and scattered inselbergs. The middle and lower basin lies mainly on the Post-African erosion surface. The Post-African surface is polycyclic and more hilly and incised than the African surface.

The relationship of well-defined catchment zones to the erosion surfaces is indicated in Table 2.

Table 2 Geomorphological zoning of the Kafue basin

Catchment zone	DWA Station	Area (km²)	Elevation range (m asl)	Erosion surface	Geology
Upper Kafue Upper Mutanda Upper Lunga	4-200 4-500 4-560	11655 1706 620	1370-1132	Cretaceous (post-Gondwana) & African (early Tertiary)	Folded Kundulungu limestones and shales & Basement Complex culminations
Middle - Kafue Hook	4-670 (4-669)	94920	1132-1073	African (mid- Tertiary)	Kundulungu shales, siltstones & sandstones with Karoo & alluvium
Middle - Itezhi-Tezhi	4-710	105620	1073-980	Post African	Hook granites & Karoo
Lower - Kafue Flats	4-980	151000	980-966	Post African (late Tertiary)	Karoo deposits & alluvium
Lower - Gorge	*Confluence*	155000	966-370	Congo (Pleistocene)	Karoo deposits & alluvium

The brief examination of the physical background of the Kafue basin confirms the need to consider a geological/geomorphological framework in which to discuss water resource development and management. At this stage it is only possible to outline the framework: Figure 4 depicts the major subcatchments and storage components of the system and the relationship to the hydrogeological sub-systems and centres of irrigation demand. It is a first attempt at dividing and distributing the catchment into discrete basin sub-systems, which will doubtless require modifying and updating as more data become available and more research is carried out, but it does offer a template upon which water resource optimisation and planning can be tested realistically.

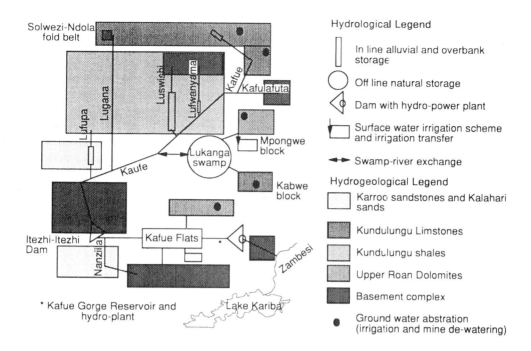

Figure 4 Kafue Basin hydrological elements and hydrogeological sub-systems

WATER RESOURCE ISSUES

The economic development of the basin commenced in the 19th century. A brief outline of the basin's development is best summarised by a table of events that have, directly or indirectly, impinged upon the water resources (Table 3). The main development of the Kafue has involved the construction of the Kafue Gorge dam and power plant completed in 1971 and the subsequent construction of the regulation dam at Itezhi-Tezhi which was completed in 1978.

This sequence of development may be considered typical of the 20th century evolution of Zambesi sub-basins and others in southern Africa. The principal development has been driven by hydro-electric considerations, and the power generation authorities hold the principal water rights. There are burgeoning irrigation demands within the sub-catchments and many towns regularly deal with water shortages and load shedding. The impact of mining activities in the Copper Belt has not been strictly regulated and deterioration of water quality is locally significant. This characteristic cycle points to poor overall water resource management with very little attempt to optimise existing resources, let alone plan for future developments.

Table 3   Economic chronology of the Kafue basin

Date	Event
1890-1930	Establishment of mining activities and railway line in the Copper Belt
1930	Establishment of capital, Lusaka: groundwater development for Lusaka water supply
1936	Water Affairs and Irrigation Dept. established. Monitoring networks initiated and large-scale commercial irrigation commences
1940	Mine de-watering becomes significant
1964-5	Independence and Unilateral Declaration of Independence (UDI)
1958	Impoundment at Kariba and resettlement in Gwembe
1960	Establishment of the Nchanga Tailings Leach Plant near Chingola
1966	Expansion of irrigation schemes, Nkambala, Chilaga, Chisamba, Chirundu and significant population growth
1968	FAO multipurpose study published
1970-1	Kafue intake and Kafue-Lusaka pipeline constructed (4.5 m^3 s^{-1}). Impoundment at Kafue Gorge
1975	Kafue industrial developments, Kasempa rice scheme
1978	Impoundment at Itezhi-Itezhi
1992	Completion of Konkola mine feasibility study. Mine dewatering at Konkola set to continue at 7.0 m^3 s^{-1}
1993	Lusaka water supply requirement for a further 4.0 m^3 s^{-1} from the Kafue

Catchment resources and demands

From a mean annual flow of only 366 m^3 s^{-1} at Kafue Hook and 314 m^3 s^{-1} at Kafue Gorge, and assuming a population in the catchment of 3 000 000 with 65% urban and 35% rural, the current and projected uses are tabulated in Table 4.

Options for water resource development

Options for further large-scale impoundment in the upper and middle catchment are severely limited if mean annual flows in the lower catchment are to be sustained and the environmental integrity of the basin preserved. Despite the limitations, there are plans to abstract a further 4.0 m^3 s^{-1} from the Kafue at Kafue Town to supply Lusaka. These plans are likely to conflict with hydroelectric and irrigation demands upon the baseflow of the Kafue in the lower catchment. The demands for freshet releases to simulate an annual flood in the Kafue Flats exacerbates the conflict.

Bearing these issues in mind, two observations can be made with regard to surface water development options. First, the Lunga sub-catchment contribution is crucial and the sub-catchment needs to be protected to conserve its contribution to the overall Kafue yield. Second, options for minimising existing losses from permanently inundated areas, principally the Lukanga swamp, may become increasingly attractive if further consumptive and non-consumptive use is required downstream. On purely hydrological criteria, the environmental demands of the Kafue Flats are difficult to justify, unless the Flats are less efficient at evaporating water than the Itezhi-Tezhi

Table 4 Summary of water use in the Kafue basin.

Category of use/loss	Use ($m^3 s^{-1}$)	Assumptions
NON-CONSUMPTIVE USE		
Hydropower	168.0	ZESCO water right
Kafue wetlands simulated flood release in dry years	(117)	Optional freshet release to sustain 300 $m^3 s^{-1}$ discharge over 4-week period in March in dry years
Fish farming	–	Negligible
Sub total	**168.0**	
CONSUMPTIVE USE		
Rural Water Supply	0.8	1 050 000 rural population @ 60 l/day/head
Urban Water Supply	4.0 (8.0 projected)	1 950 000 urban population @ 100 l/day/head
Stock watering	0.3	500 000 @ 50 l/day/head cattle
Irrigation	35.0	35 000 ha @ 1 l s^{-1} ha^{-1}
Industrial Use	3.0	Assumed
Sub total	**43.1**	

Even using these simplified assumptions, it is evident that the main competition for water is between power generation and irrigation. Compared to these demands the requirements for rural and urban water supplies are small. However, discussion with water supply engineers in the basin have indicated that there are difficulties in securing reliable sources for raw water. This relates primarily to the design of intake structures on inadequately researched sub-catchments and the lack of up-to-date research into local groundwater occurrences. More recent demands from environmentalists have resulted in occasional freshet releases from Itezhi-Itezhi to simulate the annual flooding regime in the Kafue Flats.

reservoir and the actual bankfull discharge of the Flats channel system is known. The need to sustain a largely unproductive wetland has always been questioned on account of the mosquito and tsetse fly populations which still prevent human and livestock colonisation.

In addition to surface water options, the groundwater resource development potential of the basin is also significant. The productive limestone and dolomite aquifers possess considerable natural storage capability. Opportunities for large-scale groundwater abstraction exist on these outcrops to develop the natural storage before much of the annual recharge is lost to unproductive open-water evaporation and evapotranspiration at the discharge points, which are the springs and the seepage zones. The widespread distribution of the limestone and dolomite aquifers offers an advantage over the linear distribution that is associated with surface water development. Furthermore, the undeveloped state of many of the aquifer blocks merits attention for future infrastructural and agricultural development plans (Wimpey Labs, 1983).

Fortunately, the present stage of groundwater abstraction from these aquifers is such that there is still time to bring existing groundwater development under control, even if that development is accelerated. Ultimately, however, when such developments approach the sustainable yield of the aquifer block, control of development and authority to prohibit further drilling will have to be enforced. In addition, the often discrete configuration of the karstic groundwater catchments needs to be determined before interference effects can be assumed. Under these circumstances regional groundwater monitoring networks need to be established across each outcrop and registration and auditing of all abstractions should be enforced. The karstic nature of the aquifers presents high pollution risks: aquifer protection zones are required to reduce them.

Data analysis
There is currently no analysis of hydrometric records. The Department of Water Affairs does not have the resources to do anything other than data collection and compilation. This state of affairs is typical of the regional situation with water resource organisations (World Bank/UNDP, 1990b). Previous support for the hydrometric network has acted purely as a data rescue operation (JICA, 1990). Ongoing support from the SADC Hydroelectric Hydrological Assistance Project (Phase 2) is concentrating on maintaining monitoring of inflows to the Itezhi-Tezhi reservoir.

However, with present regular budget constraints the DWA network is too extensive to maintain. The network could be rationalised to concentrate on fewer high quality stations in the major subcatchments. Such rationalisation should be based on present and future needs and should involve monitoring the main tributaries. Many water level stations could be discontinued and resources conserved until specific monitoring sites are required for development projects. It is also time that groundwater monitoring networks are established for the principal limestone and dolomite aquifers, ideally under the auspices of local water supply utilities.

Existing water policy and legislation
At present an outmoded Water Act promulgated in 1964 sits beside a recent Environmental Protection and Pollution Control Act of 1990. Whilst the latter defines the aquatic environment as comprising both surface and groundwater, the Water Act only makes provision for surface water regulation. The implicit water policy within the present Water Act was inherited from a previous administrative system whose aims were clearly to promote the rapid development of surface water resources to increase the economic growth of Zambia. Lacking concise knowledge of the groundwater resources in the country, the water policy included recognising private ownership of groundwater and allowed unregulated development.

After Independence, economic and social development placed increased demands of the surface and groundwater resources. The creation of the industrial township at Kafue and urban growth elsewhere required comprehensive surface and groundwater studies. The ready availability of surface water data and the lack of understanding of the vast potential for groundwater development led consultants to recommend a strategy of surface water development for many urban water supplies.

Clearly changes in Zambia's water policy and enabling legislation are long overdue. Elsewhere it has been observed that increased understanding of both surface and groundwater interactions can drive changes in water policy that improve society's use of limited resources (Dunbar, 1977). In the Zambian context it is also prudent to consider how legislation can be used as a tool for development and not merely as a regulatory device. However, there may be reasons for delaying or inhibiting change where prior water rights stand to be questioned or transferred.

APPROACHES TO INTEGRATED DEVELOPMENT AND MANAGEMENT

Against a background of a naturally limited system, deficiencies in data and hydrological understanding of the catchment, and an uncoordinated water policy and legislative framework, the reconciliation of apparently conflicting economic, social and environmental demands may seem insurmountable. However, if the single sector approach to basin planning is perpetuated, intersectoral competition for water will continue, together with its consequences: environmental degradation, urban water shortages, load shedding and declines in overall welfare and productivity. Under these circumstances, some of the options for initiating change to ensure sustainable development are discussed below.

Institutional innovation

According to institutional theory (Livingstone, 1993) the discrepancy or 'stress' between social goals and the complex of environment, technology and institutions that constitute the Kafue's actual circumstances should drive institutional innovation (water policy, enabling legislation, institutional development, etc.) Such innovation should allow society to adapt and remove the stress. Unfortunately, institutional innovation is the last thing Zambia has seen since people started applying technology to the Kafue. Moreover, it could be argued that the technology that has been applied has simply exacerbated the stress by enhancing evaporative losses from the basin. Fortunately the most recent drought in the basin, during 1992/3, seems to be driving some institutional reform, with comprehensive reviews of the functions of the Department of Water Affairs.

Institutional theory would also assert that with the expansion of the catchment database over time, technical innovations in data processing and complex system modelling, the range of institutional options should open up to allow a wider range of choices about development goals and resource allocation. Considering all the hydrological and hydraulic work carried out in the Kafue basin, and the advances in hydrological modelling, multi-objective analysis and computing power in recent years, it is unfortunate that the Kafue still has to rely on a lumped monthly rainfall/runoff model as its principal management tool (Shawinigan, 1990).

Linking hydrology and water resource policy

So far the hydrology of the basin remains poorly explained. Whilst there are many problems with data quality, both rainfall and runoff, very little analysis of the existing time series data has been carried out since the 1970s, and what has been done failed to interpret the distribution of daily rainfall and the form of annual hydrographs in physical terms. This is ironic, since the geology and hydrogeology of the catchment are generally well established. It is evident that integrated development and management of the Kafue sub-basin must appreciate the role of the hydrological and hydrogeological processes operating in the catchment. Moreover, optimal use of the sub-basin's resources will require, at the very least, a distributed model of interlinked hydrological systems and hydrogeological subsystems.

However, the real issue lies not with hydrology or modelling accuracy, but with finding a way of resolving conflicting demands upon the resource. This will depend on a thorough understanding of basin processes, the ability to simulate development alternatives such as new irrigation and hydropower schemes, and the ability to evaluate their impact on key social, economic and environmental indicators. These simulations and evaluations can be presented to government as a range of options or scenarios, each with its particular set of benefits and trade-offs.

Towards an integrated approach

It is possible to develop a physical framework for the Kafue system based on geological and geomorphological considerations that respects the spatial distribution of hydrological process in a set of sub-catchments and associated hydrogeological sub-systems. Such models have been used for more complex systems such as the Rio Colorado in Argentina (Major & Lenton, 1979) and in northern China (UNDESD, 1993).

Furthermore, such a framework could be used as a basis for a set of linked environmental and economic models to develop multi-objective/goal programming methods (North, 1993) to examine the range of environmental and economic options for the Kafue basin in a truly cross-sectoral vision of the system. Multi-objective models are well suited to integrated basin management, since a single resource (the river basin) has to supply multiple demands. Under these circumstances optimisation methods seek to allocate the resource in a satisfying manner, rather than the maximising manner associated with more conventional economic modelling.

The whole approach is not necessarily complex and can be carried out using simplified models until the processes and formulation techniques associated with multi-objective analysis are learnt, but the approach does require continuous collection and analysis of data, and the decision maker must interact with the computational processes in identifying goals, values and priorities. To this extent the multi-objective approach is no more than an integrated and transparent tool for supporting decision making. Under present circumstances the Kafue basin would be a good candidate for the application of multi-objective methods and whilst it would take time to establish and develop the methodology, it may offer the only sensible way to reconcile future water resource allocation in the basin.

On the basis of work carried out by the United Nations in northern China, (UNDESD, 1993) an outline of such a scheme could involve the following set of linked databases and models:

- A water resources management information system using key hydrometric station data;
- Subcatchment rainfall/runoff modelling with both soil moisture and groundwater components using daily data wherever possible;
- A channel routing model with conveyance losses attributable to overbank spillage and alluvial storage;
- A reservoir operation model;
- Agricultural and industrial water demand models;
- An engineering economic analysis model;
- A macroeconomic model;
- A water resources simulation model;
- A multi-objective analysis/goal programming model (special case of linear programming model).

This outline is suggested as an alternative to the present planning/resource allocation system operating in the Kafue basin which is unaware of the physical limitations of the catchment and makes no attempt to integrate the physical, social and economic circumstances. It is not proposed as a remedy to the water resource problems the basin is facing, but rather as a method for examining development options and trade-offs.

CONCLUSIONS

Very little attention has been paid to the establishment of a comprehensive and coherent framework for the hydrology of the Kafue basin. Despite advances made in information management and modelling capability in the past decade, the 1990s still sees the use of lumped models with monthly data to manage the basin's water resources. Moreover, such planning as has been carried out comprises isolated sectoral studies for hydropower, irrigation, water supply and environment. This has resulted in sub-optimal use of the system's resources and disputes between government agencies during dry season hydrograph recession. Groundwater resources in particular have been poorly understood, yet their potential is significant in terms of urban and irrigated agriculture supply. However, whilst there are attendant institutional problems to be resolved, it is possible to see a way of resolving current resource allocation problems using simple multi-objective/goal programming techniques set in a sound geological/geomorphological framework.

REFERENCES

Balek, J. 1970. *An analysis of the hydrometeorological sequences observed in the Kafue River Basin.* Water Resources Research Report No. 5, NCSR, Lusaka, Zambia.

BCP 1975. *Ndola Water Supply: Report on Phase 1 Studies.* Vol. 3 Geology and Hydrology. Brian Colquhoun & Partners, London.

BCP 1980. Groundwater Study. Interim Report. Kabwe. Vol. I Text; Vol II Figures. Brian Colquhoun & Partners, London.

Chenov, C.D. 1978. Groundwater Resources Inventory. NCSR/UNESCO/NORAD, Lusaka.

DHV 1980. *Kafue Flats Hydrological Studies.* Final report. DHV Consulting Engineers, The Netherlands.

Duff, C.E. 1959. First Report on a Regional Survey of the Copper Belt. Chapters IX-XII by The Special Commissioner for the Western Province. Government of Northern Rhodesia.

Dunbar, R.G. 1977. The adaptation of groundwater control institutions to the arid West. *Agric. Hist.*, **51**, 662-680

FAO 1968. *Multipurpose Survey of the Kafue River Basin, Zambia.* Final Report. Vol. III Climate and Hydrology; Vol. IV Ecology of the Kafue Flats. UNDP/FAO, Rome.

FIGNR 1978. *Groundwater and Management Studies for Lusaka Water Supply*: Part I Groundwater Study; Part II Data. Federal Institute for Geosciences and Natural Resources, Hanover, Germany.

Gibb 1971. Copper Belt Water Resources Survey. Preliminary Report. Alexander Gibb & Partners, London.

Hadwen, P. 1972. The Groundwater Resources of the Greater Ndola Area. Department of Water Affairs, Lusaka

Hawes, C.G. 1951. Report on the possibilities of development on the Kafue River and the organisation of the Water Development and Irrigation Department. Government of Northern Rhodesia, Lusaka.

IUCN 1985. The National Conservation Strategy for Zambia. IUCN, Geneva.

JICA 1992. The master plan study of hydrologic observation systems of the major river basins in Zambia. Japan International Cooperation Agency, Final Report.

Jones, M.J. 1970. *Hydrogeology of Chilanga Area and Central Province.* Department of Water Affairs, Ministry of Lands and Natural Resources, Lusaka.

Jones. M.J. 1971. *The Old Pump Station, Lusaka: a hydrogeological study.* Department of Water Affairs, Lusaka.

Jones, M.J. & Topfer, K.D. 1972. The Groundwater Resources of the Kabwe Area, Zambia. Groundwater Resources Paper No. 2, Department of Water Affairs, Lusaka

Lambert, H.H.J. 1965. The Groundwater Resources of Zambia. Department of Water Affairs. Ministry of Lands and Natural Resources, Lusaka.

Livingstone, M.L. 1993. Normative and positive aspects of institutional economics: the implications for water policy. *Wat. Resour. Res.* **29**, 815-821.

Major, D.C. & Lenton, R.L. 1979, *Applied Water Resource Systems Planning.* Prentice-Hall, Englewood Cliffs, New Jersey, USA.

Mander R.J. 1968. Kafue River Basin: the characteristics of minimum groundwater discharges. Department of Water Affairs, Lusaka.

NCSR 1971. Luano Catchments Research Project Book 1: Introduction. Report WR 10. Water Resources Research Unit. National Council for Scientific Research, Lusaka.

North, R.M. 1993. Application of multiple objective models to water resources planning and management. *Natural Resources Forum*, **17**, 216-227.

Sharma, T.C. 1984. Characteristics of runoff processes in the Upper Kafue Basin. Water Resources Research Report WR25. National Council for Scientific Research, Lusaka.

Shawinigan 1990. Application of the Pitman Model to the Kafue sub-catchment above Kafue Hook bridge. Review of River Basin Models in Use in the SADCC Region and recommendations for models and studies appropriate to the Zambezi Basin. Appendix A. SADCC AAA.3.4 Hydroelectric

Hydrological Assistance Project — Phase 1. Shawinigan Engineering, Lusaka.

SWECO 1968. Kafue River Regulation. Project report on main storage reservoir. Swedish Engineering Corporation, Stockholm, Sweden.

Tague, M. 1965. The Groundwater Reserves of the City of Lusaka, Zambia. Department of Water Affairs. Ministry of Lands and Natural Resources, Lusaka, Zambia.

Torrance, J.D. & Lineham, S. 1963. Meteorological Report on the Kafue Basin. Federal Meterological Department, Salisbury. Rhodesia and Nyasaland Meteorological Services.

UNDESD 1992. Diagnostic Assessment of the Water Sector in Zambia. Executive Summary and 4 Technical Appendices. Water Resources Branch, United Nations Department of Economic and Social Development, New York.

UNDESD 1993. Water Resource Management in North China. Water Resources Branch, United Nations Department of Economic and Social Development, New York.

Warren, G.D. 1972. Aspects of watershed leakage in the Upper Kafue Basin. Hydrological Survey of Zambia. Ministry of Rural Development. Department of Water Affairs, Lusaka.

Wimpey Labs 1983. Mpongwe Development Project. Review of the Groundwater Development Potential. Wimpey Laboratories Ltd., Hayes, Middlesex, UK.

World Bank/UNDP 1990a. Sub-Saharan Africa Hydrological Assessment SADCC Countries. Country Report: Zambia. World Bank/UNDP, Washington.

World Bank/UNDP 1990b. Sub-Saharan Africa Hydrological Assessment SADCC Countries. Regional Report. World Bank/UNDP, Washington.

37 The use of models in river basin flood control

DAVID RAMSBOTTOM
HR Wallingford, Howbery Park, Wallingford, UK

CATCHMENT FLOOD MODELLING

Selection of method

When selecting suitable techniques for assessment on a catchment wide basis it is important to balance simplicity against quality and reliability of results. A technique with intensive data requirements may produce detailed results but might be impractical to apply for a whole catchment. A technique which requires less data may be practicable for catchment wide application but may not provide reliable results because the analysis is over-simplified.

In the case of a catchment flood model a technique is required which is able to represent flood propagation in the catchment, and may also provide predictions of the following:

- Flood flows of different return periods
- Flood water levels
- Flooded areas
- Effects on flooding of changes in the catchment.

A number of techniques are briefly considered here, ranging from rainfall-runoff modelling to hydrodynamic modelling. The first step in these approaches is to divide the catchment into sub-catchments, for which runoff from rainfall can be estimated using standard hydrological techniques.

Rainfall-runoff modelling can be used to predict flood hydrographs for sub-catchments. The Flood Studies Report method which has been developed for flood prediction in the UK recommends that the area of sub-catchment which can be modelled using the method should not exceed 1000 km^2 (NERC, 1975).

The catchment can be divided into suitable sub-catchments, and hydrographs predicted for each sub-catchment. These could be combined to produce flood discharge hydrographs for the whole catchment taking into account the different response times of each sub-catchment. This method of combining hydrographs is, however, crude as it makes simple assumptions about the flood wave speed and does not take account of attenuation of the flood wave.

Flow routing is a technique for modelling the movement of a flood hydrograph down a river, and takes account of the variable speed of the flood wave, attenuation and flood storage. If combined with rainfall-runoff modelling, a catchment flood model can be developed which gives predictions of flood flows throughout the catchment. A catchment model of this type has been developed (Price, 1973, 1978) initially for UK rivers, but the technique is generally applicable to any river catchment.

Hydrodynamic modelling and flood plain mapping . The catchment model referred to above predicts flood discharges but does not predict water levels. To do this a hydrodynamic model is necessary, which requires relatively detailed survey information for the river channels and flood plains. Several models of this type have been developed (for example Price & Samuels, 1980; Samuels, 1983).

In order to predict flooded areas, detailed contour maps are required for the flood plains. Flooded areas are then estimated by comparing flood water levels predicted by the hydrodynamic model with ground levels. The process is data intensive, and is generally too detailed to be suitable for overall catchment management and planning.

The recommended technique for catchment flood modelling is therefore a catchment model incorporating rainfall-runoff modelling and flow routing. The model described by Price does not take specific account of the effects of flood plains which are separated from the river by flood embankments. This is an important enhancement, as many major river catchments have embanked flood plains in their lower reaches. In addition, user-defined rating curves can be used to obtain flood water level predictions at key locations.

The recommended technique for catchment flood modelling is already in existence, although some enhancements are desirable to improve the applicability of the model to catchments worldwide. The main purpose of this paper is therefore not to present new techniques, but to highlight the development and use of an existing technique as a basic component of integrated river basin management and planning.

The catchment flood model will provide a means of predicting flood propagation throughout a catchment in terms of discharge hydrographs. Changes to the catchment can then be modelled to provide predictions of their effects on flooding. Such changes might include climate changes resulting in changes to rainfall, changes in land use and therefore runoff characteristics, and flood alleviation measures.

Techniques for predicting flood water levels and flooded areas should be used for the parts of the catchment where this detailed information is required. Boundary conditions for these more detailed models in terms of flood hydrographs would be provided by the catchment model. In this way a relatively simple technique requiring limited data can be used for the whole catchment, with detailed information only being necessary for relatively small areas.

APPLICATION OF THE MODEL

The catchment flood model described by Price and subsequently developed as a software package (HR Wallingford, 1987) has been used to demonstrate the

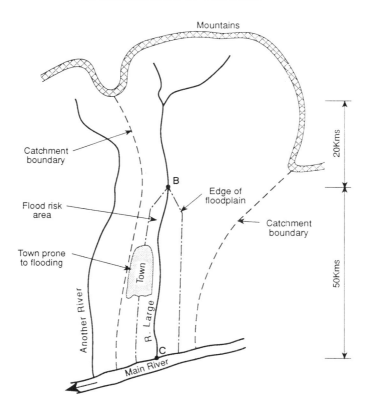

Figure 1 Existing example catchment

development of a flood control strategy for the imaginary catchment of the River Large, see Figure 1. The upper catchment is mountainous, with steep streams and rapid runoff. The lower catchment and the main flood risk area are relatively flat. The main flood risk area is 50 km long by about 1 km wide, and includes part of a large town.

Before constructing the model, it is first necessary to identify the flood problem. This can be done by mapping the known extent of flooding on a base map of the catchment, as shown in Figure 1. Having established the approximate extent of the flood risk area, methods of alleviating flooding can then be considered using the catchment model.

Model construction and calibrations

Characteristics of sub-catchments, including area, shape factors, slope, etc. These are required to determine runoff hydrographs from rainfall.

Data for routing reaches, including length, storage characteristics and hydraulic information. These are used to calculate flood wave speed curves and attenuation parameters.

It is desirable to calibrate the model, and this is done using observed rainfall and

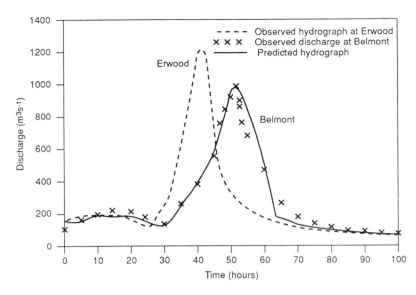

Figure 2   Predicted and observed hydrographs for floods of December 1960, River Wye

gauged river flow data. Observed rainfall data are applied to the model, and adjustments are made to the runoff and routing characteristics in order to obtain an acceptable match between predicted and observed discharge hydrographs at gauging sites. An example of the calibration of a routing reach is shown in Figure 2, where the effects of attenuation of the flood wave are apparent.

Design flood events
Having calibrated the model, it is then necessary to derive rainfall events which correspond to design floods of specified return periods. For example, the estimated 1-in-100 year flood event is commonly used in the UK for urban flood defence design in non-tidal reaches. The procedure is well documented for individual sub-catchments (for example, NERC 1975 for UK catchments).

For large catchments, where significant differences in aerial rainfall exist across the catchment, the situation is more complicated. There are a wide range of possible combinations of flows from different sub-catchments and a joint probability analysis is required to derive the required flood events.

For river basin planning purposes a relatively simple procedure is required, and a method involving the use of areal reduction factors is recommended (Bell, 1977). This is relatively easy to apply, but simplifies the flood generation mechanism. It is therefore recommended that design floods predicted by the model at gauging sites are compared with gauged discharge hydrographs for observed major flood events to ensure that the predictions are consistent with observed events.

When using gauged flow data both for calibrating the model and checking design flood hydrographs, care should be taken to ensure that the gauged flows are reasonably accurate. Gauging of flood flows is subject to large inaccuracies and

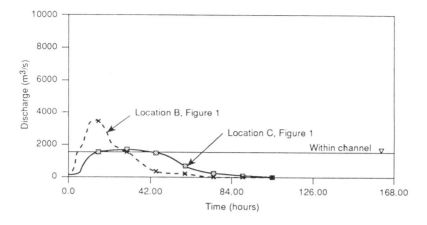

Figure 3 River Large catchment: flood hydrographs for existing conditions

uncertainties, particularly at sites where flood plain flow or significant bed movement occurs (Tagg & Hollinrake, 1987).

The first objective of the catchment flood model is to predict flood propagation in the catchment. A design flood has been generated for the catchment which has a peak flood discharge of 3,500 m³ s⁻¹ at location B on Figure 1, upstream of the main flood risk area. The shape of the hydrograph at location C on Figure 1 has changed significantly owing to attenuation and the peak discharge is reduced to about 1,800 m³ s⁻¹, as shown on the existing condition hydrographs on Figure 3. The approximate bankfull capacity of the river channel between locations B and C is 1,500 m³ s⁻¹.

River basin flood control planning
A catchment model of this type can form the basis of a flood warning system, and can be used to determine the flows at gauging sites in the upper catchment which will cause flooding downstream. Using a rating curve for the gauging site, these critical flows can be converted to critical water levels which are easily measured on a continuous basis. The model will also provide estimates of the time taken for the flood wave to pass from the gauging site to the flood risk area (for example, Dobson & Davies, 1990).

The catchment model can also be used to produce a flood control and alleviation strategy for the river basin. The river valley is represented in the model by a series of widely spaced cross sections which are sufficient to describe the approximate storage and hydraulic characteristics of the river. By making adjustments to these sections the impact on flood hydrographs of such works as embankment construction can be predicted.

For catchment planning purposes it is important to model the effects of reservoirs and storage ponds. Routines have been included in the model for both on-line and off-line storage ponds together with a range of possible control structures. Features such as flood relief channels and diversion channels can be investigated by changing the

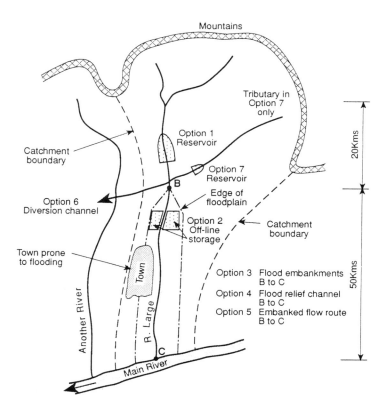

Figure 4 Flood alleviation options

configuration of the model.

The model of the sample catchment was used to investigate a range of flood control options intended to reduce flooding in the main flood risk area, and predict the changes these will cause to the design flood hydrograph. These options include an on-line reservoir, embankments, flood relief channels, a diversion channel and an off-line storage area. In addition, a tributary was added to the model and the effect of an on-line reservoir on the tributary was also investigated. These options are indicated on Figure 4, as follows:

Option 1 On-line storage reservoir upstream of location B
Option 2 Off-line storage area on the flood plain near location B
Option 3 Flood embankments 200m from the river in the main flood risk area
Option 4 Flood relief channel between locations B and C
Option 5 Embanked flow route on the flood plain between locations B and C, and embankments on the river
Option 6 Diversion channel to another catchment from location B
Option 7 Reservoir on tributary
It is clearly possible to prevent flooding in the flood risk area using an on-line

reservoir if it is big enough. In Option 1 the catchment model was used to determine the maximum permissible outflow, the required storage volume and the size of the outlet control structures. The model can also be used to demonstrate the effect of a flood which exceeds the capacity of the reservoir. Reservoirs often have other constraints including power generation and water supply, and the model can be used in the determination of operating rules.

Option 2 consists of diverting flood water into a 5 km² area of flood plain, and the effect of this on the design flood is illustrated on Figure 5a which shows flood hydrographs upstream and downstream of the storage area. It should be noted that the flood plain is below river bank level, and water begins to enter the storage area before bankfull flow is reached. Once the storage area is full the entire flood flow passes downstream, and the reduction in peak flood discharge and therefore level is small.

Option 2 illustrates the effect of a storage area or reservoir which is not big enough to accommodate a flood. The small change in peak discharge downstream means that the storage area causes only a minimal reduction in flood risk downstream and is of very limited use in the control of the flood. Very large areas of land are required for off-line storage areas on flood plains because of the shallow depth of water in the storage area.

The effect of embankments is to reduce the attenuation of the flood wave, resulting in higher flows and therefore water levels downstream. In the example given (Option 3), the effect of the embankments was to increase the peak discharge of the design flood by 300 m³ s⁻¹ at location C, as shown on Figure 5b.

A flood relief channel (Option 4) can contain the flood flow above the bankfull capacity of the river providing that it is big enough. The reduction in attenuation of the flood wave will however cause an increase in discharge where the flood relief channel rejoins the river.

The effect of the flood plain flow route (Option 5) on the discharge in the river downstream location B is shown on Figure 5c. Whilst the flood flow in the river is reduced it is still above bankfull level, and flooding will still occur unless flood

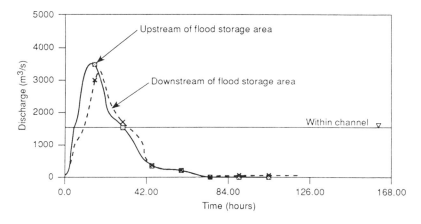

Figure 5a River Large catchment: flood hydrographs for off-stream storage area

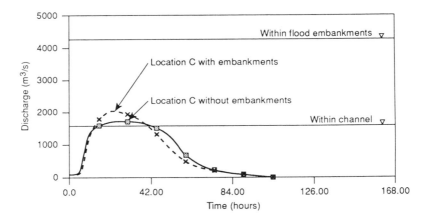

Figure 5b River Large catchment: effect of flood embankments on flood hydrographs

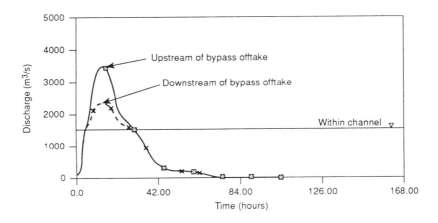

Figure 5c River Large catchment: effect of bypass flow route on flood hydrographs at Location B

embankments are provided. Figure 5d shows flood hydrographs at location C with and without the flood plain flow route. The reduction in attenuation of the flood hydrograph has changed the shape of the hydrograph and increased the peak flow by about 600 m³ s⁻¹.

A diversion channel to another catchment can eliminate the risk of flooding for the design flood if the diversion is big enough. The opportunities for diversion channels are very limited, and they may simply transfer the flooding problem elsewhere. They do however have an important advantage over storage schemes. When a larger than design flood occurs it is only the surplus flow above the design flood peak discharge which causes flooding.

For the final option a tributary has been introduced into the model. The tributary has a faster response to rainfall than the main catchment, as shown on the modified flood

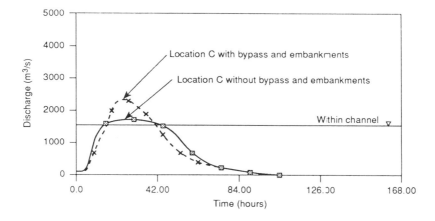

Figure 5d River Large catchment: effect of bypass flow route on flood hydrographs at Location C

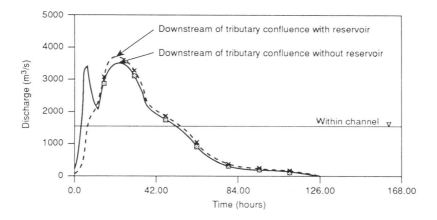

Figure 5e River Large catchment: flood hydrographs with reservoir on tributary

hydrograph for existing conditions at location B on Figure 5e. A reservoir was then introduced on the tributary to control flooding in the lower reaches of the tributary. The effect of the reservoir on the flood hydrograph at location B is also shown on Figure 5e; it can be seen that the peak flood flow is increased by about 250 m^3 s^{-1}.

This is because flood water from the tributary passed location B before the main flood hydrograph under existing conditions. The reservoir discharged the flood water over a longer period which coincided with the peak of the main flood hydrograph, thus increasing the peak flow. This example illustrates the importance of using a catchment model in the design and operation of storage facilities.

In summary, these examples illustrate how a relatively simple catchment flood model can be used to predict design flood discharge hydrographs throughout a river basin, and identify the effects of flood control works on flood flows. Peak flood flows can be converted to peak flood water levels at key locations using suitable rating curves. The model can be used to develop an overall flood control strategy for a river basin, and provide design flood flows for more detailed engineering studies of flood control options.

INTEGRATED RIVER BASIN PLANNING

Integrated river basin plans must consider all features of the river basin including river flows and water levels, groundwater, water resources, water quality, flood propagation, land drainage, erosion and deposition, ecology and fisheries.

These features are not independent: water resources depend on both river flows and groundwater; land drainage depends on river water levels; and both ecology and fisheries depend on river flows and water levels. For example, proposals to lower river water levels to improve land drainage may affect the ecology of the river corridor. The link between ecology and river water levels must therefore be understood.

For each feature a description is required in which significant aspects are identified. Suitable information must therefore be collected and techniques are then required which use the information to describe the behaviour of the catchment, and enable the impact of changes to be assessed. These must take account of interactions between the different features.

During a seminar on catchment management at HR Wallingford in 1993 the catchment flood model was described together with simple methods of assessing water resources, water quality, ecology and fisheries on a catchment wide basis. It was concluded that these form an essential basis for river basin management and planning (HR Wallingford, 1993).

These methods include modelling and other forms of assessment. Such techniques provide the basic technical information required to undertake integrated river basin planning. These fundamental "building blocks" might include modelling of surface water flows, river water levels, water resources, water quality, and flooding. The links

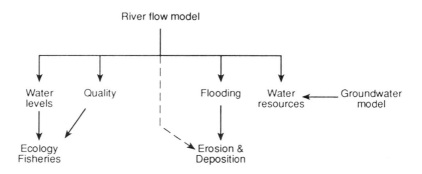

Figure 6 Techniques for integrated river basin planning

between these models and databases containing information on other features of the catchment are shown on Figure 6.

When changes to a river basin are proposed the effect on each feature of the basin should be investigated. The process will be to some extent iterative because of interactions between the various features.

Some progress has been made in the development of catchment models which combine surface and groundwater flow. However, the more complex the techniques become the more data intensive they are likely to be. Techniques for river basin planning should ideally be kept as simple as possible in order to enable integrated catchment plans to be developed within reasonable budgets.

CONCLUSIONS

A catchment flood model which includes rainfall-runoff modelling and flow routing is recommended for river basin flood control management and planning.

The effects of flood alleviation and control measures on flood flows within a river basin have been demonstrated using a catchment flood model.

To undertake integrated river basin planning and management it is recommended that a range of relatively simple techniques are used to provide basic technical information on different features of the river basin. These aspects include river flows and water levels, groundwater, water quality, flooding, land drainage, erosion and deposition, ecology, and fisheries.

REFERENCES

Bell, F. 1977. The areal reduction factor in rainfall frequency estimation. *IH Report No. 35*. Institute of Hydrology, Wallingford, UK.

Dobson, C. & Davies, G.P. 1990. Integrated real-time data retrieval and flood forecasting using conceptual models. *International Conference on River Flood Hydraulics*. Hydraulics Research Ltd. Paper M5. 469-479.

HR Wallingford 1987. RBM-DOGGS User Manual. Hydraulics Research Limited.

HR Wallingford 1993. *River Catchment Management and Planning*. Seminar at HR Wallingford Ltd, 24 June.

NERC 1975. *The Flood Studies Report*. Natural Environment Research Council, 5 volumes.

Price, R.K. 1973. Flood routing methods for British rivers. *Proc. Instn Civ. Engrs*, **55** 913-930.

Price, R.K. 1978. A river catchment flood model. *Proc. Instn Civ Engrs, Part 2*, **65**, 655-668.

Price, R.K. & Samuels, P.G. 1980. A computational hydraulic model for rivers. *Proc. Instn Civ. Engrs*, Part 2, **69**, 87-96.

Samuels, P.G. 1983. Computational modelling of flood flows in embanked rivers. Int. conf. on hydraulic aspects of flood and flood control. BHRA Fluid Engineering, Cranfield, UK, Paper H1, 257-270.

Tagg, A.F. & Hollinrake, P.G. 1987. *Flood Discharge Assessment: Current UK Practice*. Hydraulics Research Ltd, Report SR 111.

38 Strategies for handling uncertainty in integrated river basin planning

MICHAEL J CLARK
Geography Department and GeoData Institute, University of Southampton, UK
JOHN GARDINER
National Rivers Authority, Thames Region, Reading, UK

UNCERTAINTY AS A BACKDROP TO RIVER BASIN PLANNING

Integrated river basin management and water resource planning both take place within a context of uncertainty which derives from the fact that suppositions repeatedly have to be made (and justified) about many of the attributes of the systems concerned. While attention has traditionally focused on the physical and chemical systems which drive aspects of water quality, quantity and hydraulics, the same observation could be made about the biological or even the socio-economic aspects of river basin or channel management. It follows that effective planning must take account of any major shifts in the assumptions being made about any of the systems involved. Any uncertainty concerning the assumed nature or rate of the processes involved will therefore be reflected in related uncertainties of management or planning.

In the past, the tradition has arisen for elements of uncertainty in science and engineering to be regarded primarily as products of model simplicity, limited data availability and inadequate measurement resolution. Since any inherent uncertainty has in effect thus been explained as being the result of technical constraints, the implications have been handled through such correspondingly technical devices as the use of error margins, confidence limits and factors of safety. In essence, uncertainty has tended to be regarded as a temporary deficiency in the planning procedure, and the primary aim of that procedure has thus been to remove or find ways of safely ignoring the uncertainty.

Several recent developments have combined to suggest that these traditional approaches will need significant supplementation before the end of the century. Growing awareness of the likelihood (though not necessarily of the nature or pace) of global environmental change, an increased appreciation of the complex interlinkage within and between systems, and a widespread acceptance that some key elements or externalities of system behaviour are non-linear (and thus non-predictable), all combine to necessitate a new attitude to medium-term river basin planning.

At the same time, the growing social expectation that development and environmental exploitation should be achieved on a sustainable basis further adds to the pressure for creating strategies which are not vulnerable to failure due to system or predictive uncertainty. Since the social and economic cost of such strategies is also increasing, there is an incentive to project return on investment across longer periods so as to establish the associated business case, and this too generates an increased sensitivity to uncertainty. Government in the UK has previously provided guidance on environmental appraisal, and economic appraisal is commonplace. By now moving towards a deeper consideration of social appraisal, we seek to achieve the combination of environmental, economic and social principles which together underpin sustainable development.

BUILDING UNCERTAINTY INTO THE PLANNING MODEL

In practice, river basin and channel management have tended to be based on sufficing compromise perhaps based loosely on functional strategy and objectives rather than on optimising ideals for sustainable development. While this sub-optimal achievement may relate to management deficiencies or limited resources, it is more likely to be the product of conflicting planning aims. Until the aims are resolved, management is bound to under-perform in the satisfaction of one or more of the interest groups involved. Should a river channel be engineered to maximise its conductance of flood flows, managed to enhance environmental quality, or stabilised to control land-owner losses? The management strategies appropriate to each of these three aims may differ to greater or lesser extent, and while very effective and efficient compromises can be designed, ultimately their performance will have to be judged against a prioritised list of goals. This daily challenge of catchment and channel management is greatly exacerbated if significant uncertainty in system prediction is acknowledged.

Uncertainty concerning externally-imposed future states and operations of the systems that we wish to control poses limitations on our ability to target and monitor the natural resource management goals of development planning and environmental management. This is an unavoidable challenge to catchment planning, and cannot be neutralised as a matter either of political convenience or of financial conservatism. The effects of such uncertainty can, however, only be gauged if we have some concept of what the "metagoals" of sustainable development planning and environmental management (to fulfil Local Agenda 21: LGMB, 1992) actually are. This is no easy task. Far from the literature being mute on such matters, it overwhelms us with possibilities —all of which are the delight of those who produce corporate mission statements, but are often the despair of the more pragmatic day-to-day catchment managers. Table 1 is a list of design criteria for the planning goals set by society suggested by Lonergan (1985).

This list is helpful, but could hardly be regarded as comprehensive. For example, Cocklin (1989) adds several important elements which are of great relevance to integrated river basin planning (Table 2).

Table 1	Design criteria for the planning goals set by society suggested by Lonergan (1985)
	adaptivity
	flexibility
	freedom of action
	reliability
	resilience
	robustness
	stability
	inertia

Table 2	Further important elements suggested by Cocklin (1989)
	efficiency
	equity
	sustainability

The last item listed in Table 2 is of particular significance: it is currently a high priority social goal, yet it appears to be particularly sensitive to uncertainty in system definition and prediction. It is tempting to ignore such issues on the grounds that they are too philosophical or academic for practising catchment planners and engineers, but such comforting myopia overlooks the fact that ultimately these *are* the social and economic imperatives which should drive our institutions and ultimately our companies. Moreover, their challenge is increased by the fact that as aims they almost certainly conflict with one another, as Cocklin (1989) recognises in questions such as:

Is a system that is sustainable also one that is resilient and robust? Are sustainable systems necessarily efficient? Does a sustainable system necessarily imply freedom of action?

It is clear that in practice the aims which our planning system seeks to achieve are potentially conflicting, and they are also threatened to varying degrees by management uncertainty. It has been shown above that there are good reasons for suggesting that environmental systems are, and will remain, *inherently* uncertain across timescales appropriate to planning and management. If this is the case, then it may be that the idealistic list of social metagoals for environmental planning require fundamental review. One way of achieving this revision is to take the goals that are commonly promulgated as social priorities, and recast them as a linked hierarchy of reducing practicability such that emphasis is placed upon a small number of all-embracing but achievable aims, with lesser priorities and lower levels of achievability being separately recognised:

High-order goals	*Supporting aims*	*Less achievable ideals*
sustainability	reliability	stability
adaptability	flexibility	efficiency
resilience	robustness	inertia
consensus	equity	freedom of action

Far from being an exercise in semantics, this approach actually increases the practical value of goal-setting by reducing the number of targets, by resolving many of the internal conflicts between them, and by controlling the vulnerability to predictive uncertainty. The high order goals can be viewed as absolutes (of intention if not of achievement), while the lower levels in the hierarchy remain as aims or ideals, but it is accepted that they will be compromised in practice.

Any such hierarchy will inevitably be highly contentious, since it relegates some cherished ideals to the realm of the "less achievable". Few may argue with assigning *stability* and *inertia* to this category, since these are system attributes which accord ever less comfortably with current thinking on the functioning of the environmental system. To abandon the maintenance of *equity* (the total stock of environmental physical/biological attributes) and *efficiency* to the same fate, however, is certainly greatly more subjective and probably far less acceptable. However, such a possibility remains a valid consideration if system uncertainty is accepted as having significant and fundamental effect on practical management. It is for this reason that *consensus* has been added to the list of high order social metagoals. If some social priorities are to be regarded as beyond reasonable achievement (at least in the short term), then there must be a general social acceptance that this is the case and that the goals that have been substituted represent a viable and respectable compromise. Goal "substitution" may be seen as an echo of the equity transfers inherent in concepts of soft or weak sustainability, and it is for this reason that such apparently philosophical issues **must** be debated by practising planners and managers, not just by academics.

UNCERTAINTY AND SUSTAINABILITY

The point has been made that there is potential conflict between even the members of the reduced list of social metagoals for environmental or river basin planning, and that central to this conflict lies the added difficulty of achieving sustainable development when system attributes and behaviour are subject to significant uncertainty. Over the past decade, environmental sustainability has moved from being a philosophical ideal to becoming a core commitment of governments (central and local) and agencies. Typical of this transformation is the reformulation of the sustainable development philosophy inherent in the Agenda 21 Document agreed at the "Earth Summit" (the UN Conference on Environment and Development, Rio de Janeiro, 1992) by the UK Local Government Management Board (LGMB, 1992) to provide a practical guide for UK local authorities. Two key paragraphs are:

- *Water resources must be planned and managed in an integrated and holistic way to prevent shortage of water, or pollution of water sources, from impeding development. Satisfaction of basic human needs and preservation of ecosystems must be the priorities; after these, water users should be charged appropriately.*
- *By the year 2000 all states should have national action programmes for water management, based on catchment basins or sub-basins, and efficient*

water use programmes. These could include integration of water resource planning with land use planning and other development and conservation activities, demand management through pricing or regulation, conservation, reuse and recycling of water.

The aims of sustainable development and the threats of uncertainty are both clear throughout these paragraphs. The task of establishing for practical planning purposes the links between sustainability and uncertainty is rendered problematic by the wide range of current definitions of sustainable development. The concept was launched most powerfully into the public arena by the Brundtland Report, which saw sustainability as development which meets the need of the present generation without compromising the ability of future generations to meet their own needs (WCED, 1987). Agencies such as IUCN and UNEP adopted an approach based on the notion that sustainable development should improve the quality of human life while being within the carrying capacity of supporting ecosystems, while Pezzey (1992) argued that natural capital assets should not decline through time.

Sustainability is now a fashionable concept most often applied to resource use and environmental management. In these contexts it is generally perceived as having overtones of a long-term as opposed to a short-term perspective, and often embodies ideas of renewability, resilience, and recovery. *Renewability* signifies the rate at which a resource can be replaced, so that sustainability is achieved by restricting the level of use to something at or below the rate of replacement. *Resilience* signifies ability to withstand stress without long-term or irreversible damage — and may be seen in notions such as the carrying capacity of grazing land or the erosion-resistance of a footpath. Sustainability is achieved by restraining use to a level at or below that which exceeds the system's resilience. *Recoverability* is a concept which accepts that detrimental impact may take place, but concentrates on the rate or frequency of impact in relation to the inherent (or artificially supported) rate of recovery. A sustainable system is one in which impact intensity and frequency are small in relation to recovery rate — as is the case with the entirely sustainable traditional slash-and-burn cultures of the tropical humid zone. All of these concepts have great relevance to catchment planning, and all involve elements which are vulnerable to deficiencies due to uncertainty in the definition of present or future system characteristics.

In each of these approaches to sustainability, holistic attitudes are found to be central: the system is viewed as inter-connected, and short-term decisions are seen to have long-term direct and indirect consequences. Against such a background, it can be seen that the traditional view of sustainability is one which emphasises the need to maintain an equilibrium (albeit a dynamic equilibrium) within complex natural and anthropomorphic systems. A single system may have several (or many) possible equilibrium states — each sustainable in its own right. Such a view of sustainability emphasises that the concept may be effectively meaningless when restricted to a single domain (e.g. forestry). Yet the achievement of holistic sustainability (covering the interactions between subsystems) would require a holistic system understanding that would greatly multiply the uncertainty about the present or future state of the system. Thus, the move towards sustainable development redoubles the need to identify approaches which are robust and effective in the context of the overall system.

COPING WITH UNCERTAINTY: STRATEGIES FOR RESPONSE

The challenge of handling uncertainty as an accepted part of the catchment planning procedure rather than as a flaw in it can be tackled from three points of view. *Technical adjustment* focuses on the traditional devices for coping with error, and involves a growing use of risk assessment. *Adjustment of management strategy* offers a series of possibilities relating to optimising the scheduling of remedial investment and modifying the balance between capital works (which require a fixed design specification) and maintenance works (which can more easily adopt a rolling specification). *Attitudinal adjustment* requires a major re-evaluation of procedures so as to manage public and institutional perceptions and expectations in line with changing views on system behaviour.

Adaptability and flexibility are often the most effective working responses to predictive uncertainty. It follows that the most suitable strategies will be those that permit the greatest degree of reformulation as goals and parameters change. Of the many possible outcomes of this suggestion, the one with perhaps the most far-reaching implications is the preference for a shift wherever appropriate from capital projects to maintenance works. Such a bald claim requires very careful qualification in practice, but there is an underlying significance in the fact that capital projects tend to require a detailed and broad-ranging prior design specification which can incorporate uncertainty only through the relatively coarse control of increased nominal safety margins. On-going maintenance works and soft engineering approaches, on the other hand, can more readily be adjusted to a changing specification during implementation, and thus provide a highly adaptive approach which accords well with one of the primary social goals of environmental planning. This advantage notwithstanding, it must be acknowledged that the preparation of any initial business case or associated cost-benefit analysis, and the undertaking of the necessary internal and external consultations, will both require at least a nominal set of working specifications. It may be that both business case evaluation and environmental impact assessment should henceforth routinely consider the extent to which the project/works concerned have incorporated a satisfactory degree of flexibility to absorb system reformulation during the project lifespan. Such flexibility will need to provide a capability for change at least as great as the accumulated effects of uncertainty.

Not least amongst the problems faced by the uncertainty-conscious river basin planner is the awareness that inbuilt project flexibility makes it ever more difficult to demonstrate convincingly that effective target levels of system resiliency can be achieved. Indeed, one of the major hurdles to be overcome in designing planning and management procedures which are resilient in the face of uncertainty is to achieve effective target setting and performance monitoring despite the flexibilities of prediction. It is already apparent that one intuitively-attractive but inherently dangerous reaction is to reduce the timescale across which detailed targets are set. This harks back to the notion that uncertainty is a temporary deficiency which will be removed or reduced with the passage of time, and fails to grasp the management necessity of periodic routine retargetting. Only the latter can avoid still further drift towards short-term strategy setting, which already bedevils some aspects of river and basin planning. In its most extreme form, such "short-termism" develops into an excuse for total inactivity pending the removal of uncertainty —an approach which appears attractive to many politicians, but which should be regarded as absolutely unacceptable by those with the responsibility (often the statutory obligation) to plan and manage.

Perhaps surprisingly, it appears that the key uncertainties that we face in environmental management are socioeconomic not physical. The latter category is unlikely to be more than 100% in error; the former is almost certain to be in areas such as the future structure of EC agricultural subsidies or the base lending rate in the year 2050 AD. This places great demands on the need for re-education of those who make decisions so that they come to terms with uncertainty in general, and that they appreciate that physical science is not the weakest link in the chain.

One response is to recognise that much emphasis should be given to incremental investment or the phasing of decisions, so as to optimise the extent that we are able to reduce the negative impact of uncertainties and at the same time minimise the increasing costs of inactivity if we wait for a certainty that never comes. If the balance of *activity* is to shift to one driven by uncertainty by the end of the Century, then attitudinal and procedural revision must start changing now, since with other comparable institutional/NRA changes of this magnitude the lead time has been in excess of five years. Once again, we see the importance of re-education as a trigger to improved long-term management.

IMPLICATIONS OF UNCERTAINTY

The main challenge of coping with uncertainty is that we can never be sure which strategy is best, nor when to adopt it, thereby rendering it all the more important to agree on multi-functional principles and on an assessment methodology from which to derive a "best guess". In practice, the demands of uncertainty may reverse the trend towards a "mechanistic" technology-driven decision process, and restore judgement to the managers, planners or their supporting teams. Alternatively, the decision technology itself must become more flexible, as with expert systems and neural network approaches. Such techniques may have very different social implications, with the former reinforcing the sense of ownership and involvement, while the latter may reduce direct human input to decision making. Above all, however, uncertainty must now be regarded as an inherent part of the system rather than as a deficiency or aberration which can be removed in time—and this does imply a total change of perspective if not of working practice.

The outcome of the above review is relevant to all of those involved in data acquisition/manipulation, the formulation of the prevailing working culture within catchment management, and the design of professional training and management development programmes. The implications and the responsibilities are wide ranging but at risk of oversimplification, we can identify a priority list of actions that could generate a significant and rapid improvement in the ability of catchment management to cope effectively with the challenge of uncertainty:

- Identify and achieve agreement on principles of sustainability, and on methodology for assessment of policies and strategy in the light of sustainable development. The foregoing discussion has pointed towards social criteria of uncertainty to accompany the emerging environmental and economic criteria within an overall framework of sustainability.
- Increase project adaptability by encouraging a shift from capital to maintenance investment, and an associated focus on soft rather than hard engineering practice, achieving resilience through the self-renewing capability of natural systems.
- Concentrate on modelling investment phasing so as to achieve the ideal balance between waiting for further information and implementing immediate action. Incremental investment will allow adaptation to unforeseen circumstances at minimum cost.
- Reduce the negative impacts of uncertainty through enhanced communication within and between teams, and through a renewed emphasis on public involvement and consensus to focus on sustainable values —guided by the application of the precautionary principle to (modelled) information.
- Adjust the perceived values to be sought through environmental projects by the application of environmental economics, thereby providing a new (and perhaps more robust) value framework for decision makers.
- Adjust corporate mission statements to accord more closely with the practicalities of environmental good practice rather than lofty ideals.
- Make an immediate start on the task of laying a foundation for revision of attitudes (re-education) both within the existing environmental management and environmental engineering professions, and in the training of the next generation of professionals. A parallel exercise in the education of the public will also be required.

In themselves, such actions will not remove uncertainty. It is possible that in practical terms they will do little to reduce uncertainty. But they do offer a genuine contribution towards the task of making catchment management processes better understood, so that catchment managers are better able to provide optimum decisions in the highly suboptimal decision-making circumstances which are likely to prevail in the future.

REFERENCES

Cocklin, C.R. 1989. Methodological problems in evaluating sustainability. *Environ. Conserv.* **16**, 343-351.

LGMB 1992. *Agenda 21: a guide for local authorities in the UK*. The Local Government Management Board, England.

Lonergan, S. 1985. Characterizing sub-optimal solutions in regional planning policy. *Socio-Economic Planning Sciences*, **19**, 167-177.

Pezzey, J. 1992. Sustainability: an interdisciplinary guide. *Environmental Values*, **1**.

WCED 1987. *Our Common Future*. The Brundtland Report, World Committee on Environment and Development. Oxford University Press, Oxford, UK.

39 Water management in Danubian floodplain forest ecosystems

JÁN LINDTNER, ROMAN KOSTÁL and RASTISLAV SLOTA
Water Research Institute, Bratislava, Slovakia

INTRODUCTION

Danubian floodplain ecosystems in southwestern Slovakia represent a unique phenomenon in Central European geobiocœnosis, not only for the species and productivity, but also for their specificity of feedbacks (Miklós *et al.*, 1989). Water and its dynamics are the most important factors influencing the interaction between abiotic and biotic elements in this territory. The existence of floodplain forest is dependent on geomorphological and hydrological settings, above all on there being sufficient water storage in the soil profile during the first half of the growing season.

The natural water regime in this area has been altered by hydraulic engineering measures: some of these have induced river bed erosion and a further lowering of the groundwater table. At the end of October 1992, the hydrological settings in the floodplain were changed fundamentally by the damming of the Danube River, when a substantial part of the surface water was led from the original river into the new diversion canal.

This paper describes the water regime in the Danubian floodplain before the Gabcíkovo waterworks came into operation, bringing with it the possibility of conservation of the floodplain forest ecosystems by suitable water management.

SITE DESCRIPTION

The territory under investigation is located on the left floodplain of the Danube River, about 40-50 km downstream of Bratislava, 1812 to 1840 km from the source, bounded by the river itself and by the old levee. Practically the whole area is covered by floodplain forests, associations of *Querceto-Fraxinetum* with introduced poplars (*Populus* spp.) being the most frequent.

The geomorphology of the Danubian floodplain area has been formed by accumulation of sediments after floods, and consequently agradation mounds have developed along the riverside. The territory is interwoven by a system of branches, creating a unique inland delta. The morphological structure of the ground surface is highly varied, with inconspicuous extensive flat elevations and many depressions. Every small difference in the microrelief of the ground surface induces considerable changes in the height and duration of floods, in the position of the groundwater table and in the distribution of soil moisture.

The geological structure of the alluvial aquifer consists of unconsolidated deposits composed of gravel, sand, silt and clay with various hydrophysical properties. Throughout the whole territory, these fluvial deposits are covered by flood deposits with loamy, sandy-loamy and fine-sandy granulotypes. The clay content obviously decreases with depth, but the sand content increases. The thickness of flood deposits varies from 1.5 to 4.0 m, most frequently 2 to 3 m.

HYDROLOGICAL SETTINGS BEFORE THE DAMMING OF THE DANUBE

The water regime in the Danubian floodplain forests is the result of interactions among surface water, groundwater, soil moisture, atmosphere and vegetation. The driving force in the hydrological phenomena is the flow regime in the Danube River channel, which is clearly alpine in character. The highest discharges usually occur in summer, when the average monthly discharge reaches its maximum in June/July. Minimum discharges are recorded in winter. Before the Gabcíkovo waterworks came into operation, the left-hand branch system was supplied by infiltration from the Danube. Direct water exchange between the river and branches was possible at discharges over 3000 $m^3 s^{-1}$, which occur on approximately 35 days per year. At higher water stages (over 4300 $m^3 s^{-1}$), the area had flooded periodically, approximately 10 days per year.

Groundwater is closely connected with surface water. Recharge of the aquifer by infiltration from the Danube and its branches, as a result of the geomorphological structure of the territory, took place over practically the whole year. The annual precipitation in this region is about 580 mm and, compared with the influence of the Danube, does not play an important role in the groundwater regime. The fluctuations in groundwater level were synchronous with the fluctuations of surface water level, with an amplitude of approximately 2-3 m. The maximum water table, observed from May to July, can reach the ground surface. In autumn, at the low water stage, the minimum water table lay below the sandy-loam/gravel-sand interface.

The saturated and unsaturated zones are not independent units but are parts of a continuous flow system: the position of the groundwater table affects the moisture distribution above it. The soil profile was wetted in the first half of the growing season, with values for the soil water content never lower than field capacity (approximately 30-40 volume per cent). The water exchange between groundwater and soil moisture obviously took place through the capillary fringe. At the low water stage, when the water table fell below the sandy-loamy deposits, this capillary rise was interrupted and the moisture content decreased as a result of evapotranspiration and percolation. The minimum soil moisture content was about ten per cent.

BASIC DESCRIPTION OF GABCÍKOVO WATERWORKS

The waterworks has been constructed for the multipurpose exploitation of the Danube downstream of Bratislava, including the improvement of the navigation, the generation of hydroelectricity, water supply and recreation. Furthermore, it provides suitable conditions for flood protection and nature conservation on the adjacent territory.

The construction consists of two main parts: (i) the Hrusov reservoir and (ii) the diversion canal. The location of the waterworks is shown in Figure 1. The Hrusov reservoir is situated in the inundation area and extends to Bratislava. Its total volume is 143 million m^3, with a surface water area of 30 km^2. The water level in the reservoir is elevated by weirs at Cunovo.

The diversion canal, serving navigation, hydropower and flood control, is able to deliver 4000 m^3 s^{-1} at a mean velocity of 1.0 m s^{-1}. This canal is split by the Gabcíkovo power plant into head-race and tail-race sections. The hydroelectric power plant is located at a distance of 17.0 km along a head-race canal on the right side, with two navigation locks on the left side. The tail-race canal, 8.2 km long, returns to the Danube at Palkovicovo, 1811 km along the river from its source.

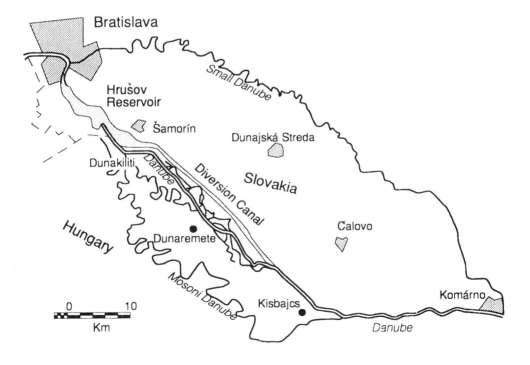

Figure 1 Layout of the Gabcíkovo waterworks

ENVIRONMENTAL REQUIREMENTS FOR WATER REGIME

Any major modification of the hydrological system results in some retardation in response by the ecosystem trying to adapt to the new conditions. The harmonising of the influence of the Gabcíkovo waterworks with the natural system of the Danube, including the floodplain forests, has therefore been considered.

The Danubian floodplain forests result from the physical forces, particularly from hydrological and geomorphological settings; hence water and its dynamics are very important factors in this territory. Because the precipitation deficiency is considerable in comparison with evapotranspiration, the frequency and the duration of flooding, the position of the groundwater table and the subsequent soil moisture are among the main determinants in the functioning of floodplain forests. In order to design an environmentally sound water management system for conservation of floodplain forests, it is necessary to understand the role of water in these ecosystems.

Because the hydrological system is a continuum, any modification of one component has an effect on the contiguous abiotic and biotic components. For example, the degradation of the Danube River bed caused by some hydroengineering measures and the gravel excavation were immediately reflected by a successive lowering of the groundwater table, with soil moisture content consequently decreasing biomass productivity.

To minimise the negative impacts of the Gabcíkovo waterworks on floodplain forest ecosystems, it is necessary to fulfil the following conditions:

- to control the water regime in the old river channel and branch system;
- to achieve such conditions for infiltration of surface water to the aquifer as will elevate the fallen groundwater table back up to the sandy-loam/gravel-sand interface;
- to secure permanent water flow in the branch system;
- to improve the soil moisture balance in the root zone;
- to secure artificial flooding periodically in forested territory in the growing season, and
- to monitor groundwater table fluctuations and soil moisture distribution.

WATER MANAGEMENT SCHEME

During the construction of the large Gabcíkovo waterworks, a substantial part of the river water was led from the original Danube River course to a new diversion canal and, as a consequence, the inundation frequency and height of the groundwater table have changed considerably. Therefore, on the basis of knowledge from other catchments (e.g. Dister *et al.*, 1990; Pautou & Decamps, 1985; Rood & Mahoney, 1990), environmentally sound hydroengineering measures for the conservation of the floodplain forests were integrated into a waterworks project. There are two alternatives for water management in this area: (i) by weirs in the old river channel and (ii) by the artificially-filled branch system. By means of mathematical models,

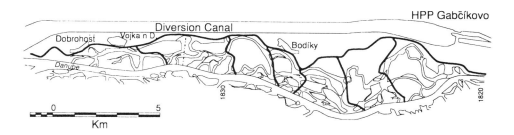

Figure 2 Layout of the branch system

it was found that water management by the branch system is more effective.

In order to supply the branch system it was necessary to build the inlet structure 1.7 km along the diversion canal, reaching the branch system 1838.8 km along the river. The inlet structure consists of three flap gates with total discharge capacity of 250 m³ s⁻¹. The branch system is split into seven relatively independent sections by dividing lines traced on the existing forest roads and small dams (Figure 2). Dividing lines include weirs and spillways to make the interaction of water between sections possible. The outlets of branches to the old Danube channel were closed and the left-hand river bank may be used as an access route to the sections, in order to maintain and repair the structures.

The base discharge of 28.2 m³ s⁻¹ through the inlet structure fills up the single sections to the level that corresponds to the discharge capacity of the hydraulic structures situated on the dividing lines. In consequence of the losses in the system, the minimum outlet of 5.0 m³ s⁻¹ is assumed to be at the end of the system. By increasing the discharge from the inlet structure, it is possible to supply the system during extremely dry periods, as well as to simulate natural floods, whenever the forest associations require it.

It can be seen that, according to this design, water exchange between the old channel and the river branches system will take place only during high natural discharges in the Danube or by artificial flood waves through the Cunovo weirs to the old Danube. A more intensive exchange will occur if the small weirs or other hydraulic structures in the old river channel are built, but this depends upon negotiations with the Hungarian authorities.

In order to investigate the hydraulic regime of the branch system, the three-dimensional physical model (1:250/50) was built. The aim of the hydraulic research (Sikora & Slota, 1992) was to check the capability of the designed solution for the water supply system. To verify the discharge capacity of the structures, to establish the relationships between the discharge through the inlet structure and total flooded area and to estimate the water demand for the flood simulation, it was necessary to carry out a number of experiments with discharges ranging from 28.8 m³ s⁻¹ (basic discharge) to 234 m³ s⁻¹ (flood discharge). For a more detailed investigation of the unsteady regime, which would be quite difficult to achieve with the physical model, the mathematical model was used to optimise the water management system (Sumbal *et al.*, 1992).

CONSEQUENCES OF WATER MANAGEMENT

The water management system in the floodplain area has been in operation since the beginning of May 1993 when the branch system was filled with water from the diversion canal through the inlet structure. During the growing season the discharge through the branches was about 60 m³ s⁻¹, and in winter approximately 35 m³ s⁻¹.

The impact that the damming of the Danube River and the subsequent water management has had on the groundwater regime in the floodplain area is shown in Figures 3 and 4. It is obvious that the fluctuations in groundwater level have decreased in comparison with pre-dam conditions (Figure 4), but that it is possible to re-establish the mean groundwater levels in a substantial part of the floodplain forests. This means that the root zone in the soil profile can be supplied via capillary rise from groundwater.

As Figure 5 shows, the soil moisture balance after the filling of the branch system is equal to, or better than, pre-dam conditions. Above all, an increase in the soil moisture content was observed in the lower soil profile, with sufficient water storage of soil water to support the stability of forest ecosystems in the left-hand part of the Danube inland delta.

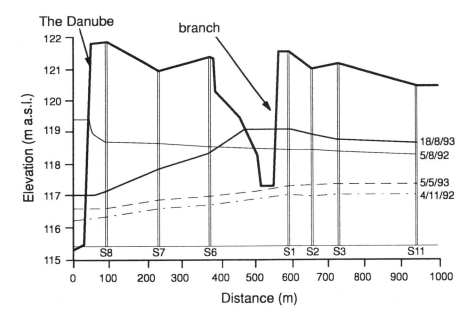

Figure 3 Groundwater level progress: cross-section in the river at 1834.5 km

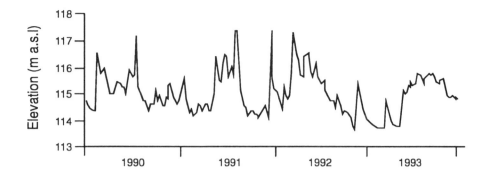

Figure 4   Fluctuations of groundwater level in observation well K14

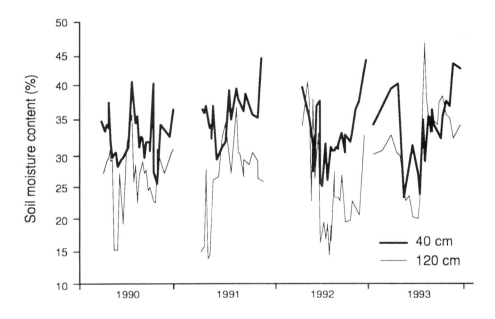

Figure 5   Soil moisture content profiles at 40 and 120 cm depths

CONCLUSION

The water management system presented in this paper was developed in the context of the design of the Gabcíkovo waterworks for conservation of the Danubian floodplain forests. It has been shown that this system is able to raise the groundwater table and thus to provide suitable conditions for capillary rise from the groundwater to the root zone of a soil profile. Furthermore, since the implemented system provides for the simulation of natural floods via existing branches, more than 60 per cent of the inundation area can be flooded several times per year.

ACKNOWLEDGEMENT

The authors thank Mrs Daniela Dzupová for the kind provision of her soil moisture data.

REFERENCES

Dister, E., Gomer, D., Obrdlik, P., Petermann, P. & Schneider, E. 1990. Water management and ecological perspectives of the Upper Rhine's floodplains. *Regulated Rivers: Research and Management* **5**, 1-15.

Miklós, L., Lisický, M. & Kozová, M. 1989. Ecological evaluation of the territory of the Gabcíkovo - Nagymaros dam project. *Ecology (CSSR)* **8**, 167-177.

Pautou, G. & Décamps, H. 1985. Ecological interactions between the alluvial forests and hydrology of the Upper Rhone. *Archive für Hydrobiologie* **104**, 13-37.

Rood, S. B. & Mahoney, J. M. 1990. Collapse of riparian poplar forests downstream from dams in western prairies: probable causes and prospects for mitigation. *Environ. Manage.* 14, 451-464.

Sikora, A. & Slota, R. 1992. *The discharge and level regime of the left-side branch system of the Danube from the physical model, river km 1840 - 1820.* Technical report. WRI, Bratislava.

Sumbal, J., Brezinová, E. & Topolská, J. 1992. *The discharge and level regime of left-side branch system of the Danube, river km 1805-1842, computations on the mathematical model.* Technical report. WRI, Bratislava.

40 The role of integrated catchment studies in the management of water resources in South Africa

F A STOFFBERG and F C VAN ZYL
Department of Water Affairs & Forestry, Pretoria, South Africa
B J MIDDLETON
Steffen, Robertson & Kirsten, Northlands, South Africa

INTRODUCTION

Integrated catchment studies in South Africa (also referred to as basin studies) were initiated by the Department of Water Affairs and Forestry in 1985. To date, some sixteen basins have been investigated or are currently under investigation, with the multi-national basins with known water problems being tackled first. There has been at least one catchment study in each broad hydrological region in South Africa.

Basin studies were initiated to provide information to better enable the Department of Water Affairs and Forestry, and others, to manage South Africa's scarce water resources. These studies are already proving to be extremely useful in giving information on the current situation in selected basins and potential future developments. The key role that catchment studies can play in the management of water resources in a fast changing South Africa cannot be over-emphasised.

In this paper, the reasons for undertaking the studies will be explored in greater detail, the contents of a typical catchment study will be described, and the methodology used to undertake the studies will be outlined. Management objectives will be described, and three essential aspects of the study, *viz* reasons, content and methodology, will be assessed in terms of the management objectives. All these components will then be synthesised to place in perspective the role of the integrated catchment study in water resources management.

WATER MANAGEMENT

The management of water in any country is a difficult task, but water management in a semi-arid, semi-industrialised, multi-cultural country with developed and developing populations such as South Africa is even more complex. The manager must be placed in a position to see the whole picture, to assess the current needs and future aspirations of all users and to make wise decisions in the interests of all.

The major changes which have occurred in South Africa, and those which are still taking place, have considerable impact on the approach to water management. These changes render the traditional concepts of management incomplete. Among the main factors which have necessitated a new management approach are:
- the increasing number of social, political, commercial, environmental and legal consequences associated with water management decisions;
- the growing complexity of the information and assumptions that need to be considered before decisions are taken;
- the escalating rate of change which has resulted in managers making more and more decisions without necessarily having all the relevant information at hand, and
- the proven need for community participation, integration and coordination of actions, education and auditing.

For continued successful management these factors demand the adoption of a dynamic and integrated approach. *Ad hoc* decisions related to one user sector or sub-region are no longer acceptable. On the other hand to properly manage the whole it needs to be divided up into more manageable and controllable units. Because natural systems are dependent upon the hydrological cycle, it makes good sense to take river catchments as those units.

For effective management it is also necessary to have clear goals and objectives. Short, medium and long-term planning is essential as well as efficient organisation, guidance, motivation and control. To achieve this, the roles and responsibilities of all parties involved should be clearly defined and understood.

Some of the main management issues are the setting of objectives and strategies and the establishment of a reference framework or platform against which proposals and impacts, daily management decisions, existing and new developments and development potential can be compared and evaluated.

For the establishment of a reference framework for decision support and impact evaluation, it is essential to have adequate information about a number of water related factors, such as:
- information on users and associated demands;
- knowledge of the physical system in terms of natural characteristics, water availability and development potential;
- knowledge of land use development and related water use and pollution impacts;
- information on institutional structures, management objectives of a user sector or region, management approaches, evaluation, implementation, funding, control and efficiency;
- information on the role of driving forces such as political, social, legal, technical, environmental and economic forces.

These main elements of water resources management should be encompassed in a proper management information system. Catchment studies are prerequisites for, and essential elements of, effective water resources planning and management, as will be discussed further in the following sections.

WHY SHOULD INTEGRATED CATCHMENT STUDIES BE DONE?

Water is both the sustaining factor for life and the limiting resource for optimal development of South Africa. The growth in South Africa's population, coupled with changing priorities and the expected rapidly expanding needs of these people for household water, food, commodities and employment, is going to put ever increasing pressure on the water resources. This growth is taking place at various rates in all sectors of the population.

There are of course competing water-use sectors. Agriculture uses about 70% of all surface water and forestry is a growing water-use sector in its impact on streamflow. Commerce and industry, including the mines, are also major users, and so are power stations, especially the older coal-fired ones.

In addition, all the water-use sectors in the numerous independent states and self governing territories in South Africa must share fairly in the resource. Neighbouring countries also have the right to share in the rivers which cross or form international boundaries.

Within a water-use sector such as agriculture there may be a number of organisations which could speak on its behalf. There are also many management and organisational levels within each sector, ranging from top management to on-the-ground operation. Both public and private influences have an effect on the demand. Legislators at governmental, regional and local level simultaneously apply numerous and sometimes conflicting laws to competing water users.

The growth in water use is certainly not constant in space or time. The current socio-political changes in South Africa may lead to appreciable change in demographic patterns, with increased migration within basins and from one basin to another. The driving forces to human activities need to be properly understood to make better projections into the future so that timely provision for the water needs of those people can be made.

There is also the growth of water-use sectors which may not have been adequately recognised and provided for in the past. Notable among these are the requirements of nature, both in terms of consumptive and in-stream water requirements. As water source development increases the legitimate demands of the environment are becoming more and more important. Eco-tourism has the potential to rank among the major earners of foreign exchange and the largest employers in South Africa.

As regards the water supply side the situation is also complex. South Africa is a semi-arid country, beset by water supply problems associated with the vagaries of the weather. The variability of river flow, with severe droughts and floods, make the provision of a low risk surface water supply system extremely expensive and difficult to implement. The total surface runoff in the country has been estimated at some 50 000 million m^3 per annum, which is only about 9% of the total annual rainfall. Because of the hydrological extremes experienced in South Africa only about 34 000 million m^3 per annum of the streamflow can be economically exploited. South Africa is also not well endowed with groundwater resources. It has few major aquifers, but even so about 75% of domestic demand in the rural areas is supplied from boreholes. Injudicious management of water resources will add to the high risks in sustaining water supply.

Another important factor in water utilisation and water resource management in South Africa is water quality. High salinity in both ground and surface water is a common feature in the country. Water of inferior quality, even if in abundant supply, is virtually valueless if it cannot be economically treated to make it acceptable for domestic, industrial or agricultural use. Each user group has its own set of quality compliance criteria. The more development that takes place in a catchment, the more the return flows to the stream will impact upon downstream users, and the more seepage of pollutants into the ground will render the groundwater increasingly unfit for use.

It is apparent from the preceding paragraphs that the water picture in South Africa is exceedingly complex. A method of handling this complexity is the use of integrated catchment studies. Both strategic and integrated thinking should be applied to water resources planning and management. Catchment studies which simultaneously consider the requirements of all water-use sectors and the total water resource in a true system analysis, are the logical building blocks towards achieving this aim.

THE CONTENT OF A CATCHMENT STUDY

The content of a typical South African catchment and the inter-relationship between the different components of the study have been described in various papers and presentations in the last few years.

A basin study is usually broken down into component parts for proper execution and ease of control of the study. Typical components include:

- *Physical description.* This includes descriptions of the geography (topography, location, size, shape), soils, vegetation, and river system. Climate, although not strictly physical, is also usually included.
- *Land use.* Here the many different land users in the catchment are described, such as urban, semi-urban and rural developments, with their associated commercial and industrial water users, agriculture, forestry and nature conservation.
- *Water use.* The water uses associated with each land-use sector would be described, namely domestic, industrial, agricultural, infrastructure (mainly losses from dams and water supply and treatment systems) and ecological. Future demands are also estimated. This needs socio-economic and socio-political input and consideration of various other factors to fully understand the driving forces for development.
- *Water resources.* The water resources of the catchment are described and evaluated, usually separated into surface water and groundwater, but also including unconventional resources where appropriate.
- *Factors influencing development.* There are many factors which impact on

development either positively or negatively. These include environmental issues, water quality, erosion and sedimentation of impoundments. Of increasing importance is provision of water for the consumptive and in-stream requirements of nature.

- *Institutional and legal aspects.* The legal and management processes which modify the behaviour and the use within the catchment surround and impact on all the physical and natural factors of the catchment, including the water use. These are normally described, and if applicable, future changes in legislation and policy explored and provided for in recommendations.

A catchment study normally comprises two major sections, the first being the assessment of the present situation within the catchment, and the second being an evaluation of the potential for development of the catchment to meet current and estimated future requirements.

METHODOLOGY

The method used in the majority of catchment studies is to assemble a multi-disciplinary team to undertake the study in a number of steps:
- data collection;
- data analysis;
- synthesis;
- reporting.

The data collection phase employs significant manpower resources and will usually be most time consuming. Data collection is done by using questionnaires, personal interviews, research, data files and reference books, aerial photographs, digital processing of satellite imagery, maps of varying scale and other sources. Data analysis, synthesis and reporting are performed by interdisciplinary teams using tried and new techniques and methods.

The day-to-day management of the study is carried out by a small study executive team or Liaison Committee comprising representatives of the client and consultant. Overall direction is given by a Steering Committee made up of members representing all the water-use sectors and all the local, regional and state authorities, who have an interest in and impact on the catchment. The Steering Committee acts as a coordinating body between all these parties and provides direction to the study on a continuous basis. The Committee is also responsible for verification of data, monitoring of progress and approval of study reports as well as providing a forum for the distribution of information and findings of the broader community. The Steering Committee could then evolve into a committee responsible for the integrated management of water-related matters in a catchment or region.

In most of the basin studies undertaken to date the overall responsibility was taken by South Africa through the Department of Water Affairs and Forestry.

COST AND BENEFITS OF CATCHMENT STUDIES

Catchment studies are inexpensive when compared to the immediate and potential benefits to be derived therefrom. First level catchment studies are done as an initial step in an extended planning exercise and, as always in the planning cycle, problems or changes which are identified early in the process cost the least and bring the greatest benefits. The cost of a catchment study would normally be only a fraction of the capital cost of any major resource development undertaken as a result of the investigation. Effort expended in the early stages to address all issues and ensure, as far as possible, the correct solution to the problem is obviously worthwhile.

In the South African experience catchment studies bring many short-term and long-term, tangible and intangible benefits. Tangible benefits include coordinated planning of water resource development, resolution of conflicts and socio-economic issues, development of system operation techniques, reduction of operational and development costs, establishment of an extensive data base, identification of shortcomings in data collection and monitoring systems, fair apportionment of water, liaison between key players and the identification of research needs. The development and use of new techniques such as geographical information systems (GIS), remote sensing, more sophisticated hydrological evaluation and communication techniques are important subsidiary benefits. Other benefits include the collation of useful knowledge and information and the development of new disciplines by combining the needs of one and the techniques of another, e.g. ecohydrology. Some intangible benefits are general education, dissemination of knowledge, integration of requirements of users, and the better understanding by participants in the study of the various other disciplines involved, e.g. those practised by hydrologists, agriculturalists, economists, demographers, ecologists and engineers.

Possibly the greatest advantage of the catchment study is that it enables the planners or managers to see the whole picture and to gain holistic perspective of the problems and the pressures that are being placed on the limited water resource.

THE ROLE OF CATCHMENT STUDIES IN WATER RESOURCE MANAGEMENT

The management of water resources is multi-faceted and is continually being developed and improved. The key to good management and decision-making is information. The more detailed and extensive the information base the more time-consuming and expensive it is to set up. Relevant and timely information must be collected, analysed and made available for decision-making. This includes current and future requirements of users and the current state of the water resource.

The water manager's aim is to meet all justifiable water requirements in the natural and built-up environment at varying levels of assurance and with water of adequate quality. These requirements are usually interrelated and have reciprocal effects.

The development of one sector or user could have a multiplier effect in that it stimulates the development, and consequently the water requirements, of other users. The reciprocal effect is particularly evident on the quality side where pollution caused by one user's effluent could render downstream sources unfit for use by other consumers.

The requirements must be met from local or outside sources. Conflicting and competing demands by the various sectors and regions must somehow be reconciled through fair allocations and optimal utilisation.

The water manager's task is thus to orchestrate the whole water utilisation cycle from demand management through development to operation and quality control.

Catchment studies form the very basis upon which all further actions such as detailed analyses, development decisions, institutional arrangements, international joint utilisation, sectoral division of water, management structures and indeed the actual integrated management of water resources in a catchment are founded. A catchment, or a larger area consisting of two or more adjacent catchments, forms the natural and logical unit for effective management. All the interactive and interrelated activities in a catchment are drawn together in such an integrated exercise, with water being the golden thread.

The Steering Committee of a basin study provides the initial forum for contact between the main role players in a catchment. The study itself provides them with mutual understanding of each other's situation. This drawing together of people with a common purpose creates momentum towards the formation of other institutional structures which will be entrusted with the operation and management of water resources in a catchment.

A catchment study is obviously not just aimed at data collection, but serves as a vehicle for community participation, team building, education of users and managers, preparation for identified roles to be fulfilled and the understanding of relationships which determine the optimal application and use of water.

And what of the future? A number of studies are now complete, forming bodies of knowledge of a whole region. This knowledge is already being applied on a wide front. It is vital that the basin studies be used as a starting point and not as an end product. Regular updating of the information base is essential for the continued beneficial use of study results. The ongoing application of the results will confirm their usefulness.

CONCLUSION

Catchment studies in South Africa have proved to be extremely useful in the integration of dispersed knowledge. There are many good reasons for undertaking such studies. The nature of the studies and the methodology used add to further synergies.

In addition to their use as a source of knowledge, it has been shown that basin studies can play an extremely important role in the effective management of the water utilisation in a basin. All the elements of a study add to the overall efficiency of this function.

There is no doubt that catchment studies as promoted in South Africa, are essential tools to be used by the water resources manager.

ACKNOWLEDGEMENTS

The authors wish to acknowledge the permission of the Director-General of the Department of Water Affairs and Forestry to present this paper. The views expressed are those of the authors and not necessarily those of the Department.

REFERENCES

Middleton, B.J., Van Zyl, F.C. & Botha, M.L.J. 1991. Water for nature: Letaba River hydrology. 5th South African National Hydrological Symposium, Stellenbosch.

Van Niekerk, P.H. 1989. Opvanggebiedstudies vir Deelnemende Waterbronbeplanning. *The Civil Engineer*, October 1989.

Van Zyl, F.C. & Pullen, R.A. 1989. An Integrated Resource Management Approach to River Basin Management. SAICE Conference, Pretoria.

Van Zyl, F.C. 1991. Making it happen — The Sabie experience. Seminar on Water for the Environment, University of New England, USA.

41 Catchment management plans: current successes and future opportunities

CRAIG WOOLHOUSE

National Rivers Authority, Thames Region, Waltham Cross, UK

INTERNATIONAL PERSPECTIVE

The United Nations Conference on Environment and Development, held in Rio de Janeiro in June 1992, ensured that the debate on the integration of environmental issues into economic and developmental decision-making received widespread recognition and consideration.

One of the key outputs from the conference was Agenda 21 (United Nations, 1992) which is described as "an action plan for the 1990s and 21st century, elaborating strategies and integrated programme measures to halt and reverse the effects of environmental degradation and to promote environmentally sound and sustainable development in all countries". The UK Strategy on Sustainable Development (Department of the Environment, 1994) identifies several principles to guide action including:

- Action should be based on the best scientific information available;
- Where potential damage to the environment is uncertain and significant we should adopt a precautionary approach;
- The carrying capacity of ecosystems and habitats or the capacity of the environment to absorb pollution should be recognised;
- Human wealth should be measured in terms of natural environmental capital as well as economic measures;
- Development decisions must make proper allowance for the interests of future generations and the pressures put on the global environment;
- The "polluter pays" principle should be applied.

Local government in the UK (Local Government Management Board, 1993) is playing a leading role in taking Agenda 21 forward by preparing environmental audits and applying relevant concepts to the production of land-use development plans. Water issues are invariably covered in detail in environmental audits and have increasingly become a component of statutory land use development plans. In the latter case this has been driven in part by the NRA who have prepared guidance (1994a) for local planning authorities on such issues. In addition voluntary organisations, including the Council for the Protection of Rural England who have considered the implications of adopting a sustainable approach to the management of water resources (1993), are also working hard to translate Agenda 21 into action.

Action to deliver Agenda 21 and the 27 principles of the Rio Declaration on Environment and Development is, therefore, under way. Several of these principles, relevant to the NRA's catchment management plans, are paraphrased below:

- **Principle 4**   Environmental protection must be an integral part of the development process
- **Principle 10** Environmental issues are best handled with the participation of local communities. Individuals should have access to information on the environment and activities which may affect it, and be able to participate in the decision-making processes
- **Principle 11** Environmental standards, management objectives and priorities should reflect the local environmental and developmental context
- **Principle 15** The precautionary approach should be widely applied.

The International Conference on Water and the Environment held in Dublin in January 1992 provided the major input to Chapter 18 of Agenda 21 which specifically addresses freshwater issues. In promoting the integrated approach Section 18.9 states that "Integrated water resources management, including the integration of land- and water-related aspects, should be carried out at the level of the catchment basin or sub-basin." Encouragement is given for plans prepared in support of this objective to be completed by the year 2000.

Although the above should not be taken out of the wider international and environmental, social and economic context pertinent to sustainable development, it does form a backdrop to gauge how the NRA's plans match up to the relevant themes and principles. It also shows that the NRA cannot sit to one side but must be central to the debate on water and sustainable development. A fuller consideration of how the principles behind sustainable development could influence our approach to water management is given by Gardiner (1994).

INTEGRATED WATER MANAGEMENT IN ENGLAND AND WALES BY THE NRA

The establishment of the NRA in September 1989 created a number of opportunities for the improved planning and management of the natural water environment.

Initially the focus of the Authority was on the creation of a single unified independent regulator. However, the NRA's 1990/91 Corporate Plan (NRA, 1990) reflected a growing realisation within the organisation that to be effective 'Guardians of the Water Environment' it was necessary to integrate the activities of functional groups such as water quality, water resources, flood defence, fisheries, recreation, conservation and navigation. This integration would flow across the functional management chimneys and ensure that the focus of the Authority was on delivering sustainable improvements to catchments. Integration would be achieved both by integrated planning and reconsidering how NRA services were to be delivered.

The basis for this change lay in pioneering work by several of the Authority's regions, which will be discussed later, and general changes in the culture of public services in the UK as typified by the introduction of Citizen's Charters (NRA, 1994a).

Current approach of the NRA

Before considering the background to the adoption of catchment management plans, which are an integral part of the Authority's corporate planning process, the current approach (NRA, 1993) to plan preparation will be described.

Catchment management planning in the NRA is the process by which the problems and opportunities resulting from water-related catchment uses are assessed and action is proposed to optimise the overall future well-being of the water environment. A catchment use is defined as a direct use of the water environment (e.g. ecology, water abstraction) or an activity which impacts upon it (e.g. mineral extraction, housing development). Catchments are defined as discrete geographical units with boundaries derived primarily from surface water considerations and comprising one or more hydrometric sub-catchments.

The plans consider the various water users' interests and develop a long-term vision and medium-term strategies and actions through consultations with local communities and organisations. It is intended to produce each plan within a twelve-month period (see Figure 1). The initial step is to prepare a Consultation Report. After a period of open consultation, a Final Plan is produced.

The purpose of the Consultation Report is to describe the resources, uses and activities relevant to the water environment, set environmental objectives, present issues and suggest potential solutions as well as a catchment vision. Public consultation then takes place over a period of eight to thirteen weeks through public meetings, press releases and direct mailing.

The Final Plan subsequently sets down the agreed vision, strategy and action plans for the catchment. It reflects the results of the consultation process. The vision looks to the long-term (ten years plus) whereas the strategy and action plans have a three- to five-year time horizon. The action plans reflect the economic and political constraints that apply to delivering environmental improvements.

Annual monitoring of commitments is undertaken and reported on by the NRA. The relevance of the plan is also reviewed at this stage so that the need to formally update the plan can be considered. A range of organisations and groups is likely to be responsible for undertaking actions. Informal liaison with key partners (e.g. local authorities, water companies) is an essential element of the plan making process.

It can be seen therefore that the plans adopt many of the principles and themes related to Agenda 21 outlined earlier. They involve local communities, reconcile conflicts between competing uses of water, integrate water and land uses, set standards or limits related to the carrying capacities of water systems and consider issues within both short-term (i.e. the action plans) and long-term (i.e. the visions)

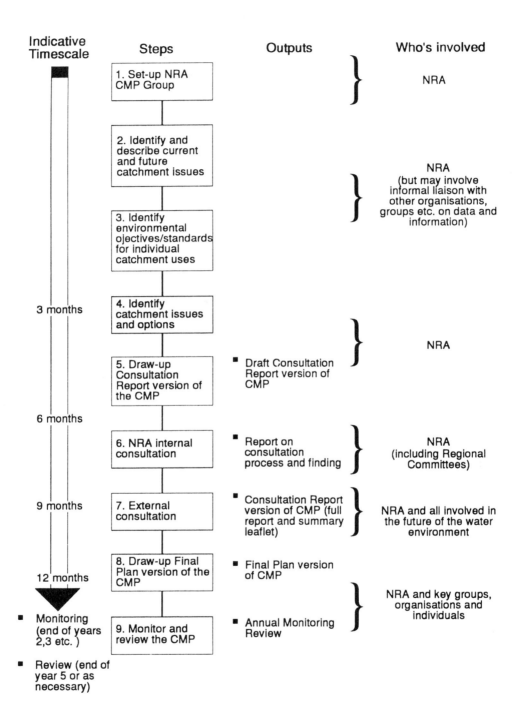

Figure 1 The Catchment Management Plan process

time horizons.

The process has been refined over the last four years by NRA staff working together to share experiences and ideas.

EARLY EXPERIENCE OF INTEGRATED PLANNING

Early, but by no means unique, investigations were undertaken in Thames region in 1988 and 1989. A set of 'Implementation Guidelines' (NRA, 1989) were then prepared for plans which were designed to 'ensure that all the problems caused by demands within a catchment relating to the water cycle are presented within a management framework which aims to maximise the overall well-being of the catchment'. Trial plans were published in mid-1991. Experience from overseas, notably Australia and New Zealand (Allan, 1989), informed the regional discussions.

The impetus for this work had its roots in the difficulties Thames region was having in promoting environmentally sensitive and economically sound flood defence works. The emphasis of the plans was therefore very much on flood defence and conservation. They reflected water quality and resources issues, but only in so much as these areas interacted with flood defence needs. At the time these plans failed to achieve unanimous internal support. A failure to sell the process and obtain broad support for the initiative limited many of the potential benefits and the plans were not successfully implemented.

The plans also suffered from the fact that they were based on new and thorough surveys of conservation and flood defence interests. To collect and analyse this information was both time-consuming and expensive. The gestation period for the plans was therefore considerable which meant that it was difficult to sustain interest across a broad range of staff whose political as well as technical support was required to make the process work. Key lessons learnt from this period of practical research were that:

- The planning process must be as short as possible;
- Planning on the basis of what we already know is, in the UK context, an acceptable risk provided subsequent action is subject to detailed evaluation;
- Integrated planning was feasible but all-round management support was needed to translate it into practice.

Different experiences were found in the Southern (Chandler, 1994) and Welsh regions. The process initiated in 1986 by Welsh Water had as its starting point a fully multi-functional perspective. This enabled plans to be more readily translated into action and contrasts with the experience in Thames region.

The move towards a national approach to catchment management planning
To help reconcile some of the differences of approach being used by NRA regions and to broaden the application of shared principles, a National Working Party on Catchment Management Planning was established in May 1990. An important output of this group was a set of draft internal guidelines (NRA, 1991) on how integrated water management plans should be taken forward by the Authority. These guidelines

established integrated water management as a basic building block of future NRA activity.

The process of producing the plans drew heavily on existing approaches to the management of water quality. These are to set objectives and measurable quality standards for proposed uses of water and to promote action to either sustain or achieve standards if there is adequate justification. Uses defined in these early guidelines included water abstraction, recreation and ecology. Activities such as urban development, which might impact upon the water environment, were less precisely dealt with. The use-based approach adopted for catchment planning recognised that environmental objectives (and standards) were necessary for all uses in terms of quality, quantity (or resources) and physical characteristics.

This approach raised a number of concerns. First, the term 'use' was considered pejorative by many working in the natural environment, as ecology and landscape are often seen as fundamental resources. In addition, although there was considerable experience of setting standards in relation to quality, virtually no objective standards were (or are now) available for water quantity and physical characteristics. The approach also marginalised the importance of land use activities which can have a significant impact on the water environment.

However, the purpose of the guidelines was to initiate the testing of the process in the real world and then to revise them in the light of experience. The key lessons learnt were reviewed in late 1992 and can be summarised as follows:

● The process, rather than the size of the catchment, determined overall resource needs. Minor catchments should be grouped to minimise costs;
● The expected cost of producing a plan was £ 90 000 (this has now fallen to £72 000);
● The process required all-round senior management support in order to be effective;
● A national target of March 1998 should be set by which time plans for all catchments should be ready;
● Plans should not raise expectations beyond what is achievable. However, they should include a long-term vision for the catchment;
● Local communities responded very positively to being involved in the planning process;
● Inter-functional co-operation and understanding within the Authority was improved;
● The catchments had their own particular needs but a nationally-consistent approach had to be fostered;
● The term 'use' should embrace 'resources, uses and activities' to recognise the full range of interests in a catchment;
● Research was required to develop objectives and standards for water quantity and physical characteristics.

These lessons were reflected in the revised national guidelines summarised earlier which are freely available to community interests, reflecting the openness inherent in the NRA approach to catchment planning.

Current situation
A national programme of activity coordinated by regional and head office staff is now in place to prepare 164 catchment management plans for the whole of England and Wales by March 1998 (thereby meeting the objective in Agenda 21 to prepare integrated plans by the year 2000). Preparation of plans is a shared activity between area and regional staff but the responsibility for implementing the plans lies with the 26 NRA multi-functional operational areas. Each area, operating within national policy guidance, has the responsibility across all the NRA areas of activity to identify the key issues facing the local water environment and to tackle them directly or in partnership with other organisations.

The move towards planning for and delivering services in an integrated way (both technically and geographically) inevitably meant that the structure of the organisation had to be reconsidered. During 1992 and 1993 the change to multi-functional areas was implemented. It is unlikely that integrated water planning and service delivery would have been possible without this change.

To date over 30 Consultation Reports and 10 Final Plans have been prepared (see Figure 2).

Benefits of catchment management plans
The concept of sustainable development is founded on the integrated management of natural resources. A catchment management plan enables the NRA and the local community to contribute to this goal in respect of water. The plans aim to provide long-lasting solutions and are the essence of the integrated approach in that they consider the various water users' interests and construct future strategies through consultation with local groups and organisations.

Internally, the plans enable resources to be targeted on key issues. In addition priorities are agreed and opportunities for coordinated action identified. Externally, the plans offer local communities the opportunity to participate in planning for their local environment. Experience to date indicates that local government is finding this particularly useful for establishing joint initiatives with the NRA.

The benefits of multi-functional operational areas and integrated planning include:
- A clear remit to tackle the key local issues;
- A single point of contact for local communities;
- A common focus and interest for staff;
- Open and clear inter-functional communications;
- Close links between planning and operational activities;
- Efficiency savings through coordinated action.

Successful integrated management of catchments is thus founded on the expert knowledge and awareness of operational staff who can recognise the key issues facing the water environment and develop practical ways of dealing with them in conjunction with colleagues from other disciplines. Integrated planning (which establishes direction and priorities) and integrated operations (which deliver the actions) are therefore two sides of the same coin. Together they provide a clear means of communication with local communities.

Future direction
Development of the principles and practice of catchment management planning has
been undertaken as much in the public domain as within the offices of the NRA.
External reviewers (Slater, Marvin & Newson, 1993), as well as practitioners, have
identified important concerns (e.g. the failure of plans to link successfully with the

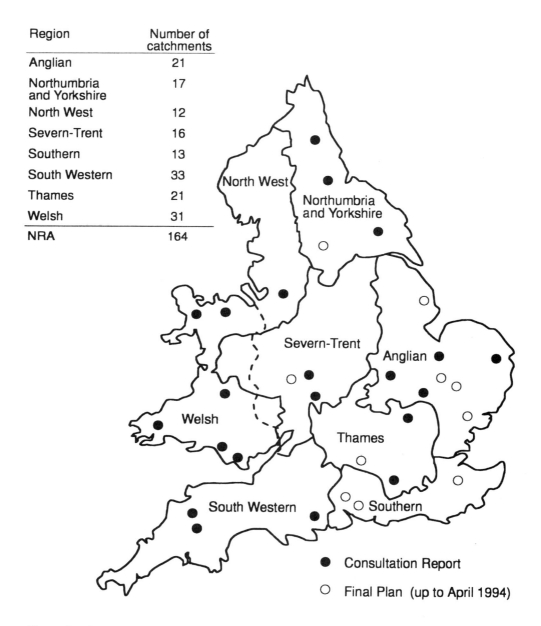

Region	Number of catchments
Anglian	21
Northumbria and Yorkshire	17
North West	12
Severn-Trent	16
Southern	13
South Western	33
Thames	21
Welsh	31
NRA	164

Figure 2   Current achievements in NRA plan production

development plans of local planning authorities) which have had to be addressed. Plans prepared today are very different from their early predecessors. The process of setting 'use-related' environmental objectives still requires further consideration and development.

Since the concept of sustainable development is still the subject of debate it is not surprising that the plans have not addressed all the possible permutations of thoughts and interests reflected by voluntary groups, industry, conservation and recreation interests, local government and individual landowners. Experience, however, is actively being fed back into the process which will continue to evolve and respond to new ideas and demands.

Indicative of the external recognition of the plans is the development of handbooks to assist staff (and members in voluntary groups) to respond to catchment management plans in a consistent and appropriate manner. This development also highlights a growing awareness amongst those interested in the natural environment that we need to share our expertise and knowledge more effectively if we are genuinely to deliver integrated natural resource planning and actions in accordance with Agenda 21. This issue was covered by the Secretary of State for the Environment, the Rt. Hon. John Gummer MP (1994) in the Laphroaig Lecture:

"In their system of catchment management plans, which have brought together so many of the river management issues, they [the NRA] are beginning to understand that the matter of public debate and consultation is not just a democratic necessity, it is actually a practical need, for it is from such debate and discussion that some of the solutions hard sought for are discovered."

Sharing of information and cross-sectoral integration of planning initiatives are areas which will also need to be considered if we are to achieve integrated planning efficiently and not over-burden those who have to deliver actions with a plethora of planning documents which may confuse, rather than clarify minds (see Figure 3). This will be a particular challenge for catchment management plans which have a very strong remit to inform operational activity within the Authority including the setting of Statutory Water Quality Objectives.

By the end of 1995 it is proposed that the Environment Agency, which will be formed by combining the activities of the NRA with those of Her Majesty's Inspectorate of Pollution and the solid waste disposal regulatory work of the Waste Regulation Authorities, will be established. The evolution of catchment management plans into 'environmental management plans' will require careful consideration and may not necessarily occur overnight. However, the experience of Agency staff in preparing statutory and non-statutory plans will ensure integrated planning continues to influence and direct action which is in tune with the wishes of local communities.

CONCLUSIONS

1.   The NRA has successfully developed and intends to complete implementation of a comprehensive programme for the preparation of 164 catchment management plans

Current Situation - "integrated sectoral planning"

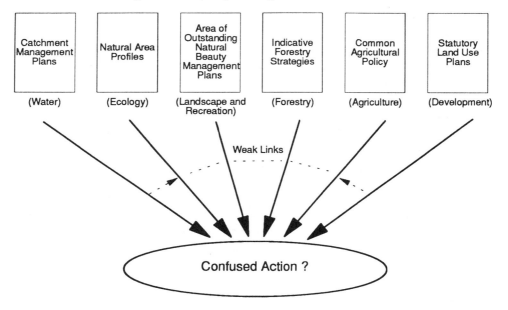

Future Situation - "integrated resource planning"

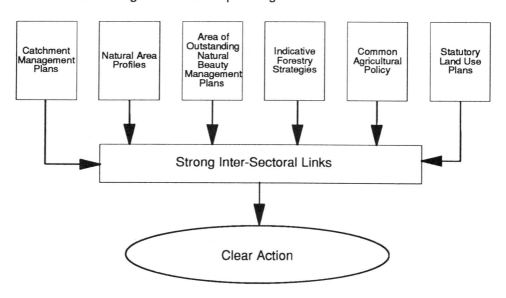

Figure 3   Integrated resource planning: current and future situations

for the whole of England and Wales in slightly under eight years (Figure 4). The cost of this programme of plan production is estimated to be less than £13 M which represents less than half of one per cent of the budget for the organisation over the implementation period.

2. By harnessing professional skills within the Authority, drawing on initiatives in overseas countries and testing ideas through live pilot projects, the NRA has ensured that catchment management plans are practical and relevant to the needs of those involved in managing the water environment. Re-organisation of the NRA into 26 multi-functional operational areas is complementary — yet fundamental — to successful catchment planning.

3. Completed catchment management plans have drawn on the skills of in-house professionals and the local knowledge of representatives of relevant communities. The pooling of knowledge and the creation of a shared approach to problem solving appears to be particularly successful and beneficial.

4. Catchment management plans have a strong basis in the principles of sustainable development and offer water environment managers a means of translating Agenda 21 objectives into action in partnership with others involved in managing natural resources.

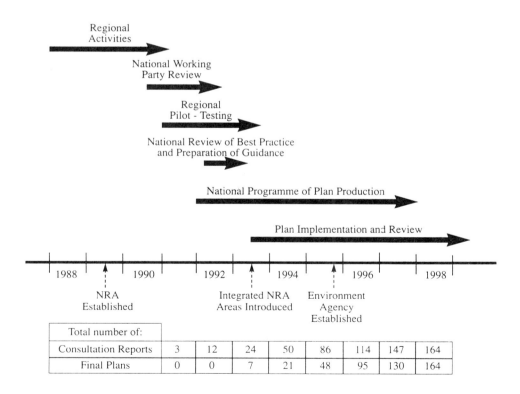

Figure 4   Integrated water management in the NRA: programme overview

5. Opportunities exist to increase the level of co-ordination between national and local agencies interested in managing the environment to reduce costs, share information, harmonise policy and simplify guidance to those who have to deliver actions on the ground.

REFERENCES

Allan, S. 1988. Water and Soil Resource Management Planning Guidelines. Ministry for the Environment, New Zealand.

Chandler, J. 1994. Integrated catchment management planning. *J. Instn Wat. Environ. Manage.* **8,** No. 1.

CPRE 1993. *Water for life.* Council for the Protection of Rural England, UK.

Department of the Environment 1994. *Sustainable development: The UK Strategy.* Cm 2426, HMSO, UK.

Gardiner, J.L. 1994. Water and the natural environment. TCPA/National Rivers Authority Conference *Planning and Water — An elemental challenge.* Town and Country Planning Association, UK.

Gummer, J.S. 1994. *Sustaining our waters into the next century.* Laphroaig Lecture, Argyll, Scotland.

LGMB 1993. Local Agenda 21: an initial statement by UK local government. Local Government Management Board, Luton, UK.

NRA 1989. River catchment plans. Flood defence and land drainage. Implementation guidelines. National Rivers Authority, Thames Region, Reading, UK.

NRA 1990. Corporate Plan 1990/91. National Rivers Authority, Bristol, UK.

NRA 1991. Catchment Management Planning Report (Internal Report). National Rivers Authority, Bristol, UK.

NRA 1993. Catchment Management Planning Guidelines. National Rivers Authority, Bristol, UK.

NRA 1994a. Guidance notes for local planning authorities on the methods of protecting the water environment through development plans. National Rivers Authority, Bristol, UK.

NRA 1994b. NRA Customer Charter. National Rivers Authority, Bristol, UK.

Slater S., Marvin, S. & Newson, M. 1993. Land Use Planning and the Water Sector: A Review of Development Plans and Catchment Management Plans. Working Paper 24, Department of Town and Country Planning, University of Newcastle upon Tyne, UK.

UN 1992. *Agenda 21: Programme of Action for Sustainable Development.* United Nations, New York, NY, USA.

42 Reconciling water regime requirements for environmental and agricultural management

D G J GOWING, J MORRIS, G SPOOR and
J A L DUNDERDALE
Silsoe College, Cranfield University, Bedford, UK

Many sites of existing or potential wetland value in Britain are used for commercial agriculture. For the most part this involves traditional, relatively low intensity farming practices. Indeed many important wetlands, such as the Somerset Levels, are themselves the product of a long period of land drainage management. Nevertheless, conflicts of interests arise between wetland conservation and agriculture. In recognition of this, various incentives have been offered recently to farmers to adapt their land use and farming practices in order to protect or enhance the environmental qualities of wetland sites. Drawing on fieldwork in the Somerset Levels of south-west England, this paper describes an approach developed to identify and reconcile the water regime requirements of natural habitats and commercial farming.

DRAINAGE MODELLING

Drainage standards can be defined in terms of the incidence of flooding and waterlogging. Flooding often occurs when flows exceed the capacity of river or ditch systems to take it away. Waterlogging arises because persistent excess water is not able to move through the soil to a drainage outfall. In flat lands, estimates of drainage standards can be derived from data on flows, drainage channel dimensions and configuration, climatic conditions, and characteristics of the land, particularly soil hydraulic conductivity and topography. Figure 1 shows the principal elements in a steady-state model (Hess *et al.*, 1989) whereby, for given site conditions, field water table levels can be estimated from the water levels in the river and ditch system. Generally, ditch levels have a greater influence on water table levels in lighter, free-draining compared to heavier, relatively impermeable soils. Furthermore, water table levels in winter tend to be higher than ditch levels due to excess of rainfall over evapotranspiration, whereas in summer the converse applies.

These principles have been incorporated into a non-steady-state model (Youngs *et al.*, 1989) to allow for variation in rainfall and evaporation during the year, and for

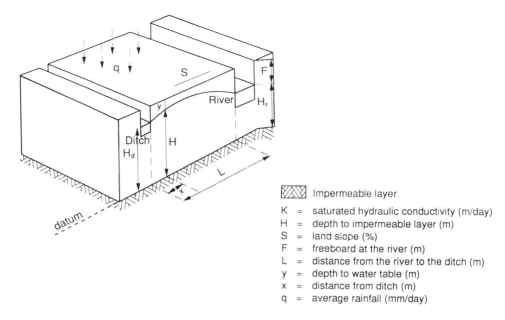

Figure 1 Steady state water table model

different field and drainage system characteristics. Estimates of field water tables using this model have been validated against dipwell observations (Mountford & Chapman, 1993). The model has the advantage of being able to consider fields with a wide range of geometries (Youngs, 1992). Use of the model allows the impact of different ditch stage-level regimes on water table behaviour to be assessed. The proposed ditch levels provide the boundary conditions for the model. The water regime at the field centre shows the greatest water-table variation in response to climate, and this regime can be modelled over a number of years (ideally about 20 if historical climate data is available) to demonstrate the effects of ditch level management. Field survey data required as input for the model include: soil conductivity, specific yields, depth of ditch base, depth to impermeable layer, field size and shape and surface elevation relative to ditch base.

WATER REGIME REQUIREMENTS

Many of the environmental and ecological interests of areas such as the Somerset Levels are dependent upon prevailing water regimes as they influence grassland flora, ornithological habitat, and invertebrate fauna.

If the existing **grassland flora** diversity is to be maintained or enhanced, then the water regime requirements of the particular plant community present must be assessed. To do this in detail, estimates are required of the movement over time of the water table relative to the soil surface. This information, described in terms of Sum

Exceedence Values (Gowing *et al.*, 1993), can be derived from the output of the hydrological model and compared to plant community tolerance information (Gowing *et al.*, 1994). In the context of the Somerset Moors, the water meadow community (MG8) excites the most conservation interest. To provide suitable conditions for this herb rich grassland, for example, it is necessary to aim for water tables deeper than 200 mm in spring, yet shallower than 600 mm in summer.

The value of the site as **ornithological habitat** will depend more directly on the water regime. For example, the availability of open water on fields to attract migrant waterfowl, and the occurrence of high water tables during spring for the successful breeding of wading birds are particularly important. To encourage such birds onto the peat sites in Somerset, the ditch levels should be managed to promote surface water in field hollows during late autumn and winter, and water tables of about 300 mm depth during the April-to-June period. High water tables are necessary to ensure that the soil micro-invertebrates are near the surface, and that the soil is soft enough to probe.

Established wet grassland usually harbours an interesting and diverse **invertebrate fauna**, both in the soil and on the surface. Plant community and vegetation management are major factors in determining the species composition at a site. In this respect, water regime influences invertebrate fauna either directly or indirectly through grassland flora or farm management practice. Its direct importance is most apparent during spring when late floods can create anoxic conditions in the soil and drastically reduce the biomass of soil invertebrates, and hence limit the food source for waders.

With respect to **agriculture,** drainage is a major determinant of land use, farming practice and performance. Standards of drainage for agriculture can be defined in terms of the wetness of soils and surface flooding as these influence the potential for crop growth, field conditions for machinery or grazing animals, and the damage to crops or livestock due to inundation. The standards of drainage service for grassland systems can be defined in terms of the water table heights required for grass growth and load bearing strengths. Research and experience suggest that water tables deeper than 400 mm from the surface in peats such as those of Somerset, do not impede farming practice. Water tables between 300 mm and 400 mm are likely to limit access by machines and animals. Water tables persistently within 300 mm of the surface, however, lead to depressed yields and restrictions on land use, especially in terms of limiting improved practices such as fertiliser use and pasture management. Surface ponding (or splashing) obviously limits grassland management. Recognising that water tables are often in a state of flux, the drainage status of a field can be defined in terms of the length of time that the water table lies within these boundaries. Table 1 shows the criteria for good, bad and very bad drainage on grassland.

The different requirements of agriculture and the environment are summarised by the boxes in Figure 2 which show the period of the year when the named aspect would be negatively affected by water tables exceeding the levels indicated. These boxes can then be compared with the estimates of water table levels generated by hydrological modelling.

Table 1 Criteria for classification of grassland drainage status

% time water table table lies at given depth		Water table depth (mm) relative to the surface	Drainage class
(i)	> 80 %	> 400	Good
(ii)	< 80 %	> 400	Bad
	< 50 %	< 300	
(iii)	> 50 %	< 300	Very Bad

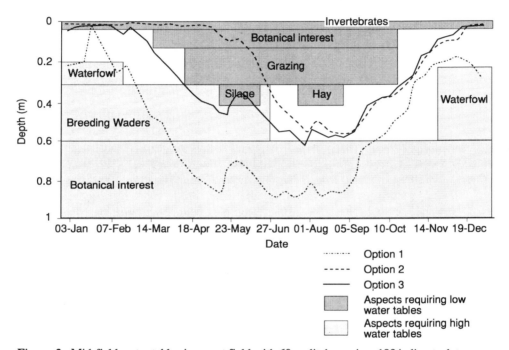

Figure 2 Mid-field water tables in a peat field with 60 m ditch spacing: 1984 climate data

APPLICATION OF METHOD TO CASE STUDY SITE

Data were derived from field work and secondary sources in order to assess the environmental and agricultural impacts of changes in drainage regime in the study area. Soils were classified by direct field investigation — supported by data from their series classification — to determine the soil's conductivity, porosity and depth. In the absence of a large scale map with 0.2 m contours, the site was surveyed to estimate mean elevations of the ditch base and the field surfaces. Existing water regimes were evaluated from historical information regarding ditch stage levels and climate records over a number of years. The hydrological model was then run both predictively and retrospectively.

In order to simplify the modelling process, fields were classified into groups according to soil type (mainly clay or peat), depth to impermeable layer (shallow or deep), ditch spacing (<50 m, 50-100 m, >100 m) and modest differences in relative elevation (low, medium and high). This classification enabled over 200 fields to be reduced to 20 categories for the purpose of modelling. The area of the land represented by each category was ascertained to allow weighting of the final estimates of environmental and agricultural impact.

With regard to farming practice, secondary data sources and informal survey methods showed that land use was predominantly permanent grass, occupied by dairy replacement stock, cattle that were part of 24 month beef systems, beef cows and their suckling calves, and lowland sheep producing fat lambs. Fertiliser rates, where applied, were commonly 75 kgN/ha for grass cut for silage and 40 to 50 kgN/ha for grazed swards. Depending on field conditions the grazing season commonly extended from mid-April to mid-October. Silage was cut about May 15, and hay in early July.

Discussions with farmers confirmed that drainage influenced sward composition, field access for nitrogen application and grazing livestock, the ability to cut for winter feed, stocking levels and overall productivity. From this analysis, productivity scenarios were formulated to represent the main combinations of drainage status, farming practice and financial and economic performance. The method was as follows (Hess & Morris, 1988; Morris & Sutherland, 1992): energy from grass production was estimated on the basis of climatic and land suitability, nitrogen use, drainage class, and grass-use (whether grazed or conserved). Stocking areas (head/ha) were derived by dividing energy available (MJ/ha) by the energy needs of animal by type (MJ per head). Estimates were derived for each animal type from the value of output, less the variable costs of purchased feed and veterinary services, to give a gross margin (£ per head). From this was deducted the costs of direct labour, machinery running, and winter housing running costs, together with the costs of growing and conserving grass and winter feed. This gave an estimate of net margin per head which can be expressed per ha according to the stocking rate. Further allowance was made for variations in the length of grazing seasons by charging a penalty per day if grazing days are lost. The final estimate was a net margin per hectare for different stock types on different drainage/productivity scenarios (Table 2).

The requirement by MAFF (1993) to reduce the financial value of output on beef systems by 35% in order to determine the economic value of extra production, has a somewhat catastrophic effect on the valuation of production at the margin. Economic net margins are negative for most beef systems. This is because extra costs are greater than 65% of the financial value of output. This suggests that poorer drainage standards, whilst detrimental to farmers, are beneficial to the nation.

The estimates were used to determine the monetary impact of changes in standards of drainage service whether in terms of the impact on farm incomes or net benefit to the nation. For example, a deterioration from good to very bad drainage results in an annual financial loss of £79/ha but an economic saving of £14/ha.

Table 2 Grassland productivity and drainage status

Drainage status		Good	Bad	Very Bad
Nitrogen	kg/ha	75	50	0
Grass use/ conservation		silage	silage/ graze	graze
Energy from grass	MJ/ha	42000	36000	24000
Stocking rate (dairy cow equivalent)	Lu/ha	1.3	1.0	0.7
Grazing penalties				
Spring	Days lost	0	14	28
Autumn	Days lost	0	14	28
Gross margin				
Financial prices	£/ha	166	139	87
Economic prices	£/ha	-46	-43	-32

STAGE LEVEL SCENARIOS AND IMPACTS

The preceding methods were used to assess the impact of drainage management options for an area within the Somerset Moors and Levels. By way of example, Figure 3 illustrates the different ditch stage-level regimes for three management regimes which are discussed below:

Option 1. This broadly reflects the existing situation in many of the Somerset Moors. Pumps are used to maintain low water levels during the winter to minimise flood risk, whilst levels are allowed to rise in the April-November period, so that the ditches act as wet fences and as irrigation channels for the grassland in summer.

Option 1 provides a water regime suitable for pastoral agriculture, with no restriction (in an average year) to grazing season (April-October) or grass conservation operations. A 'good' drainage status is achieved with an average financial gross margin of £166/ha. From an environmental perspective, there is little or no risk of anoxic soil conditions developing, and therefore no threat to the grassland flora or invertebrate fauna. The deep summer water tables, however, pose significant problems for breeding waders, because the soil becomes too hard to probe and the soil invertebrates may follow the water table down out of probing depth. The potentially drought conditions may also have an adverse effect on botanical diversity as the water tables are too deep to keep water readily available to the root zone. This may also depress grass yields and stocking rates during the summer months.

Option 2. This is based on the Tier 3 prescription of the Somerset Levels and Moors

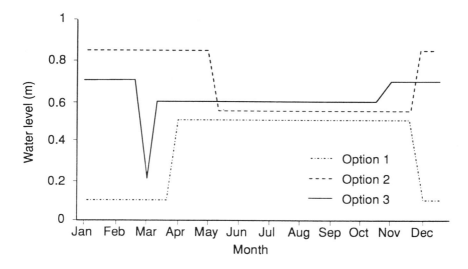

Figure 3 Ditch stage-level regimes for three management options

Environmentally Sensitive Area. Levels are maintained at mean field level during the winter for the benefit of wintering wildfowl, and lowered slightly in summer with the aim of allowing some grazing.

The increased wetness of the field results in 'very bad' drainage conditions in the spring and an average financial gross margin of £87/ha. Cattle are not able to graze without significant damage to the soil surface until the beginning of June, and silage making is precluded entirely. Conditions are suitable for winter waterfowl and, in spring, soil penetrability is adequate for wader feeding. The high levels in spring, however, are likely to result in anoxic soil conditions with significant adverse impacts on botanical diversity and invertebrate biomass, and hence on the food supply for waders.

Option 3. This is an amended version of a Tier 3 prescription whose object is to conserve the diverse wet grassland flora of the area. To do this, surface water must be removed in March, but summer water tables should be kept as high as possible. The ditches, therefore, are drawn down briefly in spring, then high stage levels are maintained throughout the rest of the year. This option is not as restrictive to grassland management as Option 2.

Drainage conditions in spring are intermediate compared to the previous options; namely 'bad' drainage with a gross margin of £139/ha. Grazing and machinery access is delayed compared to Option 1. Silage cutting is precluded in many years. Soil anoxia, however, is largely avoided, thereby circumventing the adverse environmental impacts of Option 2. Of the three options, this one provides the best field conditions for a diverse grassland flora and probably the most suitable habitat for breeding waders.

FORMULATING MANAGEMENT STRATEGIES

The preceding methodology was applied in practice to the management of a proposed core wetland site in the Somerset Levels, comparing the effects on environmental quality and agriculture of alternative pumping and ditch level regimes. The analysis accommodated the variations in soil, field and farming practice in the study area, and examined the impacts of the proposed regimes using historical climatic data over a 20-year period.

The analysis considered the impact of the water regimes associated with the existing Environmentally Sensitive Area (ESA) Tier 3 arrangements and found that these did not fully exploit opportunities for achieving the stated nature conservation objective of enhancing the ecological interest of grassland. Simultaneously, they imposed considerable restriction on traditional grazing systems. A modified regime was identified which specifically aimed to reduce the problems of anoxia due to excessive waterlogging in the spring. Compared to the existing Tier 3 arrangements, this was beneficial to both environmental and farming interests. Whilst the modified regime did result in financial losses to farmers, these deficits could be met within the range of payments available under the ESA scheme. The losses to the national economy of any reduction in agricultural output would be negligible.

REFERENCES

Gowing, D.J.G., Spoor, G. & Mountford, J.O. 1993. Determining the water-regime preferences of wet-grassland flora. Proc. Annual MAFF Conference of River and Coastal Engineers, Loughborough. Ministry of Agriculture, Fisheries and Food, London.

Gowing, D.J.G., Spoor, G. & Mountford, J.O. 1994. A field approach to determining the water-regime requirements of wetland plants. In: Wheeler, B.D. & Fojt, W. (eds). *Restoration of Temperate Wetlands*. John Wiley & Sons, Chichester.

Hess, T.M. & Morris, J. 1988. Estimating the value of flood alleviation on agricultural grassland. *Water Management* 15, 141-153.

Hess, T.M., Leeds-Harrison, P.B. & Morris, J. 1989. The evaluation of river maintenance in agricultural areas. In: Dodd & Grace (eds), *Proc. 11th Internat. Conf. on Agricultural Engineering*, Dublin, Ireland. 501-507.

MAFF 1993. Flood and Coastal Defence Project Appraisal Guidance Notes, Ministry of Agriculture, Fisheries and Food. HMSO, UK.

Morris, J. & Sutherland, D.C. 1992. The evaluation of river maintenance. Proc. Annual MAFF Conference of River and Coastal Engineers, Loughborough, UK. 8.3.1-12.

Mountford, J.O. & Chapman, J.M. 1993. Water-regime requirements of British Wetland Vegetation: characterising vegetation using the moisture classifications of Ellenberg and London. *J. Environ. Manag.,* **38**, 275-288.

Youngs, E.G. 1992. Patterns of steady groundwater movement in bounded unconfined aquifers. *J. Hydrol.* 131, 239-253.

Youngs, E.G., Leeds-Harrison, P.B. & Chapman, J.M. 1989. Modelling water-table movement in flat low-lying lands. *Hydrol. Procs* 3, 301-315.

43 The need for integrated river basin management — a case study from Eastern Europe

ZBIGNIEW W KUNDZEWICZ and MALGORZATA CHALUPKA[1]
Research Centre of Agricultural and Forest Environment, Polish Academy of Sciences, Poznan, Poland

[1] *presently at the World Meteorological Organization, Geneva, Switzerland*

INTRODUCTION

The river Warta cuts across the central and western part of Poland. With a length of 808.2 km, it is the third longest river in the country (after the Vistula, 1047 km and the Odra, 854 km). The drainage basin of the Warta (Figure 1) has an area of 54 529 km^2 (17.4 % of the area of the whole country). This drainage basin belongs to as many as 19 provinces and more than 8 million people live in it. The main population centres are the cities of Lódz, Poznan, Czestochowa and Gorzów.

The catchment of the Warta is an agricultural and industrial region. Nearly 60% of its area is arable fields, most of which are characterised by advanced agriculture. Within the basin there are also centres of lignite mining, and of power, textile and chemical industries.

The Warta basin is situated within the borders of four physico-geographical units: the South Pomeran and Wielkopolska Lake Districts, the Central Poland Lowland, the Slask-Kraków Upland and the Central Malopolska Upland.

The character of the relief in the basin was shaped by the Scandinavian glacier in the Pleistocene. In the northern part of the basin, situated within the borders of the South Pomeran and Wielkopolska lake districts, postglacial forms can be easily seen. The course of the Warta consists of mutually perpendicular reaches. The density of river networks in the Warta basin is highly variable and depends on the permeability of the subsoil.

The average annual precipitation in the Warta basin ranges from 450 mm to more than 600 mm. The highest precipitation occurs on the northern and southern edges of the basin. The lowest annual precipitation, 450-500 mm, is recorded in the central part of the catchment, where great variability is observed, both during any one year and over a long time period.

Annual nominal water resources in the Warta basin are 6974 million m^3: 3753 million m^3 can be actually used, while 2119 million m^3 are inviolable. The

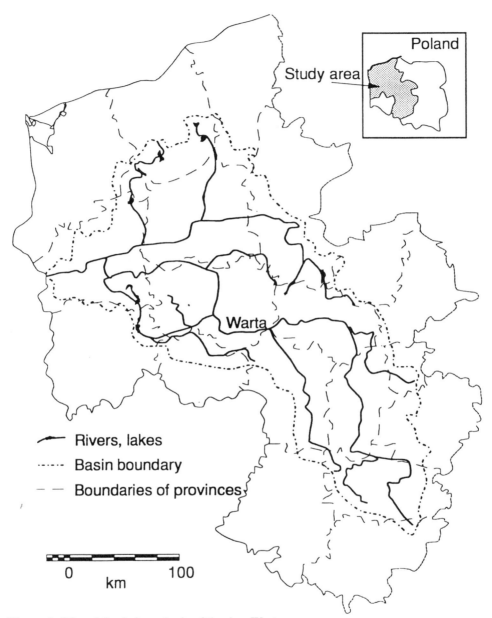

Figure 1   Map of the drainage basin of the river Warta

estimate of permissible consumptive water use during a year is 1634 million m^3 (Paslawski, 1991), i.e. 23% of the nominal resources. In various subcatchments, the value of the permissible consumptive water use fluctuates from only 2% in the catchment of the upper Notec to 39% in the catchment of the Gwda.

WATER PROBLEMS OF THE WARTA RIVER BASIN — QUANTITY AND QUALITY

Essentially all problems with water in the world concern inappropriate quantity and/or quality in a particular site and at a particular time (too much — too little — too dirty). The last two types of problem are of growing importance in the basin of the river Warta.

Much of the area of the catchment has a dry subhumid climate, with precipitation definitely lower than the national average. However, even at the national scale, the water availability per capita ranks Poland among the water-poorest of European countries. Water deficit is now, and will remain, an increasingly active constraint to economic growth at the national level, and in particular at the regional level.

Figure 2 presents a series of annual maxima and annual minima, compared with the annual mean flow of the river Warta (gauged at Poznan). Water deficit associated with low flows (e.g. annual minima for most years) causes a critical situation for navigation and different aspects of water use (including municipal water supply). An additional factor which contributes to the problem is the pollution load: there may be insufficient water to dilute the usual load of sewage. Moreover, climate change specialists consider it highly probable that hydrological extremes (dry spells and low flows) may become increasingly severe. The flows represent a supply (resistance) side of the system, whereas the demand (load) side is also likely to evolve in a direction to aggravate the problem of scarcity. There will surely be a growth in demand for water in the Warta basin, although it is difficult to assess the future

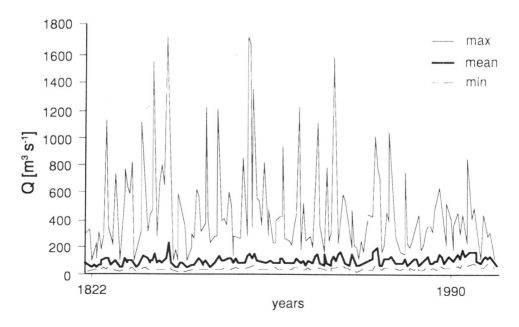

Figure 2 Annual maximum, minimum and mean flows of the River Warta (gauged at Poznan)

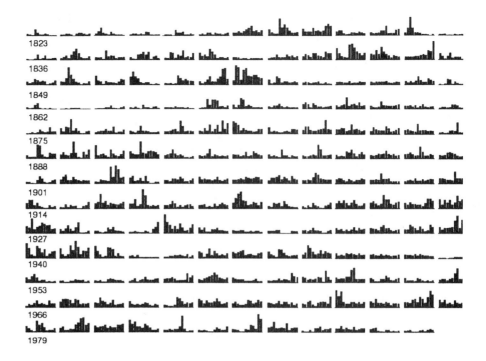

Figure 3 Monthly mean flows of the Warta at Poznan

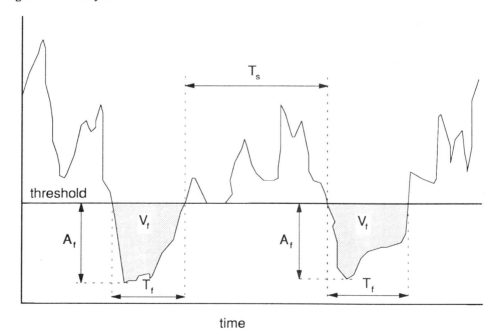

Figure 4 Definition sketch — crossing of a low flow threshold

demand quantitatively is bound to high uncertainty, as it is dependent on population dynamics and economic development.

Figure 3 presents the monthly mean flows for the time period studied. In order to analyse the low flows in a comprehensive manner, the renewal theory has been applied: this is used to study crossings of an arbitrary low flow threshold level. The characteristics (Figure 4) relevant to a particular value of a threshold are: sojourn time under the threshold, T_F; sojourn time over the threshold, T_S; amplitude of deficit measured in reference to the threshold, A_F, and volume of deficit measured in reference to the threshold, V_F. Based on these four characteristics a number of further important aggregate variables can be determined, such as reliability in the temporal or volumetric sense (R_t and R_v respectively) and resilience, measured as the inverse of the expected value of T_F. In addition to the analysis of yearly values reported above, there are several important properties of flows which need to be

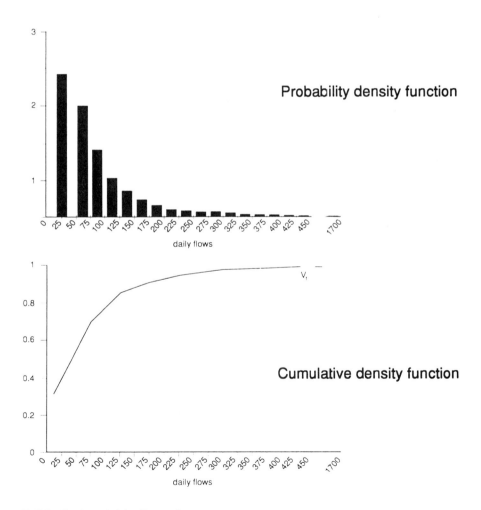

Figure 5 Distribution of daily flow values

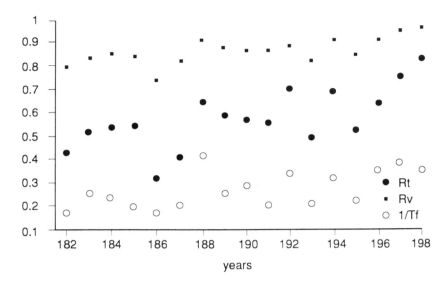

Figure 6 Temporal and volumetric reliabilities, and resilience as function of the time (decades)
for daily flows (threshold level $Q = 65$ m^3 s^{-1})

studied for shorter (e.g. daily) time intervals. Figure 5 shows the distribution
(probability density function and cumulative density function) of daily flow values,
based on a sample of 61 666 observations. Temporal and volumetric reliability and
resilience are shown in Figure 6 as functions of time (determined for decades).

The scanty water resources are effectively further reduced by water pollution. The
quality of water in the whole drainage basin of the river Warta does not actually meet
the requirements set by Polish standards for the lowest (third) water quality class.
The principal reason is a massive discharge of untreated sewage directly into watercourses: there is a serious lack of sewage treatment plants. For example, in 1990 the
town of Lódz (849 000 inhabitants) discharged 99.7% of its sewage requiring
purification (117.5×10^{-6} m^3) to a tributary of the river Warta without any treatment.

About 80% of the industrial plants do not provide adequate sewage treatment. The
scarcity of water in the river is an additional natural factor rendering the problem
of water pollution more critical.

The most common factors causes of water pollution in the area are:
- biodegradable substances (BOD5) downstream of lumped sources of pollutants
  (municipal sewage, industries);
- heavy metals, most often zinc;
- nutrients, nitrogen compounds and phosphates;
- bacteriological indicators of faecal pollution (coliform bacteria).

The contribution of agricultural pollution to the overall pollution of surface and
groundwaters of the river Warta is growing. Runoff of pollutants from agricultural
areas threatens surface waters, and initially the small tributaries of the river Warta.

This results in a considerable growth in the amount of nitrogen, phosphorus, organic compounds and the general quantity of mineral compounds in the water. The situation is aggravated by poor waste management in rural areas, by development of water supply systems without adequate sanitary sewers, and by the poor condition of existing sewerage, which results in accidental spillage of sewage into the irrigation dykes and from there into the rivers.

THE 1992 DROUGHT IN WIELKOPOLSKA

A number of consecutive snow-free winters and a series of years with warmer summers and precipitation below average aggravated the water supply shortage in Wielkopolska — a large area of central western Poland within the drainage basin of the Warta.

The most severe drought with the most far-reaching impacts was recorded in the vegetation season of 1992. After the snow-free winter of 1991-92 the precipitation recorded in all subsequent months until September was well below the long-term average. For example, in June 1992 precipitation of 3 mm was recorded at the Poznan gauge, compared to the long-term average of 61 mm. A similar tendency has been observed at other sites in the region.

Lower precipitation, combined with higher air temperatures every month in the year 1992 until August, led to a drought in every sense of the word — meteorological, hydrological, agricultural and ecological. Its extent was greater than that of any other in this region for decades.

Evapotranspiration remained higher than precipitation in every month of the vegetation period (April to September). In the summer months, the difference between the monthly precipitation and monthly potential evaporation reached values of -159, −121 and −71 mm for June, July and August 1992 respectively (Kowalczak *et al.*, 1992).

The consequences of the drought were manifold and comprised:

- Significant crop reduction on a regional scale — some municipalities claimed that crops were only of the order of 35% of the average. The mayors urged the Governor to declare local emergency status (and a tax reduction for farmers).
- Numerous and extensive wildfires of an extent never experienced before. Over the entire Poznan Province the ban on the public entering forests, issued by the Governor, remained valid throughout the summer.
- A dramatic reduction of flows in rivers and streams, destroying habitat for fish and other river fauna. Flow in smaller rivers dropped to zero. Water quality problems were aggravated because of insufficient dilution.
- The groundwater level dropped considerably. Water supply by cisterns had to be organised for a number of municipalities.
- Most ponds and smaller lakes in the region dried out completely.
- Wind erosion in early spring (bare soil with no vegetation) led to local soil impoverishment.

WATER MANAGEMENT IN THE DRAINAGE BASIN OF THE RIVER WARTA

The administrative division of the river Warta catchment has been an additional drawback in rational water management. The drainage basin of the river belongs to as many as 19 administrative units (Provinces — *województwa* in Polish). It was therefore very difficult to conduct integrated water management with so many independent players.

On 1st February 1991, the Minister of Environmental Protection, Natural Resources and Forestry of the Republic of Poland launched seven Regional Boards of Water Management in the country. The principal issues for these boards were:

- to combat pollution of surface waters and groundwaters;
- to protect the sources of drinking water, and
- to aid water users (population, agriculture, industry) in rational water management.

The Regional Boards of Water Management are executive organs of the Councils of Water Management which play the role of water parliaments. They consist of three groups of partners: state administration, local authorities and water users. It is novel feature that not only the state administration but also local authorities and water users are partners in making decisions on water management. The basic tasks of the councils are:

- setting programmes of water management in the catchment;
- distribution of financial resources in the catchment;
- pricing policy.

The Regional Boards of Water Management cut across the traditional administrative units of water management: instead they are arranged within the hydrographic boundaries of major river basins. They promote rational water management based on principles well established in Western Europe (e.g. "the polluter pays", promotion of water saving, integrated management embracing water quantity and quality, surface water and groundwater, and several components of the environment: water, air, soil). The policy for planning and investment focuses on the priority objects, granting the most advantageous cost/benefit ratio, and on simultaneous investing in the upstream parts of the catchment (affecting the whole river) and in the most critical links of the chain (e.g. the biggest polluters).

The Regional Board of Water Management in Poznan functions within the drainage basin of the river Warta. The basin embraces the entire territory of three provinces and includes parts of a further 16 provinces. There are 440 municipalities within the drainage basin.

Among the basic tasks of the Regional Board of Water Management in Poznan, responsible for the drainage basin of the river Warta, are:

- determination of principles and conditions of water use;
- development of intervention programmes aimed at the improvement of water quality, setting priorities, sequencing and scheduling;
- carrying out periodical analyses of water quality;
- proposing fees for water use and inflow of effluents;

- promotion and execution of studies and research projects, and ecological education.

Although the present environmental situation in the catchment of the river Warta is far from satisfactory, some positive changes have been already observed. First of all, environmental awareness has risen considerably and the need for an improvement in the environment has become a major issue in the perception of society. There have been several studies and reports on the state of the environment in the river Warta catchment, and in particular provinces in the catchment.

A number of projects have been carried out with foreign assistance. An example is the project called the Masterplan for the river Warta, funded within the framework of the European Community PHARE Programme. The aim of the project was to carry out a comprehensive study of the state of the environment, to identify the options and priorities and to conduct a feasibility analysis. A number of inter-regional initiatives were undertaken aimed at improving the environmental awareness of a broad range of addressees (children, youth, teachers, regional authorities). One project in the Poznan Province devoted to these issues is carried out in cooperation with (and with financial assistance from) the Province of North Brabant (The Netherlands).

CONCLUDING REMARKS

Pollution of water resources in Poland and frequent occurrence of water deficits results in the emergence of water as a barrier to the development of several regions of the country. Kindler (1987) identified several components of this barrier. It would be instructive to find out which components have remained active, despite profound political, economical and social changes in the country over the last four years.

Kindler (1987) distinguished primary and secondary natural barriers, barriers of an economic and social nature, and also the organisational-institutional barrier. The primary natural barrier is the natural water scarcity resulting from low precipitation and low water availability. The secondary natural barrier is related to the long-lasting lack of rational water management, which has led to the highly unsatisfact-ory present state where there is considerable pollution.

The economic barrier (Kindler, 1987) was related not only to scarcity of financial means, but also to the difficulties in acquisition of materials, poor investment and low capacities of specialised water engineering companies. At present, this barrier has changed its nature: the lack of adequate financial means is now definitely what is hampering rapid improvement in the environment. Although public awareness of environmental problems has undoubtedly risen in central and eastern Europe (CEE), there still exist a number of factors which limit domestic funding possibilities. As expressed by Somlyody (1993): in the public perception "other issues are so pressing: growing unemployment (close to 15% in several countries) and inflation, increasing prices and low salaries, bread-and-butter worries and so forth, which reduces willingness to pay for solving environmental and water pollution problems".

What is more, the top priorities of governments are economic restructuring and privatisation, rather than extensive investment in environmental protection. On the other hand, the annual expenditure which would be necessary in order to reach the EC water quality standards in CEE in the not-too-distant future could even exceed the gross domestic products (Somlyody, 1993). Therefore another strategy needs to be adopted, building on the existing infrastructure and its flexible extension (in stages), active demand management, regionally variable standards and cost-effective policies.

The organisational-institutional barrier mentioned by Kindler (1987) pertained to the administrative structure of water resources management, based on a division into provinces whose borders had nothing in common with catchment boundaries. It was therefore not uncommon for a set of inter-connected reservoirs to be managed by different authorities. There have been considerable attempts to remove this institutional barrier. The present basin-oriented structure of water management is undoubtedly a mark of progress.

The clean water-rich rivers of our childhood are the dream of many Poles, who remember that even in the 1960s the quality of surface water was far more satisfactory. When can this goal be realised? Several decades will be necessary, even if the problem is not unique in a technical sense. Large-scale river pollution occurred (and was remedied) in Western Europe. However, one cannot mimic the Western pattern in CEE. There is no money for installing or upgrading the many badly-needed sewage treatment plants. It is not possible to lock up the many industrial plants which are polluting the environment but cannot afford adequate treatment of their industrial waste water. The problem indeed depends on many criteria. The control of unemployment is undoubtedly an important social objective, which adds yet another dimension to the integrated perspective.

ACKNOWLEDGEMENTS

The financial support of the work by the State Committee of Scientific Research (KBN) of Poland under grant No. 4 S401 055 04 is gratefully acknowledged.

REFERENCES

Kowalczak, P., Farat, R. & Mager, P. 1992. The run of the hydrological drought in the Poznan Province in 1992. IMGW, Poznan (in Polish).

Kindler, J. 1987. Barriers to the development of water management versus national spatial arrangement policy, *Gospodarka-Administracja Panstwowa* 6, 36-39 (in Polish).

Somlyody, L. 1993. *Quo vadis* water quality management in Central and Eastern Europe, *IIASA Working Paper WP-93-68*, Laxenburg, Austria.

44 Co-operation for international river basin development: the Kosi basin

T PRASAD, S KUMAR, A VERDHEN and N PRAKASH
Centre for Water Resources Studies, Patna University, India

D GYAWALI, A DIXIT, N K LALI and B R REGMI
Royal Nepal Academy of Science and Technology, Kathmandu, Nepal

River basins which cross national boundaries are commonly characterised by a diversity of political jurisdictions which contrasts with their unity of hydrological behaviour. While the laws of hydrology and other earth sciences governing the spatial and temporal distribution of the water resources are based on the integrity of the basin, its political divisions, generally based on historical, ethnic and cultural factors, often tend to ignore this. Water is a unique natural resource which is universally necessary for economic activity — agriculture, industry, energy, etc. — as well for the sustaining of life, but it does not always serve the needs of humanity effectively and adequately in its natural regime. Thus intervention is imperative in order to optimise the use of water resources.

To ensure the measures adopted for intervention are not counterproductive, they must be planned and designed on the basis of the hydrological, hydraulic, geomorphological and hydrogeological integrity of the basin system. However, political divisions and diversity often constrain options or inhibit actions. This dichotomy usually results in poor development of an internationally shared river basin, often subjecting the peoples to suffering and depriving them of the benefits of water and related resources which occur naturally in the region. The Kosi River basin in Nepal and India is a case in point.

PRINCIPLES OF WATER RESOURCES DEVELOPMENT

Before examining the case of water resources development in the Kosi basin, it will be worthwhile to have a look at the basic scientific principles of water resources development, which may be summarised as follows (Prasad, 1974):

THE BASIN AS A UNIT

The first principle is that of basin-wide planning, where a basin is considered as a rational unit of planning for optimum development, effective management and efficient utilisation of the water resources. As the basin represents a closed boundary for various hydrological processes, it facilitates quantitative analysis of the spatial and temporal distribution of the water resources. It is only on the basis of this analysis that the best engineering measures to adjust distribution for optimal benefits with minimal adverse effects can be formulated.

COMPREHENSIVE BENEFITS

The second principle is that of water resources development for comprehensive, and not sectoral, benefits. The rationale of this principle is that while water has the potential to be put to many uses, any engineering measures taken to manage it can also serve multiple purposes. These uses and purposes may be complementary but may also be conflicting. The aim must be to resolve conflicts and maximise complementary usages. Thus, flood control, hydropower generation, irrigation, inland navigation, prevention of waterlogging and environmental enhancement can all be viewed holistically and planned for in an integrated manner. Sectoral planning often leads to a negation of this principle; for example, the land may be protected from floods but suffer from poor drainage, waterlogging and salinisation.

INTEGRATED RESOURCES

The third principle is that of integrated resources planning. Water is produced from different distinct sources, namely snow/glaciers, surface water and groundwater. Only two of these, surface water and groundwater, are easily exploitable, and there is considerable hydrological interaction between them. It is only logical that their use should be planned and developed in an integrated and coordinated manner. By taking advantage of the complementary features in their hydrology and quality, their efficient utilisation and management can be considerably enhanced and a hydrological equilibrium can be maintained in the system, so leading to sustainability. Advocacy of conjunctive use of surface water and groundwater to develop sustainably productive irrigated agriculture emanates from this principle (Prasad & Verdhen, 1990). The present and potential water resources development in the Kosi basin and the imperative for co-operation may be appraised in the light of and with reference to these principles.

THE KOSI BASIN AND ITS PRESENT DEVELOPMENTS

The Kosi basin is situated in the eastern part of the Indian subcontinent (Figure 1) with about 85% of its total catchment of 86,764 km^2 mostly in Nepal and partly in Tibet, and the remaining 15% in North Bihar (India). The river is known as Sapta Kosi in Nepal and is formed by seven tributaries; the three main ones, Sun Kosi, Arun and Tamur, meet upstream of Barahkshetra, and after flowing through a narrow gorge for about 10 km, debouch into the plains at Chatra in Nepal. Two of

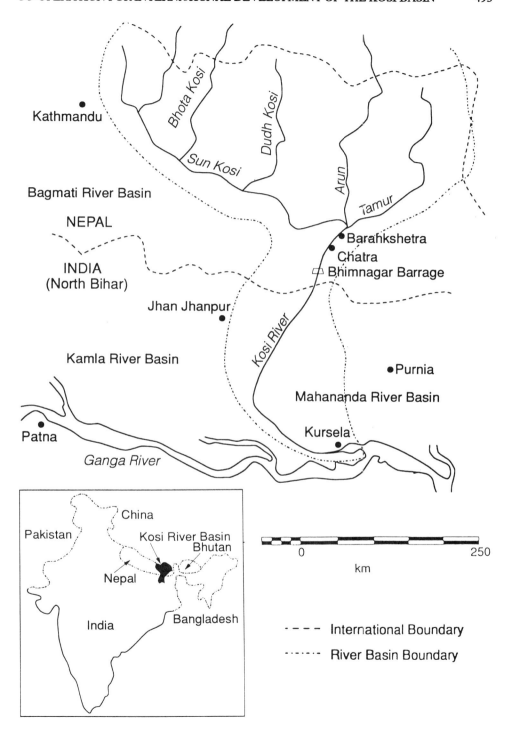

Figure 1 The Kosi River Basin

the highest mountain peaks in the world, Everest and Kanchanjenga, lie in the catchment of Arun. After flowing through another 25 km of its total length of 220 km in Nepal, it enters India, traverses a length of 248 km through the sandy plains of North Bihar, and converges with the river Ganga at Kursela. It negotiates a drop of 86 m in the 273 km from Chatra to Kursela.

High sediment load of the order of more than 2000 ppm during the monsoon flows at Chatra, and river flows varying from an average of 300 m^3 s^{-1} during the lean period to as high as 26 000 m^3 s^{-1} during floods, are problematic features which characterise the river. Earning the epithet of "Bihar's River of Sorrow" for its recurrent, devastating floods, frequent changing courses and rendering its vast flood plains infertile through deposits of coarse sand, the river Kosi is known to have shifted westwards through a distance of 112 km during the last 200 years (Carson, 1984). With uncertain and low yield agriculture as almost the sole means of livelihood, poor means of transport and communications liable to be disrupted by floods, and rampant incidence of diseases like malaria due to stagnating pools of water, millions of people living in the flood plains (where the population density is about 500 persons per km^2) suffered misery and destitution. While the plains remained ravaged by floods, people in the upper Kosi basin also lived in hardship and poverty unnoticed by the outside world.

Designed primarily to alleviate the suffering of the people in the plains, the existing Kosi Project, formulated in 1953, was aimed at containing the river Kosi between two embankments on either side with a spacing varying from 5 to 16 km, restricted upstream by a barrage located 8 km inside Nepal, which was considered a suitable gradient control measure. Secondly, irrigation was provided through a network of canals, taking water off from both sides of the barrage, as well as through an inundation canal drawing directly from the river at Chatra, upstream of the barrage. Taking advantage of a fall on the eastern main canal, an additional benefit in the form of the generation of 20 MW of hydropower was provided by the project. This project was implemented by the Government of Bihar under an agreement between the governments of India and Nepal signed in April 1954, and revised in December 1966. It has been providing flood control benefits since the late 1950s and irrigation benefits since 1964.

A proposal for a multipurpose high dam on Kosi at Barahkshetra had been under consideration by the Government of India since the late 1940s. As a project formulated by the then Central Water, Irrigation and Navigation Commission of the Government of India in 1950, a 783 ft (239 m) high dam at Barahkshetra was proposed to provide flood moderation from 850 000 cusecs (24 088 m^3 s^{-1}) to 200 000 cusecs (5668 m^3 s^{-1}), irrigation of 3.84×10^6 acres (1.554×10^6 ha) in Nepal and India, generation of 1890 MW of hydropower and a facility for inland navigation. For various reasons, this proposal, although not given up, could not be implemented. However, the possibility of taking up such a project was recognised in the intergovernmental agreement for the 1953 Kosi Project. The original 1950 Kosi high dam project was updated by the Central Water Commission in 1980, envisaging a 269 m high dam, 1.6 km upstream of Barahkshetra, another barrage and canal system at Chatra, additional irrigation of 143 000 ha in Nepal and 674 000 ha in India with enhanced intensity of irrigation in the existing Kosi command, a flood moderation from a peak of 42 500 cumecs to 14 000 cumecs and a hydropower generation of 3300 MW (Central Water Commission, 1981). This proposal is under active consideration by the two governments.

In the upper catchment of the basin in Nepal, only limited development has taken place. A 10.5 MW hydropower project on the Sun Kosi river was built with Chinese assistance in 1967. The other two projects are the 2.4 MW Panauti hydropower project built by the Soviet Union and a 240 KW small hydroelectric project at Dhankuta. Another project under consideration is the 402/201 MW Arun III hydropower scheme. Both these Arun projects are run-of-river type schemes. A master plan for the Kosi basin prepared with the assistance of Japan International Co-operation Association in 1984 estimated a hydropower potential of 22 000 MW for the basin through storage, run-of-river and inter-basin diversion schemes. In addition to these projects, implemented or planned, thousands of village level turbines and small-scale indigenous irrigation schemes cater for the day-to-day needs of the local community in the hills.

REVIEW OF THE DEVELOPMENT PROCESS

MOTIVATION

A critical look at the history of water resources development in the Kosi basin will assist in understanding the process and perspective of such development. Historically, the motivation for intervention in the Kosi basin came from the unmitigated suffering of the people in its plains in North Bihar caused by the fury and devastation of the wayward Kosi. During the British period, even large-scale intervention measures like dams were perceived as being for the benefit of only the provinces directly concerned (Gyawali, 1994).

The two projects for Kosi high dam — the original one formulated in 1950 which could not be implemented and the updated version of 1980 which is still under negotiation — are perceived as Indian projects, even though they envisage immense benefits to both India and Nepal. Similarly the projects implemented or planned in the upper Kosi catchment on Sun Kosi, Arun and Tamur are essentially Nepalese projects, although these interventions may be beneficial to India.

APPROACH

Another feature of the projects so far executed or planned in the Kosi basin is that they are based on a project-centric approach; they do not conform to the principles of water resources development, as indicated earlier, and do not aim at optimum development, effective management and efficient utilization of the water resources of the basin; neither do they provide options and alternatives to the ultimate decision makers, i.e., the peoples and their governments. This has led to unsatisfactory solutions to the problems, a poor realisation of the potential, emergence of adverse effects or counterproductive features, and distorted public perceptions on both sides. Even a cursory look at the performance of the existing projects in the Kosi basin will serve to bring out these points.

IMPACTS

The evaluation of the performance of the 1953 Kosi project, by far the most major project implemented in the Kosi basin, has been carried out in a few workshops (Carson, 1984) and several committee reports (Central Water Commission, 1981; Gyawali, 1994). The gist of these appraisals is as follows:

Flood Performance
The left embankment of about 150 km length and the right embankment of about 127 km length were planned to provide total flood protection to about 280 thousand ha, which constitutes only a part of the total flood affected area in the basin. About 500 thousand ha in the basin is still subject to floods of varying intensity (Ghose, 1982). On account of the capricious behaviour of the river between the embankments, even the flood protected area has suffered devastating damage due to breaches at various vulnerable points. As the flood plains between the embankments have considerably aggraded, the threat of overtopping as well as breaching of embankments is ever increasing and, moreover, in the hills, flash floods due to cloud bursts, glacial lake outbursts and landslides continue unabated.

Irrigation Performance
The irrigation component of the Kosi project was planned to provide irrigation benefits to 740 thousand ha from the eastern main canal, 312 thousand ha from the western main canal and 86 thousand ha from Chatra inundation canal. While the western Kosi canal system is still incomplete, according to a 1979 Government of Bihar Report on "Kosi Project and Its Irrigation Potential", the average area actually irrigated during the five-year period, 1973-78, amounted only to 26.5% of what had been planned. Subsequent figures obtained for the actual area irrigated show a distinctly declining trend. The performance of the Chatra inundation canal has also been unsatisfactory due to intake problems in the low flow season and sedimentation problems during high flows. Reasons for this include non-completion of the micro-irrigation infrastructure, silting up of water courses and field channels, reduction in carrying capacity of canals due to siltation, undulating topography of the command and poor maintenance of the system.

A more fundamental defect of the irrigation system is, however, conceptual and strategic. In view of the surface and groundwater hydrology, topography, water table and other factors prevailing in the basin, planning for irrigation exclusively from surface water sources for the Kosi command was inherently unsound. The chief conclusion of an International Workshop on "Post Fact Evaluation of a Water

Resources Project — A Case Study of The Kosi Project" held at Patna in 1979 was to this effect. It recommended that utilisation of groundwater in conjunction with surface water must be promoted to improve the irrigation performance of the project. In fact, it is the *ad hoc* emergence of extensive groundwater use through cheap, improvised bamboo tubewells in the Kosi command over the years that has "propped up" irrigated agriculture and saved the project from what would have been a catastrophic failure on the irrigation front.

Power Performance
Generation of hydropower to the tune of 20 MW through a canal fall was an incidental benefit of the Kosi project. Hardly 60% of this capacity is ever generated. Even the addition of power generated or planned to be generated by the Nepalese hydroelectric projects of Sun Kosi (10.5 MW) and Arun III (402 MW) would amount to exploitation of only a small percentage of the hydropower potential of the Kosi basin.

Ecological Effects
Several appraisal studies have highlighted the adverse ecological effects that have emerged over time as a consequence of the Kosi project. The aggradation of river bed between the embankments has encouraged seepage, and caused waterlogging and submergence of the areas beyond the embankments. This has either reduced the productivity, or has rendered uncultivable, a large area of the Kosi command. The embankments have also interfered with the drainage of the Kosi basin, necessitating a massive drainage scheme for the basin at a huge cost. A systematic rise of the water table has been observed and waterlogging conditions have been created or exacerbated in a considerable area of the Kosi command, reducing its agricultural productivity and yield. In several aspects, the Kosi project is generally considered to be counterproductive.

LESSONS

Thus, developing water resources in the Kosi basin at considerable expense of time, effort and money, has to date provided only uncertain and inadequate flood protection, created unsustainable irrigation systems and failed to realise its huge power potential. The minimal achievement on these fronts is coupled with detrimental and counterproductive ecological consequences as well as distorted and negative public perceptions. Projects emanating from Indian motivations and initiatives are perceived as possible ploys to deprive Nepal of its fair and rightful benefits, while Nepalese concern and caution, born out of perceptions of past experience, are dubbed in India as unreasonable demands, intransigent attitude and deliberate stalling.

An analysis of the process of water resources development would reveal the utter departure from scientific method. The principles of water resources development were conspicuous by their almost total absence from the formulation of any of the project plans. The most basic of these cardinal scientific principles, that of basin-wide planning, has not been observed in evolving any plan in and for the basin. No valid plan can be made unless the hydrological integrity of the basin is realized, which calls for co-operative development of the water resources. It makes co-operation between the nations sharing the basin imperative and, hopefully, inevitable.

DYNAMICS OF CO-OPERATION

Where such a vital matter as water resources development is concerned, which affects all aspects of life and living, co-operation between nations cannot be a sterile or static concept. Unfortunately, the past forty years of interactions, mainly at the nation-state level, were not supported by a commitment to co-operation, and it is no surprise that they have led neither to any major breakthroughs nor to any dazzling insights.

Given the importance of co-operation to bring about the desired process of water resources development in the basin, how to generate and sustain this co-operation is the key issue. For this, it is useful to consider the mechanics and dynamics of co-operation, particularly in the context of water resources development.

First of all, it should be understood that, while co-operation is natural between India and Nepal in all spheres of their interactions, because of their geographic affinity, historical bonds and cultural commonality, in the sphere of water resources development it is dictated, in addition and dominantly, by science. Hence, the primary step to bring about dynamic co-operation between the co-basin neighbours is to put science "in the driving seat". To achieve this transformation, the academics, who are more committed to their scientific discipline than to any narrow concept of nationality, have to come to the fore: this has been a critical missing link in the process so far.

To guard against the danger of straying into sterile scientific pursuits, it is also essential that the academics of the region assume their responsibility towards society by contributing a much-needed scientific perspective. In fact, academics have a strategic role to play in the social context to synthesise the concerns of higher level government bodies and the people at grassroots. To engineer co-operation, with good science as one of the vehicles, intermediary institutions at the horizontal contact level should be given more opportunity to plan and facilitate changes (Anon, 1992).

Another favourable factor for co-operation in development of hydrologically shared water resources of an international basin is the evolution of international law along scientific lines. From the doctrine of riparian rights, and theories of prior appropriation, territorial integrity, natural flow and equitable apportionment, international law has now come to recognise the rationality of the basin as an integral unit for water resources development in its "community of interest theory" (Chauhan, 1992). Although water resources development is too vital a matter to be left to legal instrumentality, such positive trends are greatly reassuring.

The most important factor for positive Indo-Nepalese co-operation in the development of water resources is the larger context in which such development has to be achieved. The availability of natural resources, the possibilities of technology, the dictates of science, the socio-economic milieu of the region, economic inter-dependence, demographic and cultural affinity, historical bonds, people's aspirations, responsive democratic governments, all provide the necessary pointers to co-operation: it is necessary for neighbours to accept that they share a common destiny.

INITIATIVES AND OUTLOOK

Several sporadic non-governmental efforts and initiatives have been made in the past to analyse issues and consider possible approaches to water resources development in the region. A new impetus to facilitate Indo-Nepal co-operation came in May 1992 when the Centre for Water Resources Studies (CWRS), Patna University organised a two-day workshop on "Co-operative Development of Indo-Nepal Water Resources — Prospects, Opportunities and Challenges". Attended by Indian and Nepalese water and other professionals, the workshop called for an academic approach and greater interaction among academics in the region. As a follow-up a conference was organised by Nepal Water Conservation Foundation in Kathmandu in February 1993. Entitled "Co-operative Development of Himalayan Water Resources", it provided a forum for discussing four major themes on the Himalaya-Ganga for professionals from Nepal, India, Bangladesh, USA, UK and Switzerland (Dixit & Gyawali, 1992). These two events, which have now come to be known as the "Patna Initiative", have encouraged academic interaction and helped to build confidence.

It may take several decades for any proposed major hydraulic intervention measures to become reality; in order for this to happen people need to have confidence in their ability to succeed, and the two nations must begin to see co-operation as a "win-win" situation rather than a mutual threat.

REFERENCES

Anon. 1992. Workshop on co-operative development of Indo-Nepal water resources — conclusions and recommendations. Centre for Water Resources Studies, Patna.

Carson, B. 1984. Erosion and sedimentation processes in the Himalayas. ICIMOD, Kathmandu.

Central Water Commission 1981. Feasibility report on Kosi High Dam Project. Government of India, New Delhi.

Chauhan, B. R. 1992. Settlement of international and interstate water disputes in India. The Indian Law Institute.

Dixit, A. & Gyawali, G. 1993. Building Regional Co-operation in Water Resources Development. *Water Nepal*, 3 (2/3).

Ghose, J. P. 1982. An assessment of the problems and potentials in the water resources development of North Bihar. MSc Thesis, Patna University.

Gyawali, D. 1994. The Himalaya Ganga Impasse; the need for a non-traditional approach. Paper presented in seminar "South Asia at the crossroads: conflict and co-operation", BIISS, Dhaka.

Prasad, T. 1974. Imperatives of water resources planning in the developing economy of India. *Indian J. Power and River Valley Development*, Calcutta, July 1974.

Prasad, T. & Verdhen, A. 1990. Management of conjunctive irrigation in alluvial regions—issues and approaches. Proc. Internat. Conf. on Groundwater Resources Management, Bangkok.

45 Transitional control of the River Senegal

J B C LAZENBY and J V SUTCLIFFE
Sir Alexander Gibb & Partners, Reading, UK

INTRODUCTION

The hydrology of the River Senegal has been studied since the beginning of this century, and the regime of the river is well documented (Rochette, 1974). The main feature of the upper basin above Bakel — where the river enters Senegal and Mauritania after receiving runoff within Mali and Guinea—is the extreme variability of rainfall and runoff in both time and space.

The three main tributaries, the Faleme, the Bafing and the Bakoye, drain areas where the mean annual rainfall varies between 2200 mm in the Fouta Djalon mountains of Guinea and less than 600 mm in the northern tributaries within Mali (see Figure 1). Moreover, this rainfall is highly seasonal, so that the period when the average monthly rainfall exceeds potential transpiration varies from six months in the southwestern limits of the basin at Mamou and Labe to three months at Kayes, just above Bakel.

The variations of rainfall from year to year have been similarly severe, with the Sahel drought of recent years resulting in a southward migration of the annual isohyets and a marked decrease in rainfall. The effect of these variations on river flows has been considerable, with the effect of lower rainfall exaggerated by the relatively constant potential transpiration. Thus the annual variability of runoff is as high as 43% over the whole basin to Bakel, with the variability of individual tributaries ranging from 55% on the relatively dry Bakoye to 31% on the Bafing which drains the wetter Guinea highlands. The runoff totals were complete for tributary stations for the period 1952-1984 (Sutcliffe & Lazenby, 1989) and the results are summarised in Table 1, where the contributions of individual tributaries may be seen to range from nearly 400 mm on the Bafing to about 20 mm from the drier Baoule. The mean flows of the tributaries are included in Figure 1.

However, these long-term averages disguise the dramatic effect of the Sahel drought on the available flows. This effect is best illustrated by the annual flows at Bakel for the whole period of records (Figure 2): recent flows for 1971-90 were about half the flows of the 1951-70 period, although the flows have been affected since 1987 by storage in the Manantali reservoir.

The Bafing has the highest as well as the least variable runoff of the individual

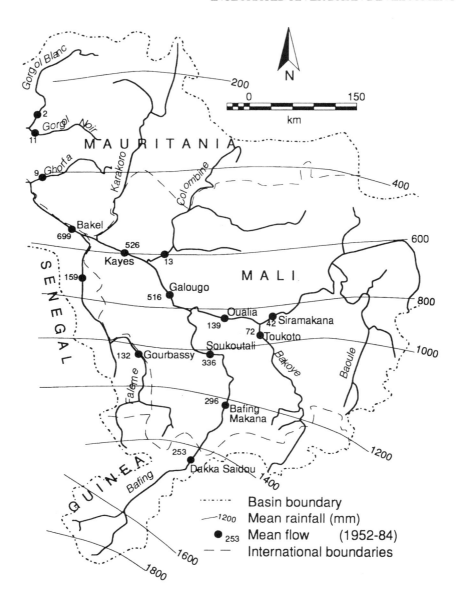

Figure 1 Map of the basin, with average rainfall (1951-80) after Sow (1984), and mean flows (1952-84)

tributaries, and therefore the reservoir at Manantali, just above the gauging station at Soukoutali, is a logical site for controlling the river flows. The lower variability of the Bafing implies that a higher proportion of the total runoff is contributed by this tributary in dry years and therefore in the recent drought. The reasonably high correlations between the flows at the different sites and the downstream flows at

Bakel are important in terms of control, since less than half the average runoff at Bakel passes the dam site.

The uncontrolled flow has been highly seasonal, with the high flows and some 80% of the annual runoff at Bakel being concentrated within the period August to October (see Table 2).

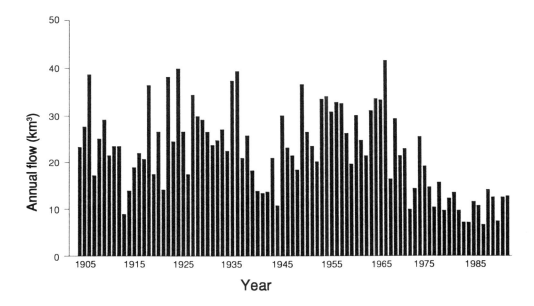

Figure 2 River Senegal at Bakel: annual flows, 1904-92. Flows since 1987 have been affected by reservoir storage.

Table 1 Mean flow statistics for the Senegal basin (1952-84)

River	Station	Area $km^2 \times 10^3$	Rainfall mm	Runoff $m^3 s^{-1}$	Runoff mm	CV %	R^2 Corr (Bakel)
Senegal	Bakel	218	850	699	101	43.2	—
Faleme	Kidira	28.9	1100	159	174	52.1	0.910
Senegal	Kayes	157.4	920	526	105	43.0	0.940
Bafing	Soukoutali	27.8	1425	336	381	31.4	0.875
Bakoye	Oualia	84.4	850	139	52	54.7	0.893
Bakoye	Toukoto	16	1150	72	141	55.5	0.834
Baoule	Siramakana	58	750	42	23	53.5	0.878
Colombine	Kabate	25	500	13	17	43.1	0.790

Table 2 Monthly mean flows of the Senegal at Bakel, m^3s^{-1} (1904-84)

Jan	Feb	Mar	Apr	May	Jun	Jul	Aug	Sep	Oct	Nov	Dec	Year
129	74	39	16	8	90	573	2281	3215	1558	536	235	732

TRADITIONAL LAND USE

For many centuries the land use of the Senegal valley below Bakel, with over half a million inhabitants, has been based on the uncontrolled flood flows of the river. The process is fairly complex as high flows pass above various thresholds through the high alluvial banks of the main river, spill into low-lying basins which are flooded for varying periods, and empty back into the river as flows recede. The result of the average annual flood, over an area about 10-20 km wide and 500 km downstream from Bakel, has been the seasonal inundation of some 5500 km^2 of the floodplain (Drijver & Marchand, 1985); however, the area flooded varies from very little in a dry year to about 8000 km^2 in a wet year.

This seasonal pattern of flooding and recession formed the basis of the local economy: it provided the opportunity for fresh water fisheries, recharged soil moisture storage in the fertile soils of the valley and also provided dry season grazing for over a million cattle before the recent drought, which were able to use the natural floodplain vegetation common to other African wetlands as well as the grazing provided by recently cultivated fields. The flooding also recharged local groundwater resources.

The local system of flood recession (Oualo) cropping of river banks and depressions has been practised over some 100 000-120 000 ha in a good year, but only 15 000-50 000 ha during the drier period of the last eighteen years. The dominant crop is

Table 3 Areas cultivated compared with flood volumes at Bakel

Year	Flood volume at Bakel (July-November, 10^6 m^3)	Area of recession cropping (ha)
1970/71	19 700	103 100
1972/73	9 200	13 700
1973/74	13 300	82 200
1976/77	13 500	29 400
1977/78	9 700	15 400
1978/79	14 900	56 900

sorghum, which is sown between mid-October and mid-December depending on the incidence of flood recession, but crops of higher value like beans and melons are grown on the land nearer the river. Detailed studies of recession agriculture carried out under UNDP auspices in the 1970s showed that the areas cultivated were very sensitive to flood volumes (Table 3).

In recent years the dry season flows have been used to support large and small irrigation schemes, including commercial irrigation of sugar cane near Richard Toll. The average rainfall of some 400 mm is inadequate to support agriculture, so without upstream control the extension of irrigation was totally dependent on the natural low river flows and some lake storage.

CONTROL OF THE RIVER

OMVS (the River Senegal Development Authority) was set up in 1972; it followed on from OERS (the Authority for the Riverine States of the Senegal). On behalf of the three Member States (Mali, Mauritania and Senegal) the authority prepared a programme for the development of the Senegal basin. The first part of this was implemented in the early 1980s with the construction of the Diama Barrage at the mouth of the estuary and the Manantali storage reservoir in the upper basin in Mali.

The Manantali Dam, which provides control over the flows in the Middle Valley, was completed in 1987. The reservoir has a live storage of 7.85 km^3, compared with a mean runoff at Bakel of 21.5 km^3, but only 9.4 km^3 over the period 1981-85; the proportion of the Bakel flow which passes the dam site has averaged 48%, but rises to nearly 70% in dry years. The Diama Barrage was constructed in 1986 near the mouth of the river to prevent saline intrusion during low flows and also to raise water levels for irrigation, but storage is limited at this site.

The mode of operation of the Manantali storage dam and the Diama barrage were the subject of studies for the OMVS from 1985 onwards. Although the Manantali dam is designed as a multi-purpose project, with benefits from hydroelectric power, improved navigation in the downstream reaches, and the provision of enhanced dry season flows, it is the use of controlled flows for irrigation which dominates the consumptive water use.

However, the conversion of an economy based on traditional recession agriculture to one based on modern methods of irrigation management would require either a controlled system like that of the Sudan Gezira, where crop planning and operations were controlled centrally, or a gradual extension of modern techniques. Because the second approach is more appropriate to the social conditions of the Senegal valley, it is important to control the river flow in such a way as to retain and as far as possible improve the benefits of the natural regime while providing increased dry season flows to enable modern irrigation to be practised. In order to combine these two objectives, the concept of an artificial flood was developed to provide a more regular and predictable source of inundation than uncontrolled flows, to be combined with other benefits.

ANALYSIS OF ARTIFICIAL FLOOD

The design of the artificial flood was based on an analysis of the natural floods from the hydrological records, including flood volumes and flooded areas, duration and timing of flooding, and the rates of rise and fall of the flood waters. The areas flooded by past natural floods have not been measured directly, but simulations have been made using a complex mathematical model, while estimates for a longer period have been made by comparing historic river profiles with floodplain topography on the assumption that the water levels on the floodplain coincide with adjacent river levels. Recent detailed studies (Hollis, 1990) have suggested that the flooding process is more complex than this assumption implies, and that it may be necessary to compare estimates of inundated areas based on satellite imagery with the hydrographs of historic floods.

However, estimates have also been made for the middle valley by comparing inflows at Bakel for the period 1951-1975 with outflows at Dagana (Sutcliffe & Parks, 1989). This study was based on water balance analysis of the floodplain, using monthly river flows with the low season outflows completed by regression, and comparing average monthly rainfall with evaporation from flooded ground; it assumes a linear relation between area and volume of flooding but uses local information to calibrate this relation. The results are limited by the quality of the input data and by the assumptions made, but the estimates of flooded areas correspond well with independent information. In Figure 3 the maximum areas flooded each year, as estimated by the water balance analysis, have been compared with the annual total flows at Bakel. The sensitivity of flooding to the river regime is evident.

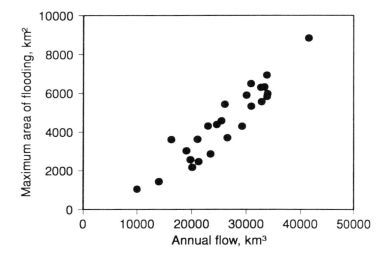

Figure 3 Estimated areas of flooding, Bakel to Dagana, compared with
annual flows at Bakel, 1951-75

In order to design the artificial flood note was taken of traditional practice. Cultivation is usually carried out on land flooded for 15-45 days, the lower limit being needed to recharge the soil moisture and the higher figure being the limit for uncovering in time for sowing. Estimated areas flooded for this period were of the same order as those of cultivated areas, though the correlation was poor. The areas flooded for more than 15 days varied from 400 000 ha in exceptionally wet years to under 50 000 ha in dry years. The flood peaks occurred at any time between 20 August and 20 September, and the rates of rise and fall were analysed. The maximum rise occurred in 1970 when the river rose from 500 to 2500 $m^3 s^{-1}$ in 11 days, while the recessions were slower.

The design of the artificial flood took agricultural requirements into account. The maximum area which could be cultivated with the available labour was estimated at about 100 000 ha; it was thought preferable that the rise in the river should be rapid to induce recharge, while the recession should be less rapid and spread over at least a month; the duration of flooding over the cultivated area should be between 15 and 30 days. Three floods were selected to allow areas of 50 000-100 000 ha to be cultivated, with the rise set at 200 $m^3 s^{-1} day^{-1}$ and the fall at 100 $m^3 s^{-1} day^{-1}$ above 1500 $m^3 s^{-1}$ and 50 $m^3 s^{-1} day^{-1}$ thereafter. The durations were based on historic floods, and the areas flooded deduced from modelling; it was assumed that half the area flooded could be cultivated. The artificial floods are summarised in Table 4: the flood volume is small compared with the historic records.

The initial impact of the artificial flood on the hydro-electric, irrigation and navigation components of the project was assessed using an operational model of the reservoir. The guaranteed regulated flow over the period of record would be reduced from 250 to 90 $m^3 s^{-1}$, and an irrigated area over 100 000 ha would be subject to water shortages in dry years; the guaranteed mean annual energy would be reduced from 1050 GWh to 910 GWh and the scope for navigation would be reduced in dry years. It is anticipated that these constraints could be removed in the longer term when the development of large-scale irrigation might allow the artificial flood to be phased out. However, Master Plans recently undertaken for development on either side of the river have emphasised the environmental importance of continued flooding in terms of regeneration of forest, grazing and fisheries.

Table 4 Characteristics of artificial floods at Bakel

Flood type	Cultivable area (ha)	Flood volume (Aug-Oct, km^3)	Peak flow ($m^3 s^{-1}$)	Duration at 2000 $m^3 s^{-1}$ (days)
A	50000	7.5	2500	10
B	75000	8.5	2750	15
C	100000	10.0	3000	20

OPERATION OF RELEASES FROM MANANTALI

In order to obtain a given flow hydrograph at Bakel, the releases from Manantali dam have to be coordinated with the natural flows from the unregulated Faleme and Bakoye tributaries. Pending a complete network of upstream transmitting stations, or the development of a flow forecasting system based on rainfall data, which is possible to achieve but with limited precision (Hardy *et al.*, 1989), it was recommended that the releases from Manantali could be based initially on the comparison of flows at Gourbassy on the Faleme and Oualia on the Bakoye with those at Bakel, as the travel times to Bakel were about three days in each case and intermediate inflows were small. The transmission of these flows to the control centre has been based on satellite transmission, but these have been supported with radio links.

It was further recommended that releases could be estimated according to simplified rules based on a study of historical flows. Releases should start when flows at Gourbassy and Oualia have increased by 350 m^3s^{-1} over three days; a rise of 200 m^3s^{-1} could be achieved by taking account of the increase at the other sites; the peak flow could be maintained by comparing the current flow at Bakel with those at the other sites, and the hydrograph recession could be maintained in a similar way. During the dry season compensation flows should be released in line with OMVS policy and to provide irrigation water at Diama. These rules were kept simple in order to test them empirically during operation. They were tested by simulation with sample years of historic data.

The operating procedure was expected to be improved by the establishment of a flow forecasting system which was being developed to give further information about the flows to be expected at Manantali and the adjacent tributaries. The success of such a system may be judged by further computer trials of the operation and by the actual flows recorded since the completion of the dam.

MULTIPLE USE MANAGEMENT OF MANANTALI DAM

In order to define optimal operating rules for the reservoir, ORSTOM were requested by OMVS to test the effect of different rules on the different objectives of control, using the historic 1904-1989 series of daily flows (Bader, 1992; Albergel *et al., 1993*). The objectives were hydroelectric production, flood control on the Bafing and Senegal, water supply for irrigation, principally below Bakel, maintenance of levels for navigation, and provision of an annual artificial flood. The priority of the objectives and the limits of the reservoir storage were taken into account in a statistical study of the minimum and maximum levels required on a given date to ensure that irrigation and other requirements, including flood control, were met with a certain level of assurance. This analysis of levels corresponding to reserves required on certain dates to meet the various objectives with a given degree of assurance could be used to define reservoir operating rules.

The trials made use of a propagation model for flows down the valley, which enabled the release required to provide a given flow at Bakel to be deduced. A reservoir trial was carried out to test the ability of the reservoir to meet the objectives over the historic period. The objective which is most difficult to combine with other objectives is that of flood control, which at certain levels of reliability requires large storage to be retained for flood attenuation. Indeed it could be argued that the reservoir is better sited for maintaining low flows than for downstream flood

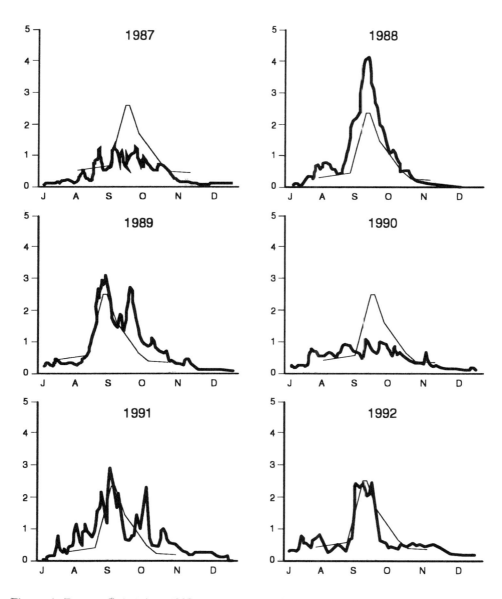

Figure 4 Flows at Bakel since 1987, showing effect of reservoir operation

control, as the reservoir commands a greater proportion of the flow at Bakel at low flows than at high flows.

A further study is planned (Bader, 1992) to test the value of flow monitoring and forecasting on the operation of the reservoir. The system would be based on transmission of flows from a network of gauging stations in the upper basin above Manantali and on adjacent tributaries, and the estimation of required releases from the Manantali reservoir.

RECENT OPERATION

The operation of the reservoir since its completion in July 1987 is illustrated by the actual flows measured at Bakel. However, the operation during this period has been affected by the filling of the reservoir, which began in 1987 and reached full supply in September 1992. Daily flows for these years are shown in Figure 4 (previous page). In 1987 filling of the reservoir took place during a dry year while the high flows of 1988 gave rise to a flood of the same shape but exceeding the artificial flood of Type A. It will be seen that the artificial flood was approximated by the actual flows in 1989 and 1991, but second peaks resulted from reservoir releases required to test hydraulic structures.

The dry year of 1992 was the first without constraints and the ability of the reservoir to provide an artificial flood during a dry year was demonstrated. This experience confirms that the reservoir can provide a reliable artificial flood during a dry year, because the flow at Manantali is a high proportion of the flow at Bakel in such a year; this has been further demonstrated by the release of a similar flood in 1993. In the reverse situation, when the requirement is to alleviate exceptional floods, the dam is less effective because a smaller proportion of the flow passes Bakel in wet years. The need for a clear operating policy and the benefit of improved flow forecasting methods are also evident.

CONCLUSIONS

The control of the Senegal through the operation of the Manantali reservoir is a complex problem which must take account of the present as well as the future requirements of the local economy. The operation must take account of the present land use and should not assume that the local economy will adapt quickly to a river regime designed to meet a drastic change in agricultural practice without a period of transition. If a transitional phase is not planned and the control of the river is not geared to this transition, the potential benefits of an expensive project will not be realised and the effects on the traditional land use may be detrimental. The example has shown that it is possible to combine the benefits of assured river flows on traditional systems with those of higher dry season flows for expanding irrigation.

REFERENCES

Albergel, J., Bader, J.C., Lamagat, J.P. & Séguis, L. 1993. Crues et sécheresses sur un grand fleuve tropical de l'Ouest Africain: application à la gestion de la crue du fleuve Sénégal, *Revue Sécheresse*, **4**, 143-152.

Bader, J.C. 1992. Consignes de gestion du barrage à vocation multiple de Manantali: détermination des côtes limites à respecter dans la retenue, *Hydrol. continent.*, **7**, 3-12.

Drijver, C.A. & Marchand, M. 1985. Taming the floods: Environmental aspects of floodplain development in Africa, Centre for Environmental Studies, Leiden, The Netherlands.

Hardy, S., Dugdale, G., Milford, J.R. & Sutcliffe, J.V. 1989. The use of satellite-derived rainfall estimates as inputs to flow prediction in the River Senegal, *IAHS Publ. No. 181*, 23-30.

Hollis, G.E. 1990. Senegal River Basin Monitoring Activity: Hydrological Issues, Parts 1 and 2, Institute for Development Anthropology, Binghamton, New York.

Rochette, C. 1974. Le Bassin du Fleuve Sénégal, Monographie Hydrologique No.1, ORSTOM, Paris.

Sow, A.A. 1984. Pluie et écoulement fluvial dans le bassin du fleuve Sénégal: contribution à l'hydrologie fluviale en domaine tropicale humide africain. Université de Nancy II, Centre International de l'Eau, Vol. V.

Sutcliffe, J.V. & Lazenby, J.B.C. 1989. Hydrology and river control on the Niger and Senegal, The Sahel Forum, Ouagadougou, Burkina Faso, 611-620.

Sutcliffe, J.V. & Parks, Y.P. 1989. Comparative water balances of selected African wetlands, *Hydrol. Sci. J.*, **34**, 49-62.

46 Integrated resource development in north-west England: a review of the Lancashire Conjunctive Use Scheme

H A SMITHERS, J M KNOWLES and K J SEYMOUR
National Rivers Authority, North West Region, Warrington, UK

INTRODUCTION

The principle of conjunctive use is that by operating different types of source as a single unit, greater yields and average supplies can be obtained than from the components operated in isolation. The Lancashire Conjunctive Use Scheme (LCUS) has been a significant water supply source in the North West since the 1970s. The National Rivers Authority, North West Region (NRA-NW) have recently carried out a review of the operation and management of the scheme for a number of reasons:

- A large body of data recording the behaviour of the system is now available;
- Environmental perceptions are changing;
- A search is in progress for additional sources of supply for the region.

Operational data availability
The accumulation of around 20 years of operational and experimental data is extremely valuable to the water resources modeller. However such information is inevitably found to require considerable care in interpretation. Despite this, it represents a significantly greater body of information than was available to the designers of the original scheme.

Changing environmental perceptions
The original licence conditions for the groundwater component of the scheme were designed principally to protect existing abstractors. Three-year total abstraction limits were designed to maintain the long-term viability of the aquifer. The need to maintain positive groundwater gradients at the aquifer boundaries to prevent saline intrusion was also identified. However it was soon recognised that groundwater abstraction was associated with reduced river flows.

With the formation of the NRA the emphasis changed considerably. In its role as "guardian of the water environment" the NRA is charged with sustaining the entire hydrological cycle (Law, 1993), and its customers include not just water supply companies and abstractors, but also wildlife and recreational users. As Skinner points out (1993), this principle has been recognised to some extent for many years in the setting of compensation requirements for reservoir systems. He suggests that it should be extended to the setting of groundwater abstraction limits, to ensure that the environment benefits in years of water stress. Downing (1993) also notes increased attention to the preservation of wetlands.

As a practical result of this change in emphasis, the NRA has identified lists of priority sites for alleviation of low flows: the LCUS is one. NRA-NW is in the process of developing a management plan which addresses the environmental problems and ensures the long-term sustainability of the aquifer. The review reported here represents the initial phase of this process.

Search for additional sources of supply
The North West Region has abundant rainfall and well-developed water supply systems. One option considered by the recent NRA national water resource development strategy was to divert water from the Lake Vyrnwy reservoir southwards out of the North West Region and use it to regulate the Severn, for possible onward transfer to the Thames. However, consideration of the regional balance between resources and demand indicated that, under a high-demand growth scenario, this diverted water would have to be replaced by new sources.

At present the LCUS licence is not fully utilised, so additional yield is theoretically available from this source. However, the environmental concerns raised above provided an additional impetus for a reappraisal of the scheme.

At a more local level, there has recently been increasing interest in smaller-scale groundwater abstractions across the aquifer, for industrial and agricultural use. An embargo has been imposed within the region on the authorisation of increased abstraction pending the outcome of the current investigations.

PHYSICAL BACKGROUND

Figure 1 shows the main features of the LCUS. The Wyre catchment drains the eastern side of the Bowland Fells in Lancashire and flows into Morecambe Bay. The major part of the catchment is agricultural, with pasture for sheep and cattle; the upper reaches are surrounded by moorland, used mainly for shooting and rough grazing.

The water quality is generally of a high standard, either Class 1A or 1B, although there are problems with high acidity in the upper reaches of the Wyre.

The groundwater component is derived from the Permo-Triassic Sherwood Sandstones of the Fylde aquifer, bounded by older Carboniferous strata to the east and overlain by low-permeability Mercia Mudstones to the west (Figure 2).

The Carboniferous strata, forming the high ground on which the River Wyre rises, consist predominantly of mudstones and sandstones, with some limestones. The presence of a faulted contact along part of the boundary was thought to imply direct

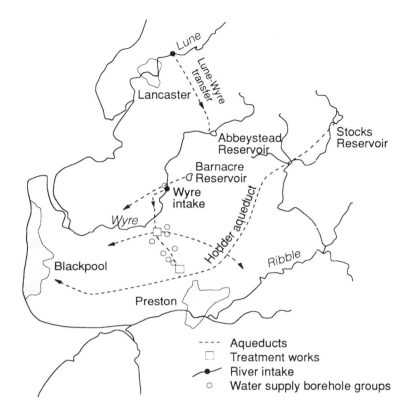

Figure 1   General features of the Lancashire Conjunctive Use Scheme (LCUS)

hydraulic continuity between the Carboniferous and the Permo-Triassic aquifer, and in the investigations in the early 1970s it was assumed that the majority of the recharge to the aquifer was derived from this source. However it is now considered likely that the presence of low permeability Manchester Marls at the base of the Permo-Triassic sequence could limit such flow, particularly south of Garstang. In the west, groundwater salinities are high, indicating that no significant groundwater flow occurs westwards under the mudstones.

The Fylde aquifer is mainly covered by drift deposits, comprising glacial silts, sands, gravels and boulder clay, the latter tending to limit rainfall infiltration. However the deposits are highly variable in composition and disposition. Furthermore, there is known to be hydraulic continuity with stretches of the various watercourses which cross the aquifer, either directly or via permeable alluvial deposits.

Groundwater movement is generally from the east to the north-west in the north and to the south-west in the southern part, with the groundwater divide occurring just south of Broughton (Figure 2). Under natural (non-pumping) conditions, outflow to the various watercourses takes place along stretches where glacial till is absent or has been incised (e.g. Wyre, Calder, Brock, Cocker and Pilling Water).

Groundwater gradients are very shallow in the north-west of the area, coinciding with the extensive tract of marine and estuarine alluvium. Piezometric heads are naturally above sea level, even at the coast.

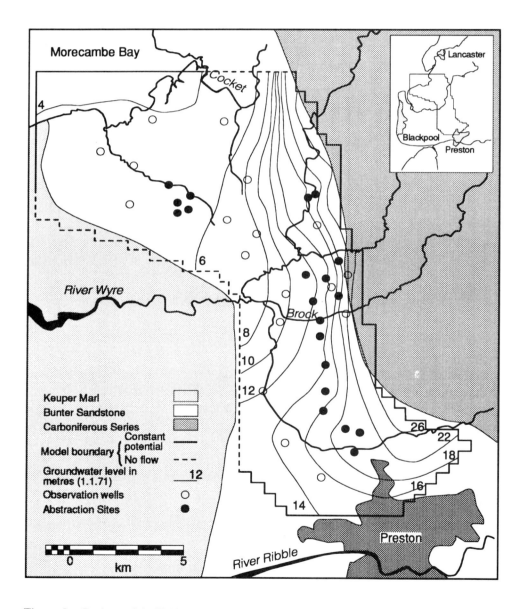

Figure 2 Geology of the Fylde area and model features

DEVELOPMENT FOR WATER SUPPLY

The earliest developments for public water supply were surface water abstractions from tributaries of the Tarnbrook Wyre above Abbeystead, where catchwaters and stream intakes gather the whole flow of the various tributaries. Abbeystead reservoir was built to supply compensation to mills on the River Wyre. These compensation rights were bought out in the 1920s, although a small flow is still released from the reservoir, in addition to a fish pass. Barnacre reservoirs were built in the late 1800s, again with no provision for compensation or bypass at low flows.

Groundwater has been abstracted from the northern part of the Fylde aquifer since the 1890s. Serious investigation of the aquifer started in the late 1950s with trial boreholes to prove the existence of the aquifer and to determine its possible yield. The quantities available for supply were limited initially by statute and later by licences.

The rest of the aquifer was virtually unexploited until investigations in the early 1960s by Fylde Water Board in the area between Garstang and Preston, where the first stage of development was commissioned in 1965, with treatment at Broughton.

The interim report of the Northern Technical Working Party (NTWP, 1967) identified future shortages in Lancashire during the 1970s. In 1969 Lancashire River Authority proposed a scheme with conjunctive use of the reservoirs and the rivers Ribble, Wyre and Lune. In 1971 Water Resources in the North Working Group II recommended a scheme which included an enlarged Stocks Reservoir, the Fylde aquifer, and abstractions from the rivers Lune and Wyre. This was developed (without enlarging Stocks reservoir) into the current LCUS (Figure 1).

In the LCUS, direct gravity supplies from the reservoirs are the cheapest sources. The boreholes can be used to back up the reservoirs, particularly for short periods of intensive use, but require expensive pumping. The river abstractions can be used to 'rest' the reservoirs early or late in the season as soon as they respond to rainfall. However it should be noted that in drought periods flows will be below the prescribed flow levels and these sources are then unavailable. Thus the licensed totals for the scheme give a misleading impression of total system yield.

In addition to the LCUS groundwater abstractions from the central portion of the Fylde aquifer, there are further groundwater uses by industrial concerns to the north and the south and by private individuals throughout. For example, the Lancaster canal has an intake on the River Calder, which can take the whole flow, without compensation.

MODELLING

Two types of model have been used to determine operating rules and licence conditions for the LCUS, with largely separate consideration of surface and groundwater components.

Groundwater modelling

The LCUS groundwater licence conditions were derived from a model prepared by the Water Resources Board (WRB) (Oakes & Skinner, 1975). Data were obtained from a series of pumping tests carried out in 1972-4 on groups of abstraction boreholes, with associated monitoring of groundwater levels and surface water flows, supplemented by historical groundwater data.

The principal source of recharge was assumed to be the Carboniferous strata to the east, because of the extensive clay drift cover over the aquifer itself, although it was recognised that certain sections of the Wyre and Brock river beds were in hydraulic continuity.

The criteria applied in determining the maximum acceptable abstractions from groundwater were:

● Demand should not exceed available recharge over a three-year period, to ensure groundwater levels are not permanently depressed;

● Positive groundwater gradients should be maintained to prevent incursion of saline groundwaters from the coast and/or from beneath the Mercia Mudstones forming the western boundary of the aquifer;

● The interest of existing abstractors should be assured.

The resulting licences attempted to encourage balanced use from north to south, by setting conjunctive conditions for groups of boreholes.

The pumping carried out in 1972-74 identified that groundwater levels in the drift and flows in certain watercourses were affected by abstraction from the main Sandstone aquifer. Reductions in head in the drift caused even by intermittent abstraction from the sandstone can result in delayed yield and drainage taking place after pumping has ceased. Threshold levels were therefore set on two observation boreholes and two drift wells to act as an early warning.

In an attempt to reduce the impact of pumping on surface waters the licences include augmentation of various river stretches, dependent on flows in the Wyre at St Michaels. Subsequent work, mainly related to the quality of water, inferred that indirect recharge occurring through the present river alluvium via the glacial sands and then into the sandstone aquifer was of comparable importance to groundwater flows across the eastern boundary from the Carboniferous (Sage & Lloyd, 1978). Their report also emphasised the complexity of the drift influence on ground/surface water interrelationships.

The recently published 1:50 000 solid and drift geological maps of the Garstang area have also revised the interpretation of the Fylde aquifer in terms of its structure, boundary conditions and drift cover, such that the WRB model assumptions may now prove to be inappropriate.

Modelling for resource planning and operation

After reviewing available methods for simulation and analysis of water resource systems Walsh (1977) constructed a simulation model of the LCUS and used it to investigate a large number of operating policies. The final order of preference for use of the various sources was:

1. Barnacre reservoirs;
2. Stocks reservoir;
3. River Wyre abstraction, supported by River Lune transfer when necessary;
4. Borehole abstractions;
5. Further calls on Stocks, then the boreholes.

The resulting policy, applied in simulation of the historic period, showed how supplies could have been maintained through both 1933-34 and 1959 droughts.

In the early 1980s, simulation modules from NWWA's Resource Planning Suite of Programs (NWWA, 1981a) were used to carry out a series of investigations into the economic operation of the LCUS (NWWA, 1981b &1983). These identified the minimum operating cost configurations for different demand levels and the detailed changes to operating policy required for each. Marginal costs ($£ Ml^{-1}$) were derived for the different system configurations appropriate to each level of demand. The report envisaged the combination of LCUS and Lake District outputs to allow a least cost allocation of water on a regional basis in the Northern Command Zone (NCZ).

This was carried out, initially using the RP suite and subsequently using MOSPA (PWSC Ltd, 1992), which provided facilities for a more detailed simulation and also advanced methods of optimisation to determine least-cost operating policies which maintained the desired reliability. Figure 3 illustrates the relationship between the LCUS and the rest of the NCZ. The Fylde boreholes are represented by two groups, at Franklaw and Broughton. This reflects differing treatment costs, and has been found to be adequate for operating policy derivation and simulation. It also allows incorporation of the overall licence limits of the system.

OPERATIONAL EXPERIENCE

Borehole observations

NWW Ltd have been operating their Fylde groundwater sources since the 1960s and using them conjunctively with surface water since 1979. During this period comprehensive groundwater level monitoring has been carried out in both the abstraction boreholes and numerous observation boreholes.

The boreholes are operated in groups, subject to three-year as well as annual licence limits, and the actual abstraction pattern and duration varies from year to year according to demand, operational and licence constraints. NWW's operating strategy of preferential use of their more efficient and high quality boreholes has altered the distribution of abstraction across the aquifer. These effects are evident in the hydrographs for the last 20 years from key abstraction sites. They also indicate more intensive use during dry years.

The boreholes in the north of the aquifer showed evidence of saline intrusion following daily abstraction rates of over 23 Ml d^{-1} in the early 1950s. A gradual reduction to about 9 Ml d^{-1} in 1960 seemed to relieve the problem, but recent intake increases have again affected groundwater levels. Although they remain above Ordnance Datum, the historical record illustrates the sensitivity of this area to

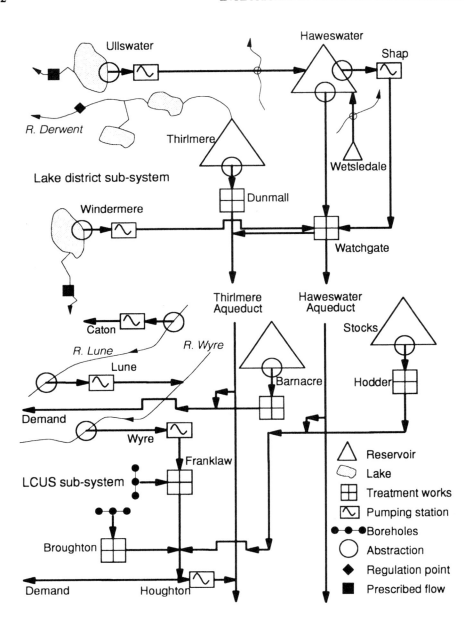

Figure 3   Representation of the LCUS in the NCZ model

changes in the operating regime.

The LCUS licence is also subject to "hands-off" level conditions relating to two observation boreholes (shown on Figure 2). These conditions were incorporated following the modelling of the aquifer, and were intended to protect the resources and groundwater users to the south-east and south-west of NWW's abstractions.

Hydrographs from these sites indicate that resources are limited in the south, and could not sustain significant increases in actual abstraction without breaching the existing licence conditions. The environmental impact of further decline in levels has not been assessed.

Artesian areas

Before the development of the aquifer there were several areas of groundwater discharge to the Wyre and Cocker catchments where heads were at, or above ground level, i.e. artesian. These high heads have supported wetland areas and spring discharges. Many of the latter were formerly used for agricultural purposes. This is also reflected in the name of the 'Running Pump' Inn at Catforth.

As abstraction increased, thus reducing the piezometric head, the springs ceased. The Water Board, recognising the effect of pumping, converted many of the spring-fed cattle troughs to mains. Lowering of the groundwater head has allowed the ground to dry out more quickly allowing earlier access to fields. Beneficial as this may be to agriculture, there are other consequences in the drying out of many conservation areas.

River flow

Figure 4 shows the stretches of rivers which are known to either dry up or have reduced flows in summer periods. As described above, for economic reasons abstraction has been concentrated on the most efficient sources. These are the ones in hydraulic continuity with the watercourses, so have the largest impact. This is thus the worst situation for the river environment. However the introduction of new operating regimes will lead to increased operating costs.

Conservation sites

There are a number of areas of conservation interest within the aquifer area where lowering of groundwater levels is likely to have an adverse effect. These are indicated on Figure 4, and include three SSSIs, and several SBIs. In addition there are many nature walks and trails, some of which are along rivers, e.g. the River Brock, and 'non-designated' sites which are still areas of significant conservation interest within the NRA's environmental purview.

Other areas of importance from a conservation viewpoint are ponds, spring lines, and wet boggy areas. Carr Green SBI and Winmarleigh Moss SSSI are drying out, as are many other non-designated ponds and small watercourses. Certain protected species, such as the Great Crested Newt, require not just ponds to breed in but also associated damp grassland.

FISHERIES

The Wyre catchment supports recreational fisheries for migratory salmonids, resident trout and coarse fish. A recent study was carried out to assess the salmonid

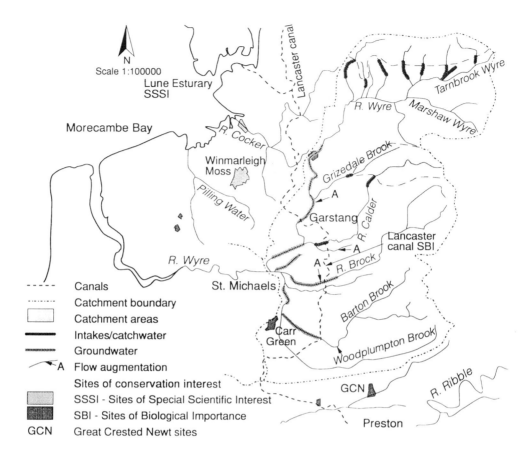

Figure 4 The Wyre catchment and Fylde district, showing sites of conservation interest and
river stretches where flows are affected by abstractions

and coarse fish population of the Wyre catchment. This also looked at distribution of
spawning and nursery areas and species distribution.

The survey indicated that natural recruitment is poor, i.e. numbers and distribution
of juveniles are limited, both for migratory salmonids and coarse fish. A number of
factors contribute to this: availability of spawning/nursery grounds, amount of wet
area, barriers to migration, and acid stress. The area of water available (wet area)
limits the population of fish. As this decreases with low flows so numbers decrease.

In particular, low flows in the Calder and Grizedale Brook cause a loss of spawning
and juvenile nursery habitat for migratory salmonids. The abstractions from these
watercourses result in the drying up of reaches during low flows, with a complete loss
of fish habitat.

CONCLUSIONS

The main conclusions of the review were as follows:

- The current regime and rates of abstraction from both surface and ground water in the Wyre catchment and Fylde area are giving rise to low flow conditions and the drying up of areas of ecological interest, in dry weather periods.

- Additional abstraction (possibly to support transfer of water to other regions), even though authorised in existing licences, may cause irreversible effects and greater environmental damage. If the scheme were to be promoted today, far greater attention would be paid to the environmental impacts of abstraction, and this would be reflected in the licence conditions, possibly reducing yields for water supply.

- The original LCUS licence constraints (threshold levels, prescribed flows, augmentation and daily, annual and three-year limits) have proved their value, and may need to be reconsidered if low flows are to be alleviated. Annual licence totals for a conjunctive system including groundwater and river abstractions give an unrealistic indication of sustainable yield.

The paper has described a recent desk review of over 20 years' operational experience of the LCUS. During this time there has been a considerable increase in awareness of the need for sustainability. In order to move towards this ideal a workable management policy for sustainable use of the Fylde water resources is essential. This must give much greater weight to the needs of the environment than has previously been the case. The policy should include short term strategies to alleviate the existing low flow problems, and reduce the risk of others developing in the future, as well as long term policies to cope with institutional and environmental changes. It should also clearly establish whether there is spare water in the Fylde for future development, or for possible transfer elsewhere as part of a national strategy.

REFERENCES

Downing, R.A. 1993. Groundwater resources, their development and management in the UK: an historical perspective. *Q. J. Engng Geol.* **26**, 335-358

Law, F.M. 1993. Defining a sustainable water cycle. Seminar on sustainable water resources, ICE, London.

NTWP 1967. Northern Technical Working Party, interim report.

NWWA 1981a. An introduction to the Resource Planning suite of programs. North West Water Authority.

NWWA 1981b & 1983. Economic operation of LCUS: 2 internal reports. North West Water Authority.

PWSC Ltd 1992. Modular optimisation/simulation package with AMORS simulation module (MOSPA): program description.

Oakes, D.B. & Skinner, A. 1975. The Lancashire Conjunctive Use Scheme Groundwater Model. WRC TR 12.

Sage, R.C. & Lloyd, J.W. 1978. Drift deposit influences on the Triassic sandstone aquifer of north-west Lancashire as inferred by hydrochemistry. *Q. J. Engng Geol.* **11**, 209-218.

Skinner, A. 1993. Sustainable use of groundwater. Seminar on sustainable water resources, ICE, London.

Walsh, P.D. 1977. The practical analysis and operation of multi-source water supply systems with particular reference to the Lancashire Conjunctive Use Scheme. PhD thesis.

WRB, 1967. *Interim Report on water resources in the North*, HMSO, London.

WRB, 1971. Report to the steering group on the Lancashire Conjunctive Use Scheme, Reading, UK.

47 An investigation of the Malmesbury Avon catchment in the Cotswolds of southern England

R GRAY
W S Atkins Water, Epsom, UK

INTRODUCTION

The Malmesbury Avon is the name generally given to the upper reaches of the Bristol Avon catchment, which in this study was defined as the catchment upstream of Great Somerford gauging station, covering an area of approximately 300 km². The catchment is located in the Cotswolds of Wiltshire in the south of England, about 40 km to the east of Bristol. The catchment location is shown in Figure 1, with a study area plan in Figure 2.

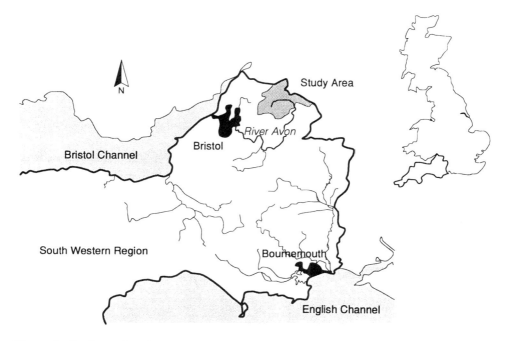

Figure 1 Catchment location

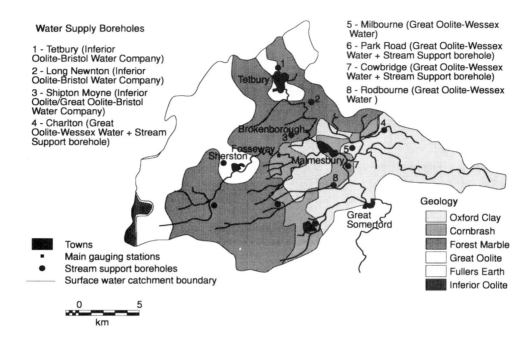

Water Supply Boreholes

1 - Tetbury (Inferior
Oolite-Bristol Water Company)
2 - Long Newnton (Inferior
Oolite-Bristol Water Company)
3 - Shipton Moyne (Inferior
Oolite/Great Oolite-Bristol
Water Company)
4 - Charlton (Great
Oolite-Wessex Water + Stream
Support borehole)

5 - Milbourne (Great Oolite-Wessex
Water)
6 - Park Road (Great Oolite-Wessex
Water + Stream Support borehole)
7 - Cowbridge (Great Oolite-Wessex
Water + Stream Support borehole)
8 - Rodbourne (Great Oolite-Wessex
Water)

Geology

Oxford Clay
Cornbrash
Forest Marble
Great Oolite
Fullers Earth
Inferior Oolite

Towns
Main gauging stations
Stream support boreholes
Surface water catchment boundary

0   5
km

Figure 2   The study area

The river consists of two main tributaries, known as the Sherston Avon and the Tetbury Avon after the towns through which they flow. These tributaries meet in Malmesbury from where the river flows south east and is joined by three smaller tributaries (Woodbridge Brook, Gauze Brook and Rodbourne Brook). The catchment topography slopes gently to the south east, reflecting the dip in the underlying geology. The river system flows within narrow valleys incised into the catchment and 20 to 30 metres below the surrounding land.

CATCHMENT GEOLOGY

The catchment geology (shown in Figure 2) consists of interbedded layers of fractured oolitic limestones, marls and clays, dipping to the south east at a slightly steeper angle than the topography. The geology can be simplified as follows:

Oxford Clay — thick low-permeability clay sequence, only found in the south-east of the catchment.

Cornbrash (3 to 6m) — alternating sequence of marl and limestone, not considered significant in catchment water supply terms.

Forest Marble (10 to 20m) — calcareous and occasionally sandy clays with lenses of limestone. Acts as a leaky aquitard and will transmit groundwater, particularly where thin or strongly faulted.

Great Oolite (20 to 40m) — flaggy and massive oolitic and reef limestones with some calcareous siltstone. Groundwater flow is through fractures and fissures. This sequence provides the main groundwater baseflow to the river system and outcrops across much of the central part of the catchment.

Fullers Earth (20 to 30m) — thick and extensive sequence of clays with some limestone. Previously considered as an effective barrier to flow between the Great and Inferior Oolite, it now appears that flows do occur across this layer.

Inferior Oolite (12 to 45m) — a relatively homogenous sequence of fine to medium grained shelly and oolitic limestones. This sequence outcrops as a thin band along the western scarp boundary and underlies the entire catchment.

GROUNDWATER RESOURCES DEVELOPMENT

The Malmesbury Avon catchment has a long history of groundwater abstraction for public supply, and wells and boreholes have been installed in both the Great Oolite and Inferior Oolite aquifers by Wessex Water and Bristol Water Companies. Conventional groundwater sources had been developed at regular intervals up until 1980 when the most recent scheme, known as the Malmesbury Groundwater Scheme, was licensed. This scheme combined increased public supply with eight new or re-allocated surface water augmentation boreholes, abstracting from the Inferior Oolite in the upper reaches of each tributary and operated at or below trigger flows on downstream gauging stations. The complete supply and augmentation system is shown on Figure 2 and the average daily yield from each aquifer in 1992 was as follows:

	Supply ($Ml\ d^{-1}$)	Augmentation ($Ml\ d^{-1}$)
Great Oolite	40	0.2
Inferior Oolite	10	2.6

Public concerns as to the impact of groundwater abstraction on the river environment grew in the 1980s. The concerns were principally related to low summer flows and focused on (i) the amenity and environmental value of the lower reaches of the Sherston and Tetbury Avon, particularly around Malmesbury and (ii) the fisheries value of the main Avon downstream of the confluence. In November 1991 the National Rivers Authority (NRA) appointed WS Atkins to undertake a detailed investigation into the catchment, the results of which are presented in this paper.

CONCEPTUAL FLOW MODEL

A conceptual understanding of the dominant hydraulic processes in the catchment was developed at an early stage of the project, the main features of which are:

- The catchment responds rapidly to winter rainfall and the baseflow index for the catchment to Great Somerford is only 0.58. This is considered to be due

rapid flow through the fissure system, and (b) low specific yield of the aquifers
resulting in annual head variations in the order of 20 metres and a rapid baseflow
response to recharge.

- The groundwater table falls to low levels in the summer and there is little or no
recharge in response to summer rainfall. Low flows in the Sherston and Tetbury
Avon through Malmesbury are commonly in the order of 5 Ml d^{-1}, compared with
mean flows in the order of 100 Ml d^{-1}.

- The Great and Inferior Oolite aquifers had been thought to be essentially discrete.
Over much of the catchment there is a significant head difference between the two
aquifers with the head in the Inferior Oolite being up to 40 metres below that in
the Great Oolite. A dye tracer test carried out in the 1970s was intended
principally to assess the flow directions in the Great Oolite. However, the Inferior
Oolite directly below the Great Oolite release point was also monitored and
recorded a dye trace around 60 days following release, indicating a distributed
downwards seepage flow between the aquifers. In addition, when the groundwater
heads in the Great and Inferior Oolite were studied closely, mounds were
identified in the Inferior Oolite where the groundwater level had risen, in some
cases up to the Great Oolite level. These mounds have been interpreted as zones
of discrete downward flow, related to fault and fissure development.

- Monthly flow accretion data revealed zones of stream flow losses on both the
Sherston and Tetbury Avon, just upstream of the confluence. This was surprising
as the Great Oolite was considered to be confined along these reaches. Boreholes

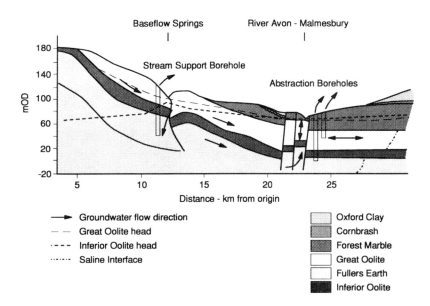

Figure 3 Conceptual section showing hydraulic interactions between aquifers and surface
system

installed by the NRA in 1991 along the lower reaches of the Sherston Avon revealed that the Great Oolite is within a few metres of the ground surface along a 250 m reach in Malmesbury. There is therefore potential for interaction between the aquifer and the river in this area.

In summary, the basic conceptual model is of a catchment where groundwater flow is dominant but which nevertheless responds rapidly to winter rainfall with very low summer flows. There is also evidence of connections both between the river and the confined Great Oolite and between the two aquifers. The section in Figure 3 illustrates this.

PUBLIC OPEN DAY

In June 1992 an Open Day for the project was held in Malmesbury. This had a dual purpose of (a) publicising the project to the public and (b) recording their recollections of the river. The main findings were as follows:

- In the years before the last war a swimming race was held in Malmesbury every August, racing along the lower reaches of the Sherston Avon. One competitor recalled a considerable variation in river temperature along the route, which was generally interpreted as due to the strong spring flows into the river. Around the same period a spring, known as Daniel's Well, was used by locals as a drinking water supply. These spring flows have not been present in recent summers.

- The river tributaries around Malmesbury used to be controlled by moveable sluice structures allied to old mills. By the 1950s these were no longer operated properly and the area was regularly flooded. The structures were progressively replaced by low level static weirs, substantially reducing the flooding risk. One interviewee had moved away from the lower Tetbury Avon in the 1930s, returning in the 1980s. He was very surprised to see, each summer, bright green algal blooms develop over the river surface. Another interviewee had recorded the movement of a lemonade bottle floating in the river, travelling a total distance of around 200 metres in three weeks.

  The causes of eutrophication and algal blooms are generally considered to be increased nutrient loads. However, in most lowland rivers algal blooms are not nutrient limited and more important requirements are residence time (at least five days), low water velocity and high water temperature. The replacement of moveable sluices, which provided a regular flushing flow to the river, with static weirs appears to have combined with reduced river flows to produce conditions favourable to algal bloom development.

- The loss or reduction of submerged aquatic plants along the river, and replacement with emergent and marginal flora, was also recorded by a number of respondees. There were other general comments on the reduction in, for example, invertebrate fauna and wild trout spawning. These features are common to many of the lowland streams suffering from reduced river flows.

RIVER DIARIES

A River Diary is a regular record of the condition of a particular river reach. In addition to general observations the diaries provide a subjective view of whether a river flow is acceptable as a summer minimum flow. Diarists recruited largely from the general public provided a weekly record of river conditions during the summers of 1992 and 1993.

A typical comparison of Diary records and river flow conditions is shown in Figure 4. The records proved very consistent, both in time and between diarists. However, in each case, the minimum flow requirements of the same diarist were significantly higher in 1993 than in 1992. The two main reasons for this are thought to be that first, whereas the summer flow recession in 1992 was gradual, in 1993 the flows reduced rapidly accompanied by a rapid deterioration in river appearance; the low flows in 1993 were therefore contrasted unfavourably with conditions only a few weeks previously. Second, the summer of 1992 was reasonably dry with sunny weather and drought conditions throughout southern England were receiving nationwide publicity. The summer of 1993 was dull and damp and, nationally, the drought was considered to be coming to an end. As noted earlier, summer rainfall has little effect on the low river flows but may nevertheless influence the acceptability of river conditions.

Algal blooms were a feature of the river in 1992 and 1993. In late summer 1992 a brief spate resulted in the flushing of these blooms from the river system. Following this spate, river flows reduced to a lower rate than recorded previously but were considered more acceptable.

These results indicate that the public can provide a coherent and reasonable assessment of the acceptable flow requirements for a river. Perceptions can be influenced however by a number of related — and sometimes unrelated — factors and should not be the sole means of defining acceptability.

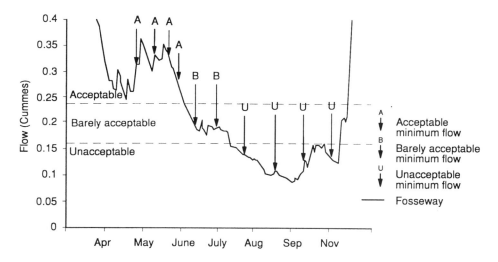

Figure 4 A typical comparison between diary records and river flow conditions

FIELDWORK

There are six public supplies from the confined Great Oolite and three from the Inferior Oolite. The fieldwork was designed to assess the impact of abstractions on groundwater levels by organising, with the co-operation of Bristol and Wessex Water Companies, to vary abstractions at each major supply source and monitor the impact on groundwater levels in the Great Oolite aquifer. Two investigations were carried out, in Summer 1992 and Spring 1993, and groundwater levels were monitored along the lower reaches of the Sherston and Tetbury Avon (the losing reaches) and upstream in both sub-catchments.

The main findings were as follows:

- The impact of Cowbridge and Milbourne sources is largely local to Malmesbury.
- Robourne and Charlton sources yield from a wider area of the confined and unconfined Great Oolite.
- Shipton Moyne yields directly from the Great and Inferior Oolite and Long Newnton yields from the Inferior Oolite only. Both sources appear to induce downward flow from the Great Oolite however, which appears to be the ultimate resource for both sources. The sources yield largely from the unconfined Great Oolite with little impact on groundwater levels around Malmesbury.

MODEL SET-UP AND CALIBRATION

The model was set up using the MIKE SHE finite difference modelling system. Separate modules were linked for:

- recharge through the soil zone and unsaturated zone flow
- groundwater flow through a four layer system
- two-dimensional overland flow
- one-dimensional river flow.

An initial model was set up at the start of the project and run in order to test the basic conceptual understanding of the flow system and to identify the fundamental areas where additional information was required. The model output was compared with actual records on the basis of temporal and spatial variations in groundwater head in the Great and Inferior Oolite and surface flow records at gauging stations, flow accretion curves and flow duration curves.

The model was developed further as the conceptual understanding improved, with the following features included in the final version:

- A four-layer groundwater model. Layers 2, 3 and 4 represent the Great Oolite, Fullers Earth and Inferior Oolite respectively. The unconfined Great Oolite was divided into two layers with Layer 1 representing an upper high conductivity layer. This was necessary to reproduce the groundwater head variation in the aquifer and is consistent with other studies on the Great Oolite (Rushton *et al.*, 1992). Elsewhere in the model, layer 1 represents the strata overlying the Great Oolite, *viz* the Forest Marble, Cornbrash and Oxford Clay.

- A soil bypass mechanism, by which up to 75% of the rainfall to the Great Oolite Catchment bypasses the soil zone through fissure systems. This mechanism is necessary to explain the rapid response to recharge in the winter period. The effectiveness of the bypass mechanism reduces to zero as the soil moisture deficit increases. This accounts for the reduced response to summer rainfall, with some surface runoff but little or no groundwater recharge.
- The specific yield in the Great Oolite was set at 0.5 to 0.7% to account for the large head variations (20 to 30 m) in response to winter recharge.
- Zones of higher conductivity were set in the Fullers Earth (Layer 3), corresponding with the location of the mounds in the Inferior Oolite groundwater head.

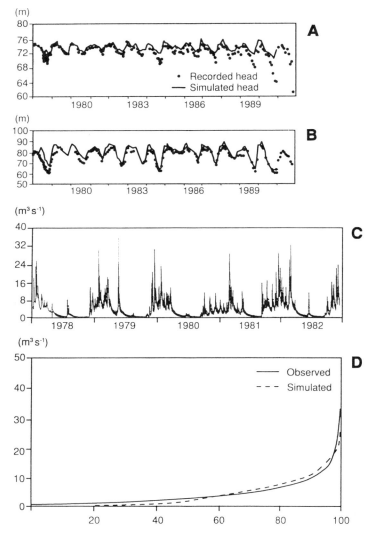

Figure 5 Final calibrated output for A: Great Oolite; B: Inferior Oolite; C: Great Somerford surface flow records; D: Great Somerford flow duration curve

- Heterogeneities were introduced into the Great Oolite to account for the spatial variation in groundwater heads and the flow accretion patterns in the river system.

Examples of the calibrated output from the model are shown in Figure 5.

MODEL SIMULATIONS

The calibrated model was used for both scoping simulations and to assess specific management options. The scoping simulations were undertaken initially to assess, in broad terms, the impact of various abstraction sources on the surface flow record. The impact of, for example, naturalising the system (no abstraction), removing or doubling augmentation, and operating only specific groups of sources was tested over the 10 year run period, comparing the modelled actual output with the output from each simulation run. Figure 6, for example, shows the difference between historical and naturalised flow at the Great Somerford gauging station. The flow differences are expressed in absolute terms and as a percentage of the historical flow and show that historical flows have been, on average, about 0.4 m^3s^{-1} below the natural flow and that

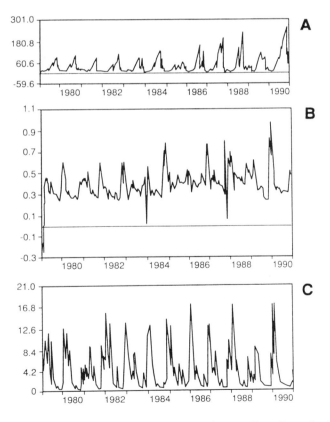

Figure 6 Comparison of historical and naturalised flow at Great Somerford.
A: in percentage terms; B: in cumecs; C: actual flow record (m^3 s^{-1})

the flow reduction was at a minimum during summer low flow periods, due to the positive impact of augmentation. Nevertheless, actual flows still show a reduction of between 30 and 70 per cent over low flow periods.

Further model runs, to test specific management options, included transferring boreholes from supply to augmentation, increasing augmentation yields and lining specific river reaches, with the main findings summarised thus:

- Under naturalised conditions there is relatively little lateral groundwater flow from the unconfined to the confined aquifers and downward flow from the Great Oolite to Inferior Oolite. Groundwater abstraction has induced (a) increased lateral flows into the confined aquifers (b) increased downwards flow to the Inferior Oolite and (c) reduced baseflow from both the unconfined aquifer and the partly confined aquifer around Malmesbury.

- Abstraction is modelled as reducing surface flows to both Brokenborough and Fosseway gauging stations, at the downstream limit of the unconfined Great Oolite, by around 5 Ml d^{-1}. During low flow periods, however, augmentation provides a small net benefit in comparison with natural conditions.

- The average flow reduction to the confluence at Malmesbury is modelled as 14 Ml d^{-1} and this decreases to around 5 Ml d^{-1} under low flow conditions. Nevertheless actual low flows are only around 40 per cent of the natural flow. In addition, augmentation provides around two thirds of the actual flow and without augmentation these reaches would virtually dry up.

- The impact of abstraction at the base of the catchment is not as severe, although minimum flows are still only around 60 per cent of the natural flow in drought years. The beneficial impact of augmentation is still evident, providing up to half the total flow in drought years.

- The augmentation scheme is very beneficial to the catchment over the summer low flow period. A net benefit of 50 per cent of borehole yield is provided. Increasing augmentation from existing sources and/or transferring groundwater from supply to augmentation was shown to provide significant additional benefits, particularly in the lower Sherston and Tetbury Avon.

SUMMARY

The Upper Avon is a very complex hydraulic system with an intimate interaction between the surface and groundwater flow regimes. Storage is low in the aquifers and groundwater levels respond rapidly to winter recharge. Conversely, summer recessions are steep and surface flows are particularly vulnerable to groundwater abstraction over the summer period.

Groundwater abstraction, from both the Great and Inferior Oolite, has had a particularly adverse impact on the lower reaches of the Sherston and Tetbury Avon around Malmesbury where, from the model results, actual minimum flows are only around 40 per cent of the natural flows. Local inhabitants recalled that these reaches used to receive year round spring flows whereas flow losses can now occur in the summer. The appearance of these reaches has been compounded by the installation of fixed weirs which, whilst they retain a high water level, also provide ideal conditions for the development of algal blooms.

The lower reaches of the Avon to Great Somerford are also adversely affected and lowest flows are reduced by around 40 per cent by abstraction. The public have kept diaries over the last two years. The diary records revealed an impressively consistent assessment of acceptable flows for each year but also showed that other related and unrelated features can influence public opinion as to the acceptability of river flows.

The augmentation scheme is providing a significant proportion of the river flow during the lowest flow periods. Various proposals are being considered by the NRA for developing the augmentation system further as part of the overall management plan for the catchment. These include (i) increasing the yield from the existing scheme and (ii) transferring a proportion of the public groundwater supply to augmentation. These and other management options are currently being developed, in consultation with all the interested parties. A definitive Action Plan for the catchment will be prepared at the end of this process.

ACKNOWLEDGEMENT

This paper is based on a project undertaken on behalf of NRA South Western Region and the author gratefully acknowledges the help and support of the NRA. The opinions in this paper however are those of the author and do not necessarily reflect those of the NRA.

REFERENCE

Rushton, K.R, Owen, M. & Tomlinson, L.M, 1992. The water resources of the Great Oolite aquifer in the Thames Basin, UK. *J. Hydrol.* **132**, 225-248.